Organisational Ethics in the Built Environment

Organisational Ethics in the Built Environment

Jason Challender
Department of Chemistry
University of the Andes
Merida, Venezuela

WILEY Blackwell

This edition first published 2023
© 2023 John Wiley & Sons Ltd

All rights reserved. No part of this publication may be reproduced, stored in a retrieval system, or transmitted, in any form or by any means, electronic, mechanical, photocopying, recording or otherwise, except as permitted by law. Advice on how to obtain permission to reuse material from this title is available at http://www.wiley.com/go/permissions.

The right of Jason Challender to be identified as the author of this work has been asserted in accordance with law.

Registered Office
John Wiley & Sons Ltd, The Atrium, Southern Gate, Chichester, West Sussex, PO19 8SQ, UK

For details of our global editorial offices, customer services, and more information about Wiley products visit us at www.wiley.com.

Wiley also publishes its books in a variety of electronic formats and by print-on-demand. Some content that appears in standard print versions of this book may not be available in other formats.

Trademarks: Wiley and the Wiley logo are trademarks or registered trademarks of John Wiley & Sons, Inc. and/or its affiliates in the United States and other countries and may not be used without written permission. All other trademarks are the property of their respective owners. John Wiley & Sons, Inc. is not associated with any product or vendor mentioned in this book.

Limit of Liability/Disclaimer of Warranty
While the publisher and authors have used their best efforts in preparing this work, they make no representations or warranties with respect to the accuracy or completeness of the contents of this work and specifically disclaim all warranties, including without limitation any implied warranties of merchantability or fitness for a particular purpose. No warranty may be created or extended by sales representatives, written sales materials or promotional statements for this work. This work is sold with the understanding that the publisher is not engaged in rendering professional services. The advice and strategies contained herein may not be suitable for your situation. You should consult with a specialist where appropriate. The fact that an organization, website, or product is referred to in this work as a citation and/or potential source of further information does not mean that the publisher and authors endorse the information or services the organization, website, or product may provide or recommendations it may make. Further, readers should be aware that websites listed in this work may have changed or disappeared between when this work was written and when it is read. Neither the publisher nor authors shall be liable for any loss of profit or any other commercial damages, including but not limited to special, incidental, consequential, or other damages.

A catalogue record for this book is available from the Library of Congress

Hardback ISBN: 9781394186242; ePub ISBN: 9781394186266; ePDF ISBN: 9781394186259; oBook ISBN: 9781394186273

Cover Design: Wiley
Cover Image: © AP911_Studio/Shutterstock

Set in 9.5/12.5pt STIXTwoText by Integra Software Services Pvt. Ltd., Pondicherry, India

Dedicated to the memory of Brenda Challender (1939 – 2023)

Contents

Author Biography *xii*
Foreword *xiii*
Acknowledgements *xv*
List of Figures *xvi*
List of Tables *xvii*
List of Appendices *xviii*

1	**Introduction** *1*	
1.1	Introduction to the Book *1*	
1.2	Overarching Rationale, Synopsis and Justification: The Aim and Objectives for Writing the Book *1*	
1.3	The Need to Rethink the Concept of Organisational Ethics *2*	
1.4	The Fragmentation and Bespoke Nature of the UK Construction Industry *3*	
1.5	Examples of How This Area Is Used or Represented in Our Society Today *4*	
1.6	The Lack of Trust in Construction Contracting and Reputational Damage for the Industry and Those Organisations Associated with It *4*	
1.7	The Book's Value Proposition and Its Unique and Outstanding Features *5*	
1.8	Audience for the Book *8*	
1.9	Structure of the Book *8*	
1.10	Summary *10*	
	References *11*	
2	**An Introduction to Organisational Ethics** *13*	
2.1	Introduction to Chapter *13*	
2.2	What Are Ethics? *14*	
2.3	The Importance and Relevance of Organisational Ethics in the Construction and Engineering Sectors *16*	
2.4	Possible Effects of Workplace Ethics on Employees and Organisational Productivity *17*	

2.5	Organisational Culture Linked to Ethics *18*	
2.6	Summary *19*	
	References *20*	

3 Environmental Ethics *22*
3.1 Introduction to Environmental Ethics *22*
3.2 Climate Change and Sustainability *23*
3.3 Context for the Discussion on Environmental Ethics *23*
3.4 What Are Environmental Ethics and Sustainability? *24*
3.5 The Importance of Sustainability and the Ethical Case for Participation *25*
3.6 The Global and National Agenda on Sustainability *26*
3.7 Environmental Social Responsibility *27*
3.8 Dilemmas around Environmental Ethics *28*
3.9 Examples of Environmental Ethics and Sustainability in Practice: Case Study One – A UK Northwest College *29*
3.10 Examples of Environmental Ethics and Sustainability in Practice: Case Study Two – A UK Northwest University *36*
3.11 Summary *41*
References *43*

4 Corporate Social Responsibility *45*
4.1 Introduction *45*
4.2 What Is Corporate Social Responsibility? *46*
4.3 The Benefits of Corporate Social Responsibility for Businesses *46*
4.4 Social Value as Part of Corporate Social Responsibility *47*
4.5 Different Types and Models of Corporate Social Responsibility *50*
4.6 The Considerate Constructor Scheme *52*
4.7 Corporate Codes and Ethical Statements *56*
4.8 Sustainable and Ethical Construction Linked to Corporate Social Responsibility *58*
4.9 Summary *60*
References *62*

5 Relationship between Ethics and Other Factors *63*
5.1 Introduction *63*
5.2 The Relationship between Risk and Ethics *64*
5.3 The Relationship between Ethics and Quality *67*
5.4 Potential Impact of Professional Ethics on Construction Quality *69*
5.5 Reputation and Ethics *69*
5.6 Ethics and Trust *70*
5.7 Ethics and Leadership *70*
5.8 Ethics and Motivation *71*
5.9 Summary *72*
References *74*

6 Ethical Human Resource Considerations 75
6.1 Introduction 75
6.2 Inclusivity, Equality and Diversity 75
6.3 Ethical Issues around Culture 81
6.4 Initiatives and Accreditations Available for Construction Organisations to Drive Positive Change for Diversity and Inclusion 83
6.5 A Toolkit for Change: Reflection on How the Construction Industry Can Improve Diversity and Inclusion 83
6.6 Ethical Supply Chain Management 85
6.7 Modern Slavery 86
6.8 Data Protection 91
6.9 Summary 93

7 Ethics around Construction Health and Safety 95
7.1 Introduction 95
7.2 Governance and Management of Health and Safety in the UK 96
7.3 The Institution of Occupational Safety (IOSH) 99
7.4 Ethical Dilemmas around Construction Health and Safety 100
7.5 The Construction Design and Management Regulations (CDM) 2015 101
7.6 Procedures and Documentation around Construction Health and Safety 105
7.7 Summary 106
Reference 107

8 Ethical Considerations around Trust and Collaborative Working 108
8.1 Introduction 108
8.2 Professional Ethics and Duties of Trust 109
8.3 The Fragility and Robustness of Trust in a Constantly Changing and Dynamic Environment 110
8.4 The Importance of Trust for Collaboration and Professional Ethics 112
8.5 The Interdependency between Trust, Collaborative Working and Professional Ethics 117
8.6 The UK Construction Industry and Recent Reforms to Encourage Trust and Collaboration 120
8.7 Traditional Procurement of Construction Projects 121
8.8 Overall Context of Collaborative Working and Partnering within the Construction Industry 123
8.9 Trust as a Collaborative Necessity 125
8.10 Different Context for Clients to Consider in Construction Procurement 127
8.11 Summary 131
References 132

9 The Importance of Ethical Leadership in Construction and Engineering Management 137
9.1 Introduction to Chapter 137
9.2 Key Components of Ethical Leadership for Construction Professionals 138

9.3	Ethical Leadership and Best Companies	*139*
9.4	Leadership Identity and Relevance for Construction Professionals	*142*
9.5	How to Think and Behave as a Responsible Leader	*143*
9.6	Leadership Challenges for Construction Professionals around Professional Ethics	*147*
9.7	Leadership Styles and Professional Ethics as Applied to Construction Professionals	*147*
9.8	The Importance of Communication for Construction Professionals in Ethical Leadership	*148*
9.9	Adaptability of Leadership Styles for Construction Professionals	*149*
9.10	The Relationship between Poor Leadership and Project Failure	*150*
9.11	Summary	*151*
	References	*153*
10	**Professional Ethics as a Means for Motivating Project Teams**	*155*
10.1	Introduction	*155*
10.2	The Relationship between Motivation, Professional Ethics and Trust in the Construction Industry?	*156*
10.3	The Bespoke and Unique Nature of the Construction Industry	*157*
10.4	Factors Which Affect Motivation Levels of Project Teams	*158*
10.5	Communication as a Motivational Factor	*159*
10.6	Training and Education Related to Motivation and Ethical Compliance	*161*
10.7	Cultural Factors Affecting Levels of Motivation for Construction-related Staff	*163*
10.8	The Use of Financial Incentives as a Motivational Management Tool	*165*
10.9	Summary and Conclusion	*167*
	References	*170*
11	**Case Study of How Ethics Can Be Influenced by an International Crisis**	*172*
11.1	Introduction	*172*
11.2	Professional Ethics in the Context of the Coronavirus (COVID-19) Pandemic	*173*
11.3	Positive Examples of Ethical Behaviours during the Pandemic	*174*
11.4	Examples of Unethical Behaviours during the Pandemic	*176*
11.5	Ethical Considerations around the UK Government Job Retention 'Furlough' Scheme	*178*
11.6	Contractual Implications Brought about by the Pandemic	*179*
11.7	Additional Measures to Ensure Health and Safety of Workforce during the Pandemic	*179*
11.8	Ethical Leadership Throughout the Pandemic	*182*
11.9	Reflections for the Future	*184*
11.10	Summary	*184*
	Reference	*186*

12 Independent Research Study *187*
12.1 Introduction *187*
12.2 Review of Existing Literature and Theory *188*
12.3 Research, Design and Methodology *199*
12.4 Conclusion and Recommendations *217*
References *219*

13 Implications in Practice for Organisational Ethics in the Built Environment *223*
13.1 Introduction *223*
13.2 The Importance of Reputation and Image for Organisations in the Built Environment *224*
13.3 The Relationship between Ethics and Quality for Organisations *225*
13.4 Initiatives and Measures to Improve Organisational Ethics *225*
13.5 Education, Training and Continued Professional Development (CPD) around Organisational Ethics *229*
13.6 Summary *230*
References *231*

14 Summary of Key Points, Reflections, Overview and Closing Remarks *232*
14.1 Introduction *232*
14.2 Summary of the Key Issues Raised throughout the Book *232*
14.3 Final Reflections, Overview and Closing Remarks *250*
References *251*

Appendix A Energy, Water and Carbon Management Plan *253*
Appendix B Environmental and Sustainability Strategy Document *269*
Appendix C Considerate Constructor Scheme Site Registration Monitors Checklist *283*
Appendix D COVID-19 Project Review of Site Operating Procedures *289*
Appendix E Supply Chain Communication and Coronavirus Business Continuity Planning *298*
Appendix F COVID-19 Management System *299*

Index *301*

Author Biography

Dr Jason Challender MSc FRICS FAPM FAHE

Dr Jason Challender has acquired thirty-three years 'client side' experience in the UK construction industry and procured numerous successful major construction programmes during this time. He is Director of Estates and Facilities at the University of Salford, member of its Senior Leadership Team and responsible for overseeing a large department of approximately 350 estates and construction-related staff. He is also a construction researcher with five books and eleven published academic journal and conference papers in recent years, all of which have been dedicated to his studies around construction and engineering management. Furthermore, he has previously participated as a book reviewer for Wiley. He has also attended many national and international construction and institutional conferences as a guest speaker over the years and is a Fellow and Board Director of the Royal Institution of Chartered Surveyors and a Board member of the Northwest Construction Hub.

Foreword
Peter Farrell

Professional ethics is one of the most challenging areas to be addressed in UK and international construction, and indeed in society as a whole. A determination in our professional and personal lives to 'do the right thing' by other people, is a value we should all hold, even if outcomes are in some way detrimental or costly to ourselves. Often 'costs' are measured in monetary terms; so be it, but perceived short-term gains can often be lost in other ways looking forward. The people we deal with can often see through inappropriate activities and will no doubt measure these as they look to award further work. They will also see acts of sincerity, honesty, and trust, and often use that as a basis to award work on a repeat basis, giving it a higher priority than lower cost. After all, none of us can work with people whom we do not trust; we enjoy working with people we do trust and derive great job satisfaction from that. If you are trustworthy and uphold a moral compass, people will reciprocate, and no doubt enjoy working with you. If we lose the trust of a client or individual work colleague, it is unlikely we will ever repair that damage. I remember a very important person historically oft said to his all-male workforce *'whatever you do in life son* (sic), *be honest'*. Whilst this rolls of the tongue easily and usefully, honesty is of course just one element of ethics.

There is a need to continue to professionalise the construction workforce, by bringing as many people as possible into membership of professional registration bodies. If practitioners are not in membership, they may not have awareness of the increasingly tough codes of conduct that we need to comply with; also, they cannot be so strongly held accountable if they do stray from what is right. Whilst the law can intervene if people commit fraudulent or criminal acts, only professional bodies will intervene if actions are taken which are judged to be immoral. Clients are increasingly insisting that they only want to deal with practitioners who act ethically and can be held to account if they do not; professional body membership is increasingly imperative. If readers are not in a professional body, I hope they are encouraged to become so.

This text by Jason is a welcome and valued addition to the literature. Jason has been able to write it from the unique perspective of being in a very senior client-side position responsible for awarding work to construction supply chains, member at fellow level and proactive involvement in the Royal Institution of Chartered Surveyors, and from his doctorate award and subsequent visiting professorship at the University of Bolton. It is important that we keep this subject area extremely prominent in our minds and in our actions. Jason will not mind that I describe him primarily as a practitioner, and it is books from practitioners that we all want to read, since they are often of most interest and easy to apply in our lives. The text admirably addresses the complexity of the subject area in a thoughtful way, using case studies and the insights of colleagues to give real-life examples and scenarios. I know Jason has written this book with sincerity; he gets no great reward other than the satisfaction that he is doing the 'right thing', and in his way hopefully mentoring some of his readers to challenge themselves, and to act in ways they would otherwise not have done, if they had not read of his experiences. It is a long journey ahead, so that we can hold our heads high in construction to say we 'do do' the right thing. Thank you, Jason, for taking us some distance on that journey.

Professor Peter Farrell MSc CEng, FRICS, FCIOB, FCABE, SFHEA
University of Bolton, UK

Acknowledgements

The author would like to thank his family, Margaret, Kristin and Bobby for their encouragement during the book.

He would also like to acknowledge Professor Peter Farrell, at the University of Bolton, whose long-term mentoring has enabled the author to achieve all his career and academic successes. This book would not have been possible without his encouragement and proactive continued support.

Finally, a specific thanks goes out to Simbarashe Robertson Kanotunga for his contribution to the independent research study in Chapter 12.

List of Figures

3.1	Environmental measures which should be incorporated into the Environmental Management System	38
3.2	Beneficial outcomes from an environmental sustainability perspective	40
4.1	Examples of the type of social value initiatives provided by contractors	49
4.2	The Corporate Social Responsibility Model (adapted from Fewings 2009)	52
4.3	Code of Considerate Practice	53
4.4	Code of considerate practice checklist questions	55
4.5	Main area of environmental and sustainability strategy; University of Salford	61
6.1	Open communications routes for inclusion	78
6.2	Typical list of documentation, policies and procedures required to be created and maintained as part of compliance with Modern Slavery legislation	89
6.3	Correct action plan	89
6.4	The criteria for whether commercial organisations are required to publish an annual modern slavery statement	90
8.1	Diagram illustrating traditional procurement structure	122
8.2	Traditional procurement route under RIBA plan of work	122
8.3	Illustrating design and build procurement structure	124
8.4	Design and build procurement route under RIBA plan of work	124
8.5	Benefits for clients in early integration of contractors	128
9.1	Benefits of staff development and training for improving ethical leadership	139
11.1	List of documents prepared and used by a national main contractor and adopted across the all-UK regions to manage the pandemic and to provide advice and support	179
12.1	Data from frequency counts from the different categories of Industry Appropriatenes	208
12.2	Chart illustrating the percentage of Industry Appropriateness calculated from the tables	209
12.3	Pie Chart illustrating the percentage of Industry Appropriateness calculated from the tables above	209

List of Tables

3.1	Potential recommendations to address the sustainability agenda	32
4.1	Categories for action and possible measures to address as part of Considerate Constructors Scheme	55
4.2	The four key stages towards carbon neutrality	59
7.1	Code of Conduct IOSH	99
7.2	Summary of duties under the Construction Design and Management Regulations (CDM) 2015	102
8.1	Examples of potential breaches of trust and their effects on contracting relationships	112
8.2	The importance, reliance and influence of trust in partnering strategies	118
12.1	Question 1: Trustworthiness	199
12.2	Question 2: Leadership Balance	200
12.3	Question 3: Accountability/Health and Safety	201
12.4	Question 4: Collusion	202
12.5	Question 5: Worker's Rights	202
12.6	Question 6: Conflict of Interest	203
12.7	Question 7: Environmental Issues	203
12.8	Question 8: Environmental Issues/Insubordination	204
12.9	Question 9: Sustainability	204
12.10	Question 10: Quality	205
12.11	Question 11: Nepotism	205
12.12	Question 12: Bribery	206
12.13	Analytical table of findings	206

List of Appendices

Appendix A Energy Water and Carbon Management Plan
Appendix B Environmental and Sustainability Strategy Document
Appendix C Considerate Constructor Scheme Site Registration Monitors Checklist
Appendix D COVID-19 Project Review of Site Operating Procedures
Appendix E Coronavirus Supply Chain Communication
Appendix F COVID-19 Management System

1

Introduction

There is no more powerful institution in society than business. The business of business should not be about money, it should be about responsibility. It should be about public good, not private greed. Stakeholders want companies to make a profit, but not at the expense of their staff and the wider community.

Anders Dahlvig, CEO of IKEA

1.1 Introduction to the Book

With reference to the above quotation there have in recent years been a stronger focus on ethics and moral responsibilities aligned to businesses. Accordingly, the subject matter of what this book refers to as '*organisational ethics*' has become a hot topic and one that businesses need to pay special attention to if they want to succeed. For this reason, the book will explore many different aspects and themes around organisational ethics, specifically within the context of the built environment. It will highlight some of the issues and challenges that are currently being experienced in the sector on a national and international level. It is hoped that this will stress the importance, relevance and significance of ethics in the workplace. The book will explore and articulate many different innovative initiatives and measures to addressing some of these challenges. It is hoped that these can be introduced within organisations in the construction and engineering sectors particularly and represent a model of good practice for the future of the built environment. The following will hopefully give some context and background to the discussions around organisational ethics, alongside the justification and aims and objectives for the book.

1.2 Overarching Rationale, Synopsis and Justification: The Aim and Objectives for Writing the Book

Organisational ethics in the built environment is concerned with how we make improvements and positively contribute to the construction industry in terms of providing a safe and sustainable environment to work. Furthermore, organisational ethics should be

---

Organisational Ethics in the Built Environment, First Edition. Jason Challender.
© 2023 John Wiley & Sons Ltd. Published 2023 by John Wiley & Sons Ltd.

focused on human relations adversities, to create a thriving and aspiring, motivational and learning workplace. For this reason, the book covers motivational management and leadership in the context of the built environment alongside areas such as social value and corporate social responsibility. The focus and relevance in these important areas is related to all those construction and engineering companies involved with the procurement or life cycle of buildings from planning through design, construction, operations and facilities management. Accordingly, it includes many different organisations including clients, consultants, contractors, suppliers and manufacturers.

The aim of the book is to provide the foundations for advance thinking and shape future practices in the rapidly emerging field of organisational ethics and articulate how there can be a vast array of methods, motives and rationales for achieving more successful outcomes on both an organisational and industry-wide basis. In this pursuit there are clearly many unexplored challenges and dilemmas to grapple with and it is recognised that we are at the start of a largely unexplored journey. Such challenges include leadership, corporate social responsibility, motivational management and collaborative working. The improvements in practice in these key areas will be draw from case studies across many other knowledge fields. These include other industries such as manufacturing and retail and involve the cross pollination of ideas from a broad range of collaborating stakeholders. In this regard, previous examples of what works best in practice will be given to pave the way forward for the advancement of applied knowledge as a new and clear agenda for change in the built environment. The cultural changes in the construction industry especially in the UK over the last thirty years, including significant improvement in health and safety can lead us to be optimistic that further reforms can be made possible around organisational ethics. This may be assisted by changes through legislative, leadership and training and education initiatives to achieve the same shifts for organisational ethics in those areas previously referred including sustainability, human resources, motivational management and collaborative working.

1.3 The Need to Rethink the Concept of Organisational Ethics

Organisational ethics, rather like corporate social responsibility, are concepts that are not new. Their roots have gone back as far as the slave trade over 300 years ago when anti-slavery groups lobbied communities and governing organisations to ethically source supplies from countries that had not enforced slave labour. In recent years, however, the world has been witnessed to many different corporate financial and sometimes environmental failures and scandals. These have included major global institutions such as BP, WorldCom and Enron. Such events which have been followed by public scorn and reputational damage for those organisations involved have heightened attention to the urgent need for corporate social responsibility and corporate governance. More recently there is the ongoing war and atrocities in Ukraine which has highlighted good and bad responses from global organisations who have historically had business dealings with Russia. Accordingly, the need to rethink organisational ethics has never been more profound and the pressure exerted from the general public and stakeholders to behave ethically is something that can no longer be ignored (Koh and Boo 2004).

Although over many years business organisations have focused mainly on strategies linked to making increased profits and creating value for shareholders (Stormer 2003). However, in recent years companies are realising that they need to focus on non-financial strategies alongside their economic goals. This involves behaving ethically which in turn can raise their reputations and in turn increase their success in the marketplace. This is supported by Orlitzky et al. (2003) and Joyner and Payne (2002) who found from their research that a positive correlation exists between corporate financial performance and environmental/social performance. Furthermore, Koh and Boo (2004) found that organisational commitment to ethical initiatives can have a positive relationship with job satisfaction and career success of employees. Companies that embraced ethical policies around social, environmental and human resources found that this raised the motivation levels of employees. In these instances, they found that staff loyalty and willingness to suggest organisational improvements were prevalent where motivation levels are high. In addition, such policies were found to reduce absenteeism and staff turnover which are very costly for organisations. The evidence would therefore suggest that ethical initiatives underpinned by moral considerations for staff and the communities they work in, could have far reaching positive effects on organisational success. It is for this reason that the book will articulate the various ethical measures that organisations in the built environment should be progressing to achieve more successful outcomes. Such measures will revolve around leadership, human resources management, environmental sustainability strategies, ethical procurement, motivation of employees and pursuing corporate social responsibility policies.

1.4 The Fragmentation and Bespoke Nature of the UK Construction Industry

In the UK nearly half of all work completed is undertaken by approximately 200,000 small contractors. Some of the workforce in this regard, especially those who undertake smaller jobs, are often individuals with limited and potentially insufficient experience. Because of such inexperience there may be cases where their knowledge and awareness of issues relating to regulations, compliance, best practice and ethical considerations is lacking. Furthermore, in the context of the built environment, which is predominantly associated with fragmented, complex and potentially confrontational practices, this can create a dilemma for the sector. It is, therefore, perhaps not surprising that there have been many reported cases in the press of examples of malpractice and contractors' breaches of regulatory standards and codes of practice. Some of these cases have resulted in litigation proceedings brought by their clients, which has resulted in reputational damage for the industry. In addition, these practices have frequently led to less than acceptable project outcomes in terms of value for money, delays and poor-quality build standards.

The book will respond to the increasing demand for practical, and industry aligned, ethical practice in the built environment, covering construction and engineering management. More specifically the book will address how existing ethical standards can be pragmatically applied to professional practice from an organisational perspective and provide case studies and example scenarios aligned to modern-day requirements. Furthermore, the book will provide coverage of those real-world situations where the minimum legal and contractual requirements continue to require practitioners to demonstrate professional judgement and ethical decision-making. It will outline how ethically problematic situations can arise from internationally geographical differences in expected behaviours, conflicts between legal and moral dictum, or the mismatch between business goals of client and contracting organisations and society's perceptions of them as social facilitators. The book will then address how decisions can and should be made to uphold ethical standards and values. To enable this, it will bring together ethical theory, existing worldwide ethical standards and the requirements of many professional bodies including the Royal Institution of Chartered Surveyors (RICS), the Royal Institute of British Architects and the Chartered Institute of Building (CIOB). Practical advice will also be provided on how the largely theoretical ethical principles and guidelines adopted can be applied to all professional practitioner disciplines at both a national and international level.

1.5 Examples of How This Area Is Used or Represented in Our Society Today

The book will explore the extent to which organisational ethics are a viable tool in delivering improved outcomes in the built environment and will provide an important insight into the influence they have on the success of construction and engineering projects and redevelopment programmes. The book is intended to be a natural extension to *Professional Ethics in Construction and Engineering* (2022) by Dr Jason Challender. Whilst this book was largely focused on ethical dilemmas and types of unethical and illegal practices and behaviours and institutional codes of conduct to govern such adversities, the new book in will deal with organisational ethics to enhance working practices from many different perspectives.

1.6 The Lack of Trust in Construction Contracting and Reputational Damage for the Industry and Those Organisations Associated with It

The lack of trust in the construction industry has been highly documented by authoritative sources over many years and the underlying reason for this would predominantly be around breaches of professional ethics. The book will seek to address this ongoing dilemma and act as a catalyst for improvements to encourage more organisations and individuals to embrace and embed professional ethical standards in everything they do. This is a deliberate attempt to improve construction practices in the UK, which have arguably not been delivering the impact, and benefits that were intended in terms of successful project outcomes. Various government reports have reinforced this dilemma over the years. These have included *Construction 2015* (HM Government 2013) which identifies those fractious and sometimes unethical practices and behaviours are embedded in the UK construction industry. The report stresses the golden thread between professional ethics, collaboration and trust across the entire supply chain which is crucial to deliver successful projects. Furthermore, findings from the *Low Carbon Construction Final Report* (HM Government 2010) confirmed the growing need for increased awareness of environmental ethics especially between clients, consultants and the supply chain, in order to make greater contributions to the pursuit of sustainable solutions and efficiencies.

In addition to the aforementioned government reports, historically the credibility of the construction sector has come under increased scrutiny with many questions being raised about the morality of the industry. One example relates to the unethical tender practices in the UK associated with blacklisting of the supply chain, bid rigging and collusion. Over one hundred national construction companies came under scrutiny in this investigation which resulted in significant fines and reputational damage for these organisations and the industry as a whole. In addition, there have been UK companies that have been found guilty of unethical and illegal conduct on a global scale. One example relates to the case of construction and professional services company Sweett Group PLC in February 2016, which was ordered to pay £2.25 m after a bribery conviction. This case was followed shortly afterwards by the demise of Carillion in 2018 with revelations of unethical practices and shortcomings. These cases have presented a poor image for the construction industry, and many have called for a radical recalibration of the industry to address such adversities. One could argue that the high-profile negative practices exhibited by some construction companies is arguably resultant of the unethical behaviour and management practices of their employees and consultants.

The book will be designed to assist practitioners in their awareness, understanding and breadth of knowledge of the issues around professional ethics to build trust between contracting parties on projects, with the overarching aim of delivering construction and engineering projects that are more successful. To achieve this aim, the book is designed to provide a practical guide for organisations in ethical compliance that can develop into a common practitioner framework. Therein, it will seek to address the significant institutional risk that lies in the lack of a clear and consistent approach to professional ethics within the construction industry.

1.7 The Book's Value Proposition and Its Unique and Outstanding Features

In previous studies into professional ethics, very little attention has been focused on the process of embedding ethics into decision-making and everyday practices. Furthermore, scant attention has been paid to the role and development of organisational ethics in the context of construction and engineering management. This book will seek to infill the literature gaps through examination of those factors already referred to which could greatly improve current practices and behaviours and achieve more successful project outcomes.

In addition, there have been few books which have been written on the specific subject of encouraging and incentivising appropriate organisational ethics in the built environment specifically through a practical guide. Other books which have been published have largely focused on theoretical studies examining different client behaviours and relational analysis of clients with construction teams. Although the component elements of professional ethics have been covered previously and are therein well-trodden ground, there has been very little to articulate how these can be incorporated into organisational strategies in the built environment. As previously referred to the only other book of this kind is J Challender's (2022) *Professional Ethics in Construction and Engineering*, which the new book in this proposal intends to lead on from, drawing on case studies from the author's experiences and interviews. It will, however, take a slightly different course and approach from its predecessor by asking some very fundamental questions:

- How can improvements to organisational ethics positively promote good ethical standards, behaviours and practices and influence the success of projects in the built environment?
- What is the importance and role of leadership, motivational management, human resource management, corporate responsibility and social value in influencing strict compliance and adherence to ethical principles, values and standards within organisations?
- How can trust, collaborative working and partnering arrangements encourage professional ethics to be embedded within construction and engineering organisations?
- Why is environmental ethics so relevant in society today and how can corporate policies and strategies linked to them influence the success of projects in the built environment?

In consideration of the above questions the book's objectives are:

- To use case studies to look at how successes, failures and key risks can be influenced by organisational ethics.
- To identify how adopting policies and delivering strategies linked to environmental ethics can assist organisations in the built environment achieve more successful outcomes.
- To articulate the importance and role of leadership, motivational management, human resource management, corporate responsibility and social value for organisational ethics in the built environment.
- To identify a set of clear guidelines, national or international, to support the adherence and regulation around organisational ethics.
- To form the basis of a practical toolkit for guidance and teaching of organisational ethics in the built environment.
- To be the standard reference for businesspeople in understanding how ethics affects organisational risk.
- To educate readers on how trust, collaborative working and partnering arrangements encourage professional ethics to be embedded within construction and engineering organisations.

The book is intended to assist academics, construction related practitioners and clients in their awareness, breadth of knowledge and comprehension of the issues around ethical considerations, with the overarching aim of delivering more successful project outcomes. This is felt to be particularly important as in previous studies into ethics in the built environment, very little attention has been focused on giving practical advice. The book has sought to infill the literature gaps through examination of traditional roles of clients, design consultants, main contractors, and subcontractors and through providing guidance on potential improvement measures. Case studies and practical examples have been included

1.7 The Book's Value Proposition and Its Unique and Outstanding Features

to assist the reader on how theoretical perspectives can be applied to real-life construction projects and scenarios which involve ethical dilemmas. There have been academic calls for greater insight into how leadership around ethics can be created, mobilised, and developed and more understanding of the resultant positive effects and impact that can be generated therein (Walker 2009). The book responds to this gap in knowledge and Chapter 9 is specifically dedicated to this area. This chapter will articulate the links between academic ethical leadership theories and professional practice as a vehicle for improving ethical behaviours and leadership skills of construction professionals. Relationship between poor leadership and project failure scrutinised to encourage best practice is explored from a lessons learnt perspective to avoid further repeated failures in the future.

There will be frequent reference to construction practitioners' views and opinions throughout the book and these have been sought through qualitative research carried out in 2020 from a small sample of semi-structured interviews. In this regard one chapter of the book (Chapter 11) has been dedicated to ethics around the global COVID-19 pandemic which emerged from China in 2019 and the impacts, behaviours and practices, good and bad, that have been experienced by many organisations and individuals through the crisis. Examples have emerged of companies using unethical and unfair practices to seek competitive advantage over other businesses during the epidemic in the name of necessity. Participants from these interviews included clients, design consultants, main contractors and subcontractors.

There have been few books which have been written on the specific subject of incentivising organisational ethics in the workplace and the benefits this can have on motivation levels of staff. Most other previous publications have largely focused on theoretical studies examining different construction practitioner behaviours and relational analysis of clients with construction teams. Chapter 10 bridges this gap in knowledge. It identifies from previous research studies which factors can affect motivation levels of project teams and the reciprocal influence that these can have for professional ethics.

Understanding the risks posed for breaches of ethical standards in the built environment is a growing area as the tolerance for project failure reduces. Contemporary books to this seek to externalise professional ethics and codes of conduct by rooting them with professional institutions that create and regulate them. This book takes a different perspective and is unique in considering organisational ethics as an integral and valuable component of the construction and engineering industries. In doing so it creates a practical baseline for awareness of ethical considerations and therein smoothing the transition between traditional ethics teaching and embedding standard, values and principles within the built environment. Accordingly, it is a starting point for standard ethical practices to be developed and integrated within organisations to support and de-risk their business activities.

The value proposition of this book is that it will hopefully be read, understood and accepted by businesspeople as their main guidance and reference tool for reducing construction-based risk associated with unethical conduct. Accordingly, this is a book written for businesspeople by businesspeople based on sound theory (how to do it) and sound practice (lessons derived from case studies). It is considered unique in that it represents a comprehensive and wide-ranging analysis of best ethical practice in the built environment as well as other industries and sectors and what can be learned from them. The book will take established and widely accepted business ethical management practices and models and align them with development/construction issues. From this perspective, it does not seek to adapt established project management systems and processes but simply use ethical best practice models to co-exist alongside them. The book also investigates the area of environmental ethics and corporate social responsibility. These have become more prominent and important over recent years with the advent of cases such as the Deepwater Horizon oil spill which had major financial implications and reputational damage for British Petroleum PLC.

Although the research was undertaken in the UK, and all findings are likely to therefore have best fit with the UK construction industry, the overall knowledge and understanding to be provided by this book will have international relevance. Other countries seeking to develop ethical guides using similar approaches to the UK will be able to utilise the book, with consideration of how the findings fit with their own understanding in practice.

Finally, it is worth acknowledging that the author has gained over 33 years' experience of construction management from both a practitioner and academic perspective. From this, the book has drawn on both academia and practice, and it seeks from both these perspectives to prove an important insight into an area which has long been problematic for the construction industry.

1.8 Audience for the Book

The book is mainly intended for construction and engineering management practitioners but could suit a wide target audience including under- and postgraduate students and academics. The research findings will be presented to prioritise use by professional practitioners and therein provide a practical guide for adherence to professional ethics, as a toolkit, for improvements in project management in the built environment.

The book will not be designed as a holistic course textbook although it could be worthy of inclusion in a recommended reading list for courses related to construction procurement. As such, it is not intended solely as a practitioner guide. Rather, the book aims to cross this divide and provide useful insight to both academics and practitioners in developing their understanding of the topic area. Although the research was undertaken in the UK, and all findings are likely to therefore have best fit with the UK construction industry, the overall knowledge and understanding to be provided by this book will have international relevance. Other countries seeking to develop strategies for improved ethical practices using similar approaches to the UK will be able to utilise the book, with consideration of how the findings fit with their own understanding in practice.

1.9 Structure of the Book

The book has been structured into 14 chapters and covers the subject of organisational ethics from many different angles and perspectives but specifically in the context of the built environment. A brief outline of the themes of each chapter is detailed below:

Chapter 1: Introduction Introduction, justification, aims and objectives, readership and brief description of the areas covered throughout the book.

Chapter 2: An introduction to organisational ethics Introduces the concept of organisational ethics and what it means for both companies and employees in the context of the built environment. Articulates how organisations can sometimes suffer reputational damage by not pursuing policies and strategies linked to ethical initiatives. The importance and relevance of organisational ethics in the construction and engineering sectors is covered and the possible effects of workplace ethics on employees. The relationships between ethics, productivity and culture within organisations will also be explored.

Chapter 3: Environmental ethics Discussions around environmental ethics and encouraging sustainability in the design and construction processes and estates strategies. Outline of what we mean by environmental ethics and definition of what sustainability is within the context of the construction and engineering sectors. Articulation of the importance of environmental sustainability policies for construction organisations and the ethical case for participation.

Chapter 4: Corporate social responsibility (CSR) What corporate social responsibility means in practice and how CSR philosophies can be embedded into organisational cultures, policies and procedures. Benefits that adopting CSR practices and conversely reputational and financial damage from breaches of corporate social responsibilities. Introduction to the different types and models of corporate social responsibility from a legal, economic, philanthropic, and volunteering perspective, and the interwoven relationship between CSR and economic, social, and environmental considerations. Examination of the Considerate Constructors Scheme from a CSR perspective.

Chapter 5: Relationship between ethics and other factors Introduces the relationship between ethics and other factors including risk, quality, reputation, trust, leadership, and motivation. Hypothetical and real-life practical examples included where risks were taken due to the unlikelihood that they would occur and the catastrophic consequences that ensued. Relationship between ethics and quality examined and various examples and scenarios will be presented wherein unethical behaviour, including dereliction of duty and 'cutting corners' of one or more parties. Resulting catastrophic failures of buildings and structures and loss of life from deficiencies in quality of construction is also explored.

Chapter 6: Ethical human resource considerations Ethical and legal aspects around employment law including inclusivity, equality, and diversity in the workplace. Discrimination and the consequences for organisations in breaching regulations around human resource management. The ways and means that organisations can improve the culture of their workplaces. Encapsulates the requirements of the Modern Slavery Act 2015 and the General Data Protection Regulations (GDPR).

Chapter 7: Ethics around construction health and safety The legal and regulatory requirements in the UK, designed to protect the health and safety of employees and members of the public. Articulation of the measures and initiatives that the construction industry has adopted over recent years to reduce accidents and fatalities on site, including near miss reporting and improved supply chain management. The role of the Health and Safety Executive in policing and managing health and safety regulations.

Chapter 8: Ethical considerations around trust and collaborative working Focus on trust and collaboration as a way and means of encouraging improved ethical practices and behaviours and improving business relationships. Discussion around duties of trust in the context of professional ethics and how trust can be gained and lost through moral and responsible actions and behaviours of individuals.

Chapter 9: The importance of ethical leadership in construction management Articulates the links between academic ethical leadership theories and professional practice as a vehicle for improving ethical behaviours and leadership skills of construction professionals. Relationship between poor leadership and project failure scrutinised to encourage best practice is explored from a lessons learnt perspective to avoid further repeated failures in the future.

Chapter 10: Professional ethics as a means for motivating project teams Considers the reciprocal relationship of how fostering a culture of professional ethics can lead to raising motivation levels within organisations. Examination of how communication as a motivational factor can maintain staff awareness of changing practices, codes of practice, standards and values from an ethical perspective.

Chapter 11: Professional ethics in the context of the Coronavirus (COVID-19) pandemic Real-life examples from interviews carried out by the author, to some of the behaviours, issues and dilemmas that emerged during the Coronavirus (COVID-19) pandemic. Reporting of both positive and negative ethical practices and actions that organisations have adopted during the crisis.

Chapter 12: Independent research study: Exploring ethics in the construction industry and understanding its impact on the built environment Independent research study to explore and evaluate the phenomenon of construction ethics within the built environment and to support the other findings and discussions in previous chapters of the book.

Chapter 13: Implications in practice for ethics in the construction industry Implications for the construction industry from unethical practices, including reputational damage and image considerations for organisations. Future improvement measures those organisations can adopt in their policies and strategies to address some of the current failings in the built environment.

Chapter 14: Summary of key points, reflections, overview and closing remarks Final chapter to summarise the book with reference to each chapter, extrapolating the key findings and issues raised. Presentation of the author's reflections and recommendations for the future of the construction and engineering industries, taking account of the issues raised in the book around organisational ethics.

1.10 Summary

There are many different aspects and themes around organisational ethics, specifically within the context of the built environment. Organisational ethics has become an increasingly important area and 'hot topic' over recent years, predominantly to address some of the challenges in the sector on a national and international level. Such challenges have revolved around leadership, corporate social responsibility, motivational management and collaborative working. *Organisational ethics in the built environment* is concerned with how we make improvements and positively contribute to the construction and engineering industries and adopt different innovative initiatives and measures to addressing some of these challenges. For this reason, the book covers human resource considerations, health and safety, motivational management and leadership in the context of the built environment alongside areas such as social value and corporate social responsibility.

In recent years companies are realising that they need to focus on non-financial strategies alongside their economic goals. This involves behaving ethically which in turn can bolster their reputations and in turn increase their success in the marketplace. In addition, organisational commitment to ethical initiatives can have a positive relationship with job satisfaction and career success of employees and provide a safe and caring environment for them to prosper. For this reason, companies that have embraced ethical policies around social, environmental and human resources found that this raised the motivation levels of employees and increased productivity and retention of the workforce. Accordingly, the evidence would therefore suggest that ethical initiatives underpinned by moral considerations for staff and the communities they work in, could have far reaching positive effects on organisational success. Notwithstanding this assertion, such positive measures are not always easy to integrate into the built environment, which is predominantly associated with fragmented, complex and potentially confrontational practices. This can create a dilemma for the sector and is evidenced in reported cases of malpractice and breaches of regulatory standards and codes of practice. The book responds to this dilemma and will address how existing ethical standards can be pragmatically applied to professional practice from an organisational perspective and provide case studies and example scenarios aligned to modern-day requirements.

The lack of trust in the construction industry has been highly documented by authoritative sources over many years and the underlying reason for this would predominantly be around breaches of professional ethics. The book will act as a catalyst for improvements to encourage more organisations and individuals to embrace and embed professional ethical standards in everything they do.

The book's objectives include the use of case studies to look at how successes, failures and key risks can be influenced by organisational ethics. Other objectives are linked to identifying how adopting policies and delivering strategies linked to organisational ethics can assist businesses in the built environment achieve more successful outcomes. These objectives are intended to assist academics, construction-related practitioners and clients in their awareness, breadth of knowledge and comprehension of the issues around ethical considerations, with the overarching aim of delivering more successful project outcomes.

Although the research was undertaken in the UK, and all findings are likely to therefore have best fit with the UK construction industry, the overall knowledge and understanding to be provided by this book will have international relevance.

References

Challender, J. (2022). *Professional Ethics in the Construction and Engineering Industry*. Chichester: Blackwell-Wiley.

HM Government (2010). *Low Carbon Construction Final Report (November 2010)*. London: HM Government. 52–62, 196–199.

HM Government (2013). *Construction 2025. Industry Strategy: Government and Industry in Partnership*. London: HM Government. 23–25, 61–71.

Joyner, B.E. and Payne, D. (2002). Evolution and implementation: a study of values, business ethics and corporate social responsibility. *Journal of Business Ethics* 41: 297–311.

Koh, H.C. and Boo, E.H.Y. (2004). Organisational Ethics and employee satisfaction and commitment. *Management Decision* 42 (5): 677–693. Emerald Publishing.

Orlitzky, M., Schmidt, F.L., and Rynes, S.L. (2003). Corporate social and financial performance: a meta-analysis. *Organizational Studies* 24: 404–441.

Stormer, F. (2003). Making the shift: moving from ethics pay to an inter-systems model of business. *Journal of Business Ethics* 44: 279–289.

Walker, A. (2009). *Project Management in Construction*, 5e. Oxford: Blackwell Publishing Ltd, 150–158.

2

An Introduction to Organisational Ethics

Six essential qualities that are key to success include sincerity, personal integrity, humility, courtesy, wisdom and charity.

Dr. William Menninger

2.1 Introduction to Chapter

This chapter of the book introduces the concept of organisational ethics and what it means for both companies and employees. It seeks to articulate the importance and relevance of organisational ethics especially within the realms of the built environment. Furthermore, it captures some ethical initiatives and explains and analyses the measures by which these can have a beneficial impact for businesses. As context it will attempt to define what ethics are but, in this pursuit, will articulate the on-going dilemma of the different definitions of ethics. Such differentiation could pose a problem in the quest to understand and interpret what constitute ethics and ethical practices.

The chapter will also attempt to identify why both the construction and engineering industries have faced many years where they have suffered from a negative general perception of the public and in some cases where their reputation has been tarnished by reports of unethical conduct. It will then outline how such initiatives such as education, personal development and teaching could address these dilemmas and embed a minimum standard of ethical compliance in the workplace. Following on from this, the possible effects of workplace ethics on employees and organisational productivity will be discussed and explored. In this sense it will examine the effects of unethical behaviour on employees' productivity and commitment and identify those ethical and integrity behaviours that positively influence productivity.

Finally, it will analyse organisational culture linked to ethics and specifically the potential of organisational culture as a means of improving ethics within organisations. In this regard it will suggest the ways and means by which organisations, specifically in the built environment, can create the right working environments, to enable staff to feel comfortable and well supported in their employment.

Organisational Ethics in the Built Environment, First Edition. Jason Challender.
© 2023 John Wiley & Sons Ltd. Published 2023 by John Wiley & Sons Ltd.

2.2 What Are Ethics?

To address the issues of professional ethics, particularly applied to the construction industry, one needs to firstly understand what ethics are and what constitutes ethical or non-ethical practices. Professionals are bound by a set of attitudes, principles and character dispositions that govern the way their profession is practiced, and this is commonly referred to as 'professional ethics'. Such ethics are not confined to clients but according to obligations are also owed to colleagues and members of the public at large.

Theories of ethics come from a philosopher's perspective and can be categorised as meta-ethics, relating to where ethical values and principles emerge from, normative ethics, relating to moral standards of conduct and applied ethics, involving examining controversial issues (Internet Encyclopaedia of Philosophy 2010).

Ethics has been described in general usage as:

> ...the philosophy of human conduct with an emphasis on moral questions of right and wrong' (Helgadottir 2008), 'the system of moral values by which the rights and wrongs of behaviour are judged' (Rosenthal and Rosnow 1991) and 'a moral philosophy that involves systematizing, defending and recommending concepts of right and wrong behaviour.

Alternatively, ethics could also be defined as:

> ...the systematic attempt to make sense of individual, group, organizational, professional, social, market and global moral experience in such a way to determine the desirable, prioritized ends that are worth pursuing, the right rules and obligations that ought to govern human conduct, the virtuous intentions and character traits that deserve development in life and to act accordingly.
> (Petrick and Quinn 2008 as cited in Helgadottir 2008)

A general definition of ethics was provided by Carey and Doherty (1968, as cited in Poon 2003) around the philosophy of human conduct with an emphasis on moral questions of right and wrong. Notwithstanding this, professional ethics has been described specifically around the expectations around responsibility, competence and willingness offer quality of services to the general public. There is sometimes confusion on what constitutes business and professional ethics. Accordingly, it would be helpful to examine the differentiation between the two. To offer clarity in this regard, business ethics are primarily focused around conduct. They revolve around ethical questions related to whether certain actions could be deemed good or bad, right or wrong, virtuous or vicious, worthy of praise or blame, reward or punishment. It seems that defining professional ethics is not an easy and straightforward task. To reinforce this position, Uff (2003) stated that 'in any event, it would be difficult today to pin down ethics to a particular definition in the current climate of change in matters of professional accountability and transparency'. Notwithstanding this challenge, professional ethics can be defined as 'a set of moral principles or values' (Trevino and Nelson 2004), whereas others have described them as a system of norms to deal with both the morality and behaviour of professionals in their daily practice (Abdul-Rahman et al. 2007).

Alternatively, Fewings (2009) described professional ethics as the application of values to society and a range of different perspectives whereas Bayles (1989, cited in Abdul Rahman et al. 2010) advocated that professional ethics revolve around a systems of norms wherein day-to-day behaviours and morality can be managed and regulated. Notwithstanding these alternative descriptions, when considering professional ethics, it is important to understand what 'profession' means. Profession in this sense could be defined as 'the possession and autonomous control of a body of specialist knowledge, which when combined with honorific status, confers power upon its holders' (Uff 2003). This is quite an arduous definition and as such it is probably more advantageous to consider professionalism from the perspective of providing a 'service' to clients. In this regard, a profession can be linked to a group of individuals who can provide specialist knowledge to their clients. Professional ethics therefore could be defined as a system of standards and norms to enable the behaviour and morality of those practicing professional to deal with their duties and job responsibilities. However, according to Hamzah et al. (2012) professional ethics does not always apply solely to individuals but can be extended to organisations and institutions. Furthermore, one could argue that they are not solely for the benefit of clients but should be extended to the public. Moreover, they encompass qualities and duties associated with responsibility, competence abide by established standards, rules and behaviours.

When considering professional ethics, Lere (2003) described them as a way and approach between professionals and experts as experts and clients as lay people. Conversely when we examine professional ethics specifically within a construction context the RICS defined them as 'a set of moral principles extending beyond a formal code of conduct' (RICS 2010). In terms of how such moral principles should be applied in the workplace the code of ethics for project managers stresses that 'it is vital that Project Managers conduct their work in an ethical manner' (Walker 2009).

According to Rogers (1911) ethics revolves around the well-being of human beings. Well-being in this sense signifies the permanent realisation of goodness enacted by individuals and takes into account:

- The nature of individual good
- The nature of social good
- The relationship between individual and social good
- The freedom of the will
- The ethical worth of positive morality
- The relationship between good and pleasure
- The nature of virtue (in antique ethics). In this sense virtue is linked with a person's character and applied to their motives or actions. In this context it is generally accepted that a morally virtuous person is one who abides and respects the moral codes laid down by honour, mercy, industry, temperance, charity. Conversely, the opposite of virtue is associated with vice.
- Duty and moral obligation (in modern ethics)
- The ethical motives that exist for people to pursue social good or to whatever is morally right

Another challenge for describing and defining ethics comes from the perspective that they have two distinct levels, namely Micro and Macro, relating to different aspects of the organisational environment. According to Chang (2005) micro ethics relates to personal issues and relationships with individuals and deals with issues such as personal integrity, honesty, trust and transparency. Alternatively, macro ethics deals with the wider aspects of actions affecting and impacting on society, the environment and the reputations of organisations (Chang 2005). Macro ethics can be related to ethical issues that have a significant and prolific high-level impact on the reputation of the industry or the public at large and predicated on the notion of doing good things for the wiser society.

Clearly the aforementioned alternative descriptions and definitions highlight that there are many differences of what ethics are which has given rise to problems of ambiguity and meaning. This view was supported by Vitell and Festervard (1987) who advocated that complications arise, and ethical dilemmas exist as there is no universally accepted definition of ethics. Perhaps this is one of the problems when considering professional ethics in the construction industry on a global scale which could be giving rise to different interpretations of ethics for construction professionals and what they mean in practice. This is arguably of particular importance when considering the construction industry as the understanding and views of many different professional bodies, association and organisations and their members are not always the same. The RICS Professional Ethics Working Party (RICS 2000) accepted that this potential ambiguity and inconsistency could create problems for the profession. Perhaps therefore a common framework for managing ethical dilemmas and thereby improving ethical standards is required which could address this problem. This opinion is supported by Liu et al. (2004) who argue that professional ethics is based on the subjective nature of principles, standards and values which vary between different sectors of the industry and a more consistent approach is required accordingly.

2.3 The Importance and Relevance of Organisational Ethics in the Construction and Engineering Sectors

In order to discuss ethics within an organisational context it is very important to understand why they are both relevant and important. Although views have changed over time, there is a consensus that ethical principles and practices within organisations are largely built upon behaviours, trust and values, in the work environment. Challender J (2022) reported that ethics has become a 'hot topic' in many sectors over recent years and the built environment is by far no exception to this agenda. This is explained from the perspectives of past performance where the construction and engineering industries have suffered reputational damage from reported acts of unethical conduct (Robson 2000). This has largely emanated from negative press coverage, particularly in the UK, and served to damage the perceptions of the public towards the sector. Accordingly, ethical teaching in applying professional behaviours, practices, cultures and standards and to the engineering and construction industries is of paramount importance to reverse this trend. Education, personal development and teaching in these sectors will seek to turn the tide around addressing unethical practices and hopefully promote policies to embed a minimum standard of ethical compliance in all areas of work life.

Arguably, ethics especially when considered from an organisational perspective can create the bedrock on which all our working relationships are built. In addition, it is about how we relate to our surroundings, and not about the connection we have but the quality of that connection (Trevino and Nelson 2004). Another important aspect, which is at the heart of ethical and moral issues for construction professionals, is ethical dilemmas and decision-making. It is recommended that there should be a need for sensitivity to ethical dilemmas to avoid actions being taken without the awareness of potential ethical issues.

One of the benefits of ethics for organisations revolves around trust, especially where firms are heavily reliant on their reputation for conducting business and gaining new work (Trevino and Nelson 2004). The aspect of trust upon organisational ethics will be discussed further in Chapter 9.

When reviewing the importance and relevance of ethics within organisations we should consider the decision-making processes. Weiss (2003) outlined that business ethics deals with three basic areas of managerial decision-making. The first is around choices about the law and whether to follow, the second relates to choices about economics and social issues outside of law and the third is the priority of self over the company's interests. Despite these assertions, Cowton and Crisp (1998) argued that economies depend on the profit motive, and the pursuit of profit need not be perceived as immoral. To aid practitioners to what is moral or immoral Cowton and Crisp (1998) concurred that one needs to consider (i) respect for core human values, (ii) respect for local traditions and (iii) the belief that context matters to decide what is right and what is wrong.

2.4 Possible Effects of Workplace Ethics on Employees and Organisational Productivity

Adeyeye et al. (2015) examined the effects of unethical behaviour on employees' productivity and commitment and found a significant relationship between not upholding ethical standards and low productivity within organisations. They also concluded that employment relations in any working organisation is important for productivity, stability, profitability and well-being of all employees and stakeholders. Employment relations in this regard were found to be influenced by such factors as proper management of employees at work including human resources (HR) policies and procedures, adherence to ethical values, standards and principles, teamwork and collaborative working approaches. Whilst this study was conducted in a non-construction-related context in Nigeria, the findings are believed to be equally applicable and relevant to the built environment in the UK and globally.

In addition to the above, a case study by Isabirye (2020) investigated the constellation of expected behaviours of organisation justice in order to identify ethical and integrity behaviours that positively influence construction productivity. The research covered both private and public sector work, civil engineering and building projects in South Africa and comprised contractors, consulting professionals, academics, local government officials, government-built environment professionals and construction workers. The findings of the study concluded that there are many ethical behaviour factors at an organisational level which have improved construction productivity. These include:

- Communicating effectively
- Being honest
- Able and willing to accept responsibility and accountability
- Working reliably, according to the best practices
- Being objective and avoiding conflicts of interest
- Being transparent at work
- Being fair in all dealings
- Having integrity and creating trust
- Continuously training to improve skill
- Being equitable in all dealings
- Receiving only a fair reward

In terms of applying some of these intrinsic and extrinsic measures, these will be discussed in later chapters of the book around such aspects as human relationship management and leadership.

2.5 Organisational Culture Linked to Ethics

Organisational culture has over many years been recognised as a positive construct in organisational science and in this sense could be regarded as the way individuals behave and approach their roles and responsibilities. For instance, managers will sometimes look to culture as both the source and the potential basis for solutions for problems when things go wrong (Sinclair 1993). Perhaps for this reason it has become the important focus for the way organisations are run and perform against their aims and ambitions. Accordingly, the potential of organisational culture as a means of improving ethics within organisations should not be understated.

According to Liu et al. (2004) culture can create an emotive sense of commitment and involvement to organisational moral codes and ethical values and therein positively support the performance of employees. They developed an organisational ethics model by using a systems approach methodology to depict the transformation of ethical behaviour. This had individual, local and cosmopolitan levels of ethical influence that were employed to allow examination between them.

The notion and application of the concept of organisational culture within working environments is an area which has received widespread interest over recent years through research. Some models have shown that it is organisational culture which shapes the mindsets of individuals within companies. It may be for this reason that there has been a concerted effort to encourage organisations to focus more in this area as part of their human resources management strategies. In addition, there has been industry encouragement to pursue such strategies and accreditation schemes that have been introduced to support these initiatives. For instance, some of the UK's most successful organisations have gained recognition through the prestigious 'Best Companies' accreditation. This helps them to be regarded as organisations which care for their employees, whilst preserving their values in doing the right things. Accordingly, some would regard the 'Best Companies' accreditation as the benchmark of what a truly engaged workforce means in practice. Furthermore, the accreditation, gained through staff surveys, should demonstrated that it is their cultures which have been helped them achieve positive results. In this way they should be able to attract and retain the best talent whilst giving them a benchmark from which to compare year on year how they compare with other organisations. This can provide them with an unrivalled marketing and public relations opportunity. Perhaps the importance of culture within an organisational context can explain why those more responsible employers have sought to invest time and resources as part of their human relations policies. In this way they strive to create the right working environments, to ensure that staff feeling comfortable and well supported and help them to achieve successful workplace outcomes. This will be discussed in more detail in Chapter 6.

2.6 Summary

Within an organisational context it is very important to understand what ethics are and why they are relevant and important. To help us in this pursuit there is a consensus that ethical principles and practices within organisations are largely built upon behaviours, trust and values, in the work environment. Notwithstanding this premise, such qualities have been compromised in the past within the built environment. Examples of bad practices which have emerged and been widely publicised have brought with them reputational damage not just for those organisations involved but the built environment sector at large. To address this dilemma, education, personal development and teaching in the construction and engineering sectors has sought to turn the tide around addressing unethical practices and hopefully promote policies to embed a minimum standard of ethical compliance for the future.

Ethics especially when considered from an organisational perspective can create the bedrock on which all our working relationships are built. When reviewing the importance and relevance of ethics within organisations we should consider the decision-making processes and there are three basic areas of managerial decision-making. The first is around choices about the law and whether to follow, the second relates to choices about economics and social issues outside of law and the third is the priority of self over the company's interests.

In this chapter a significant relationship has been established between not upholding ethical standards, around such aspects as employment relations, and low productivity. This has underpinned why employment relations in any working organisation is important for productivity, stability, profitability and the well-being of all employees and stakeholders. Employment relations in this regard were found to be influenced by such factors as proper management of employees at work including human resources (HR) policies and procedures, adherence to ethical values, standards and principles, teamwork and collaborative working approaches. In addition, there are many other ethical behaviour factors at an organisational level which have improved construction productivity, and these include honesty, effective communication, reliability, fairness, transparency and training/education.

Organisational culture has over many years been recognised as a positive construct in organisational science and in this sense could be regarded as the way individuals behave and approach their roles and responsibilities. For this reason, it has become the important focus for the way organisations are run and perform against their aims and ambitions. Accordingly, there should be a concerted effort to encourage organisations to focus more on improvements to organisational culture as part of their human resources management strategies. There has been industry encouragement to pursue such policies and strategies and accreditation schemes have been introduced to support these initiatives. The UK's Best Companies is one example of such an initiative and is predicated on staff surveys to score their employers on how they measure up against predetermined success criteria and metrics. The importance of culture within an organisational context can possibly explain why those more responsible employers have sought to invest time and resources as part of their human relations policies.

References

Abdul-Rahman, H., Karim, S., Danuri, M. et al. (2007). *Does professional ethics affect construction quality?* Retrieved from www.sciencedirect.com (accessed 24th November 2019).

Abdul-Rahman, H., Wang, C., and Yap, X.W. (2010). How professional ethics impact construction quality: perception and evidence in a fast-developing economy. *Scientific Research and Essays* 5 (23): 3742–3749.

Adeyeye, J.O., Adeniji, A.A., Osinbanjo, A.O., and Oludayo, O.A. (2015). Effects of workplace ethics on employees and organisational productivity in Nigeria. *International Conference on African Development Issues (CU-ICADI)2015: Social and Economic Models for Development Track*.

Challender, J. (2022). *Professional Ethics in the Construction and Engineering Industry*. Chichester: Blackwell-Wiley.

Chang, C.M. (2005). *Engineering Management: Challenges for the New Millennium*. Upper Saddle River, NJ: Pearson Educational.

Cowton, C. and Crisp, R. (1998). *Business Ethics: Perspective on the Practice of Theory*. Oxford: Oxford University Press.

Fewings, P. (2009). *Ethics for the Built Environment*. London: Routledge.

Hamzah, A.R., Chen, W., and Wen Yap, X. (2012). How professional ethics impact construction quality: perception and evidence in a fast-developing economy. *Scientific Research and Essays* 5 (23): 3742–3749.

Helgadottir, H. (2008). The ethical dimension of project management. *International Journal of Project Management* 26: 743–748.

Internet Encyclopaedia of Philosophy (2010, March 24th). *Ethics and Self-deception*. Retrieved from http://www.iep.utm.edu (accessed 24th November 2019).

Isabirye, A.K. (2020). Organisational justice: a matrix for ethics and integrity for improving construction productivity in South Africa. *International Journal of Business and Management Studies* 12 (2).

Lere, J.C. (2003). The impact of codes of ethics on decision making some insights from information economics. *Journal of Business Ethics* 4: 365–379.

Liu, A.M.M., Fellow, R., and Ng, J. (2004). Surveyors' perspectives on ethics in organisational culture. *Engineering, Construction and Architectural Management* 11 (6): 438–449.

Poon, J. (2003). Professional ethics for surveyors and construction project performance: what we need to know. *The RICS Foundation in Association with University of Wolverhampton* 148 (9): 232.

RICS (2000). *Guidance Notes on Professional Ethics*. London: RICS Professional Ethics Working Party.

RICS (2010). *Maintaining Professional and Ethical Standards*. London: RICS.

Robson, C. (2000). Ethics; a design responsibility. *Civil Engineering* 70 (1): 66–67.

Rogers, R.A.P. (1911). *Short History of Ethics*. Millan Books.

Rosenthal, R. and Rosnow, R.L. (1991). *Essentials of Behavioral Research Methods and Data Anlaysis*, 2e. Boston: McGraw-Hill.

Sinclair, A. (1993). Approaches to organisational culture and ethics. *Journal of Business Ethics* 12: 63–73.

Trevino, L.K. and Nelson, K.A. (2004). *Managing Business Ethics*, 3e. USA: John Whiley and Sons.

Uff, J.P. (2003). Duties at the legal fringe: ethics in construction law. *Society of Construction Law*, October 2010. Retrieved from www.scl.org.uk (accessed 24th November 2019).

Vitell, C. and Festervand, D. (1987). Business ethics: conflicts, practices and beliefs of industrial executives. *Journal of Business Ethics* 6: 111–122.

Walker, A. (2009). *Project Management in Construction*. Oxford: Blackwell Publishing Ltd.

Weiss, J.W. (2003). *Business Ethics: A Stakeholder and Issues Management Approach*, 3e. Ohio: Thomson South-Western.

3

Environmental Ethics

Environmental ethical behaviour is doing the right thing when no one else is watching, even when doing the wrong thing is legal.

Aldo Leopald

More and more companies are reaching out to their suppliers and contractors to work jointly on issues of sustainability, environmental responsibility, ethics and compliance.

Simon Mainwaring

3.1 Introduction to Environmental Ethics

The above quotations highlight both the importance of environmental ethics, coupled with the current trend for organisations in the engineering and construction industries to embrace all aspects of environmental sustainability. To respond to this, the chapter will set the scene for discussions around environmental ethics and encouraging sustainability in the design and construction processes and estates strategies. It will outline the recent trends in climate change and government pledges and targets to reduce greenhouse gases and carbon generation and the associated challenges around meeting such commitments. Furthermore, it will outline what we mean by 'environmental ethics' and define what sustainability is within the context of the construction and engineering sectors. The chapter will then explain the importance of sustainability and the ethical case for participation, articulating the benefits for engaging with the environmental agenda. In addition, the potential disadvantages and risk for organisations from non-participation in environmental sustainability reforms and improvements will be covered including the irreprovable damage that can be enacted for the planet through climate change. It will then outline the global and national agenda on sustainability and articulate world targets for issues around reducing carbon generation. This will then lead to the discussion around the UK Governments influence in environmental ethics and the many initiatives including grants and regulations that in recent years have been introduced to curb waste and emission levels specifically.

Environmental social responsibility will be covered in this chapter and the mechanisms by which organisations can make a positive impact and example for others to follow. In this regard initiates and commitments to waste, energy, carbon management and transport will be discussed and analysed. This will then lead into the many dilemmas around environmental ethics and how sometimes the financial aspects of introducing sustainability initiatives can sometimes present barriers to implementation. Awareness of the benefits of environmental commitments will be explained and how these can lead to advantageous outcomes for organisations in terms of aspects such as public relations, corporate responsibility and reputation. Finally, the chapter will conclude by looking at two case studies in the education sector where commitments and delivery strategies related to environmental sustainability plans are being implemented, which are intended to present models of good practice for other organisations to follow.

3.2   Climate Change and Sustainability

The UK government has committed to cut greenhouse gas emissions by 80% by 2050 compared to 1990 levels, and to halve them by 2025. In 2020 buildings in the UK accounted for nearly half of all the total carbon emissions. Accordingly, if the government pledge and commitments are to be achieved, buildings must become significantly more efficient, not just in the construction process but in terms of energy usage during their life cycle. This challenge has become more difficult owing to the fact that 80% of the UK building stock that will exist by 2025 has already been built, along with approximately 70% of the housing that will be occupied in 2025. Not only does the UK's building stock need to become more efficient, it will need to become more resilient if the climate changes as it is predicted. Despite this, government policies which including scrapping the Green Deal, withdrawing plans for zero-carbon homes, and substantially reducing financial incentives for renewable energy would appear to undermine these commitments.

3.3   Context for the Discussion on Environmental Ethics

Construction professionals in their respective leadership roles, when procuring projects, should be aware of the growing importance of environmental ethics, as part of their organisations' social responsibilities. Increasing awareness of environmental ethics has in recent years occurred owing to the potential environmental problems including resource depletion and climate change. Environmental ethics reflects the impact of policies and actions on environmental matters and future individuals' well-being. There is an argument that the construction industry plays an important part in sustainability, conservation and environmental management and environmental ethics should accordingly be particularly relevant to the ethos of projects (Tagawa 2005, p. 12). Owing to the above factors, there is a strong argument that construction professionals should be increasing political influence in this area, and therein drive the agenda forward. This could take the form of instructing their project teams to adopt environmental and sustainable design measures into their projects as one example. This could lead to changes on practice and have reputational benefits for their organisations. Notwithstanding the above premise, there has been much scepticism in the past about environmental ethics and whether simply enough has been done to curb the rise in carbon emissions into our atmosphere leading to global warming and potentially compromising the lives and livelihoods of the world's population. Construction and buildings in use account for a huge proportion of the problem and some would argue that the industry has ignored some of the issues with carbon emissions for many years now. Strong arguments have emerged in the past that if construction professionals continually adopt policies which encourage use of non-sustainable construction strategies, then this could have negative connotations in their environmental credentials and lead to reputational damage. Such is the growing importance of environmental ethics; over recent years, this chapter in the book will focus on organisational sustainability strategies and with particular reference to two organisation case studies.

3.4 What Are Environmental Ethics and Sustainability?

In considering what constitutes environmental ethics one definition is that:

> *Environmental ethics concerns formulation and moral obligations regarding the environment.*
>
> (Tagawa 2005, p. 12; emphasis added)

According to the Chartered Management Institute (CMI) Codes of Ethics and Professional Conduct, the field of environmental ethics covers the ethical relationship that people have on their natural environment. The role of environmental ethics is to identify the moral obligations in terms of concerns of effects on the atmosphere, climate change, depletion of natural resources, loss of wilderness, dwindling animal and plant biodiversity. These have been in the past referred to as 'green issues' and are becoming more important for economies and organisations to recognise the effects on the environment from their actions. This has emanated in policies linked to social and environmental responsibilities. From a moral standing position, it is not always straight forward to identify what obligations we have for them. Some would argue that our obligations should be focused on ensuring that future generations are not disadvantaged and prevented from having their basic needs met.

Given the increasing concern for environmental issues and the impact that our actions will have on the environments that we live in, the field of environmental ethics will become more important in the future. Political efforts will seek to limit or in some cases reverse the environmental effects of certain delirious actions that have caused harm to the environment. There may be more interventions on a global stage, such as the Kyoto Protocol, that could attempt to deal with climate change. Such measures could be focused on large polluters to the environment with targeted reduction imposed on them in terms of carbon generation and greenhouse gases. This could present a challenge for developing economies who may find it difficult to meet such targets. Environmental ethics should be focused on not simply apportioning blame for what could be regarded as failures in this regard but suggest means for resolution of such problems. This raises the issue of whether countries should hold overall responsibilities in preserving the environment or, adopting a bottom up approach seek to hold organisations responsible for putting forward solutions. This is becoming more topical in recent years and the 'court of public opinion' on many global organisations in food and beverage for packaging linked to plastics has turned the tide in some respects in forcing change.

In understanding what sustainability is, there have been many different definitions over previous years. The United Nations Brundtland Commission, as cited in Appleton (2006) defined sustainable development as 'that which meets all the needs of the present without compromising the ability of future generations to meet their own means'. Principle 3 of the Rio Declaration on Environment and Development (Harvard Institute, 1992), however, embedded a revised definition of sustainable development as 'to equitably meet developmental and environmental needs of present and future generations'.

To assist in considering these definitions, sustainability could mean different things to many people when taking into account of the wide-ranging contexts and environments which construction management operates within. For instance, a project to utilise energy for renewable low carbon sources such as timber could only be regarded as truly sustainable if the environmental costs of transport to its intended destination do not outweigh the benefits. Whichever definition we accept we need to establish how it applies to an organisation. For this we need to comprehend the scope of sustainability, what it involves and what is encompassed within it. Ding (2006, p. 463) explained that

> *It involves the efficient allocation of resources, minimum energy consumption, lower embodied energy intensity in building materials, reuse and recycling and other mechanisms to achieve effective and efficient short- and long-term use of natural resources.* (emphasis added)

3.5 The Importance of Sustainability and the Ethical Case for Participation

One could argue that it is important for all responsible organisations to uphold a sense of corporate social responsibility. This in turn will have far reaching benefits for their businesses in terms of public perception, marketing, operational efficiency, adherence to ever increasing regulations and maximising future government funding opportunities. In addition, organisations must not lose sight of the moral and ethical case for intervention on sustainability. The IPCC (Intergovernmental Panel on Climate Change) predicted that unless emissions of greenhouse gases decrease there will be a temperature increase of between 1.4 C and 5.8 C by 2100 (IEA 2011). More recently the Carbon Plan (HM Government 2011) set out the government's plans for achieving the emission reductions and set down targets of such cuts of at least 80% of 1990 levels by 2050 and revealed that

> Climate change is one of the greatest threats to both UK and global security and prosperity. There is an overwhelming scientific consensus that climate change is happening, and that it is very likely to be primarily the result of human activity.

The Carbon Plan articulated that all organisations should lead the way and be drivers of the transition to reducing emissions and they must appreciate and capitalise on the opportunities and challenges that this presents. In this regard, the Plan set out that innovation and cultural changes within organisations will be important in achieving government targets.

3.6 The Global and National Agenda on Sustainability

It could be argued with moving global agendas including the international targets set at the Rio Summit (1992), the Kyoto Protocol (1997) and the Cancun Climate Summit (2010) will continue to have little direct influence on most organisations. Furthermore, the international agenda that has not been adopted at national or local level may have no impact whatsoever. The international agenda, in particular, is often criticised for not being proactive enough or concrete enough at an operational level. A converse argument is, however, that it is the indirect effect of these that is important as national agendas and regulations are heavily influenced by international politics. An example of this is the Climate Change Act 2008 that required Government to cut emissions by at least 34% by 2020 and 80% by 2050. The Carbon Plan (2011) is designed to ensure that action at a national level addresses barrier to energy efficiency and effectively supports and complements action at a European Union (EU) level. This demonstrates that there is government pressure to rebalance organisations' traditional key objectives for corporate growth, profits and increased share prices against environmental issues and the growth and welfare of our communities.

3.6.1 Government Influence in Environmental Ethics

Government has in the past contributed to encouraging environmental ethics into the design and construction of projects in the UK. The introduction by the Department of Trade and Industry (DTI) for Sustainable Progress Reports, which cover principles such as waste, energy efficiency, carbon generation, safety, skills, climate change and costs has been used in the past for measuring how sustainable projects are. Sometimes government and international funding for construction projects stipulates minimum environmental performance criteria for buildings, which could be a *very good* or even *excellent* rating. Accordingly, any compromise on the environmental performance of buildings under such stipulations could jeopardise funding grants that may be available. In the UK it is quite common for grant funding on public buildings being conditional on gaining a minimum BREEAM rating. BREEAM is the world's leading sustainability assessment method for master-planning projects, infrastructure and buildings. Though third-party certification it assesses an asset's environmental, social and economic sustainability performance, using standards developed by the British research establishment (BRE). This means BREEAM rated developments are more sustainable environments, although that enhance the well-being of the people who live and work in the, help protect natural resources and make for more attractive property investments. Notwithstanding this premise, some designers have argued in recent years that BREEAM has not always moved with the pace of environmental technological advancements. An example of this relates to measures to achieve complete electrification of buildings to avoid the use of gas, which in 2020 is not recognised by BREEAM in the rating process.

In addition to the design and construction of commercial buildings, the UK Government have exerted pressures on reporting on the environmental ratings of buildings for sale or rent with the introduction of the compulsory Eco Home ratings. These need to form part of the buyer packs for buildings and once again this represents an intervention by Government to encourage environmental improvements.

3.6.2 The Local Agenda on Sustainability

In looking at local sustainability agenda in the UK and in particular local authority policies, there is strong emphasis on the participation of organisations to encourage reduction of waste, increased recycling and the adoption of sustainable transport policies. Councils have however, previously had to rely on enforcement of such policies through the planning system (Town and Country Planning Act 1990) for major developments and had limited participation accordingly. One example of this is the Stockport Metropolitan Borough Council Sustainable Transport Supplementary Planning Document (SMBC 2011) which forms part the Council's Local Development Framework and encourages 'more sustainable and active transport options where practicable'. Participating organisations in the area, have responded with initiatives linked with reducing the reliance that employees have on vehicular means of commuting to work. Measures will need to be taken to approach this issue not least because of the finite number of car parking spaces currently available at their workplace. It is important and productive for organisations to work more closely with local authorities in partnership, through liaison and consultation, to promote a more robust sustainability strategy for their local area as a whole. This would improve the external relations and the public perception of businesses.

3.7 Environmental Social Responsibility

Environmental social responsibility is referred to as environmental conservation social responsibility, and curtails to the long-term problem of global warming, increased waste production and contamination on an international scale. With the global focus on addressing the 'climate emergency' brought to light by high profile campaigners such as Bill Gates, Richard Attenborough and more recently Greta Thunberg, this has become front page news over many years. It followed on from Al Gore's campaign and lobbying to global powers in his climate emergency depiction 'An Inconvenient Truth' (2006). Such has been the prominence and exposure of environmental problems, and global climate change that it has become increasing important for organisations to align themselves in concerted efforts to help the situation. Accordingly, companies, especially in those larger economies to become more willing to commit to measures and initiatives linked to minimising environmental problems. These are designed to reduce their carbon footprint and therein decrease greenhouse gas emissions.

The questions that have been posed for business organisations across the world are whether they have a credible recycling programme in place, and have they programmes and technologies in place for reducing energy consumption. This could take the form of creation of cleaner energy creation through renewable technology such as solar power through photovoltaic technologies and wind power through wind turbines. Other questions business leaders are tasked with relate to waste recycling and the degree to which their waste is managed in a sustainable manner. This could include the reduction in plastic waste which has become a global dilemma and could be affecting the world's oceans and contaminating the global food chain. There are many 'greening initiatives' that companies can adopt which could make a difference and some of which carry funding opportunities. Commitment to sustainable policies such as these are not simply the responsibility of large corporation but can easily be relevant to small and medium sized companies. In the UK construction industry, there is much research being conducted in reducing embodied carbon as part of the construction processes. This has called for innovative solutions and challenges to more tradition construction methods and building materials such as concrete and steel which have been commonly associated in being carbon intensive in their production.

Those organisations that have shown responsibility in 'doing their bit' for the sustainability and environmental agenda have benefitted from positive media coverage and gaining environmental accreditation. In the UK there has been a concerted effort for public sector organisations to 'lead from the front' and make a positive impact and example for other organisation to follow. Many Universities have adopted strategies to raising their ranking in the global 'People and Planet' ratings and gaining accreditation through the Eco Campus initiates as a commitment to waste, energy, carbon management, transport and greening initiatives. All these steps van make a small but effective contribution to the wider agenda in improving the environment. It is important also for organisations to look at their supply chains and have influence on them for responsible procurement of supplies and services. This could become a deciding factor in suppliers being selected as part of the supply frameworks especially for large organisations. Accordingly, those suppliers who show limited or no commitment to environmental measures could be severely disadvantaged in selection processes for new work orders. In the past the question for businesses is 'can they afford to make pledges and commitments to environmental initiatives?' For the reasons previously referred to, this has in recent years been turned around into the overarching question; 'can businesses afford not to have such commitments in place?'

3.8 Dilemmas around Environmental Ethics

Ethical dilemmas could be generated by issues related to sustainability and environmental awareness. As part of the on-going global climate emergency, it could be reasonable to assume that when designing new or refurbishing existing buildings, the design process should take full account of renewable energies in a concerted effort to reducing their 'carbon footprint'. It is important for the whole project team including clients and designers to consider both the construction and occupation phases in this regard as part of the life cycle of buildings. In doing so environmental ethics should take into account the embodied carbon of a building. Embodied carbon is the carbon footprint of a material. It considers how many greenhouse gases are released throughout the supply chain and is often measured from cradle to grave. It therefore needs to consider the construction phase and energy efficiency measures as part of the occupation 'in use' period of buildings with a particular emphasis on energy consumption and maintenance. Accordingly, it should take account and include all the emissions from embodied carbon in the construction materials, the building process, all the internal fixtures and fitting and the deconstruction and disposal of the demolition material it at the end of the building's lifetime. Reducing embodied carbon is important for reducing resources and associated costs but also to reduce longer-term risks of resource availability. Furthermore, the embodied carbon agenda is growing more important given an awareness that almost one third of the total life cycle carbon dioxide emissions from buildings stem from the embodied carbon in materials and the construction process. Addressing these sources of carbon is regarded as critical to meeting government and global targets for carbon reduction.

Notwithstanding the global push for 'carbon friendly' buildings, the promotion and commitment to environmental and sustainability building technologies can sometimes create ethical dilemmas for designers. If their clients feel that they do not want or cannot afford such environmentally friendly technologies or system for their projects, then should the designers simply accept their position or challenge accordingly? This is not an easy question to answer and will depend on context and the relationship which exists between designers and their employers. At the very least the design consultants should make clients aware of the benefits to sustainable design in terms of reduced environmental impact and potentially lower running costs. It could simply involve the designers questioning whether a particular client requires an air conditioning solution for their internal environment when a more sustainable and cost-effective solution might be predicated on natural ventilation through openable windows. This could save on carbon emissions and lower future utility costs, therein reducing the carbon footprint of buildings. Furthermore, architects, engineers and other construction design team consultants should be articulating the implications and disadvantages of ignoring the climate emergency and how this could specifically be regarded as irresponsible, leading possibly to reputational damage. At least then, they would be able to uphold their position from an environmental ethical perspective, given they have at least tried to influence the decision-making in this regard. Clients may not be fully aware of the longer-term implications of not pursuing a policy of environmental sustainability and simply opting for solutions that generate the lowest capital costs in the construction phase. It is expected of professionals and ethical appropriate for them to educate their clients and make them aware of the future benefits in this regard. Notwithstanding this premise, it will always be clients or their senior advisors who will make the final decisions in this regard.

3.9 Examples of Environmental Ethics and Sustainability in Practice: Case Study One – A UK Northwest College

3.9.1 Introduction to Case Study

The College was operating against an agreed environmental and sustainability agenda and existing legislation. Following the College's new build construction and engineering workshop project in 2011, it had been successful in gaining an excellent BREEAM accreditation.

BREEAM represents the world's leading sustainability assessment method for master-planning projects, infrastructure and building. It recognises and reflects the value in higher performing assets across the built environment life cycle, from new construction to in-use and refurbishment. BREEAM does this through third party certification of the assessment of an asset's environmental, social and economic sustainability performance, using standards developed by The British Research Establishment (BRE). This means BREEAM rated developments are more sustainable environments that enhance the well-being of the people who live and work in them, help protect natural resources and make for more attractive property investments.

The College did seek, alongside our designers, to promote sustainable construction methods, mindful that the construction industry contributes to almost half of the world's energy consumption and that there is the possibility of running out of fossil fuels in the next 40 years (Roaf 2001).

The College's proposed master-plan redevelopment plans for further new energy-efficient buildings on the campus were cancelled following the Learning Skills Council withdrawal of funding in 2011. However, this did not preclude the College from looking at simple, cost-effective solutions to increasing building energy efficiency. These included improving heating controls, increasing thermal insulation, installing energy-efficient light fittings/bulbs, fitting door closers, managing classroom utilisation more effectively, fitting automatic controls and displaying energy meters so that students and staff can measure and monitor improvements. In addition, a waste-management strategy involving different types of waste for recycling could be introduced quickly and at relatively little cost. This would then hopefully comply with EU Directive 2006/12/EC of the European Parliament which states:

> In order to achieve a high level of environmental protection, Member States should, in addition to taking responsible action to ensure disposal and recovery of waste, take measures to restrict the production of waste particularly by promoting clean technologies and products which can be recycled and reused.

3.9.2 Implications and Impacts for the College and the Need for Change

The College considered that unless it was to adopt a sustainable waste management policy the changing agenda could present negative outcomes for the College on ethical, public relations, political and social grounds. Furthermore, on an economical basis the financial operating costs for the College could increase as time passes through the introduction of government initiatives such as the Landfill Tax Escalator (Department of Trade and Industry 2010). This taxation point is reinforced that 'further adoption of the "polluter pays principle" could see future differentiated taxes introduced by government seeking to remove or reduce emissions of carbon dioxide into the atmosphere'.

The dilemma for this organisation was not to assess the affordability and feasibility of sustainability as a choice of whether they chose to participate or not. The real question for the College was whether they could afford not to change the way they were currently operating and place sustainability high on their agenda.

3.9.3 The College's Corporate Social Responsibility

Given the importance of the sustainability agenda the College looked for solutions that present environmental and social advantages and that will benefit the College in doing so. Some would argue that this corporate social responsibility does not exist in the real world. Banerjee (2007), supported this view, presents a cynical representation of this ideology for most organisations and explained:

> This ideology is a convenient yet dangerous myth and this 'win-win' rhetoric is nothing but a mask, made up of glossy brochures, beguiling speeches and 'media-savvy' partnership initiatives.

One could, however, argue that the College is unlike most organisations as its primary function is not to make a profit but to serve the community that we work within through education and training, and this alone presents an important reason for our active participation in the sustainability agenda. On ethical grounds they considered that as a major public sector organisation they should be paving the way for cultural change and setting a good example for other institutions in the process.

3.9.4 The Need to Embed Sustainability

The College had for many years now concentrated more on economic factors (especially in light of the real reductions in funding over the years) and educational excellence in meeting government targets on student achievement and retention rather than the bigger global issues of the environment. Michael Gove, Former UK Education Secretary, conceded this point and argued that 'we need to embed sustainability into the heart of our academic establishments' (Guardian 2011). This view was underlined at the time by the Former Secretary of State for the Environment, Caroline Spelman:

> Genuine sustainability will never again be an add on or afterthought. It will be at the heart of everything we do, from making policy to running government buildings or purchasing goods and services.
>
> (DEFRA 2011)

The College accepted that a long-term cultural change at the highest level was required within the organisation rather than consider the initiatives as individual short-term projects. This view is supported by the work of Hjorth and Bagheri (2006) who opined that 'sustainability should not be perceived as a "project" that has an end point, but as an on-going process that needs to be regarded as part and parcel of everyday work'.

3.9.5 Potential Measures to Address the Sustainability Agenda

Table 3.1 presents some potential recommendations that were designed to allow the College to address the sustainability areas. It lists a combination of *dematerialisation* and *substitution* measures. Dematerialisation would include reduction in energy usage, waste, paper,

Table 3.1 Potential recommendations to address the sustainability agenda.

Activity of College	Creative Recommendations to Address Legislation and Mitigate Impacts of Environmental Risks
Environmental management	The College could:
	Promote sound environmental management by communicating procedures, guidelines and regulations via induction, staff training programmes and tutorials.
	Encourage and facilitate feedback and suggestions on ensuring good practice via a network of operational groups and team meetings across College.
	Develop a management system based on a sound understanding of current performance and the adoption of realistic targets to improve environmental performance.
Waste reduction	The College could:
	Implement sustainable resource management practices based on 'reduce, reuse, recycle' principles.
	Re-use and recycle materials wherever practicable.
	Reduce the amount of waste being disposed of in landfill sites, by segregating different types of waste.
	Develop a programme of target setting to reduce waste.
	Prevent pollution.
	Comply with the legislation and meet the requirements of the waste regulators.
	Reduce litter.
Water consumption	The College could:
	Make efficient and responsible use of water, including making opportunities for water re-use and rainwater harvesting.
Carbon management	The College could:
	Implement a carbon management strategy to ensure the efficient use of energy.
	Ensure the uptake of low-carbon technology in buildings and equipment.
	Improve the Energy Performance Certificate assessment grade of buildings from the base year, October 2008.
Transport	The College could:
	Implement sustainable transport practices across all activities, as indicated in College Travel Plan.
	Encourage the use of public transport, walking, cycling and vehicle sharing, where appropriate.
	Possible discounts to incentivise train and bus travel.
	Charging students and staff for car parking.
	Reduce the need to travel especially by car, e.g. working from home. Extension of bus routes and increasing bus frequency.
	Encouraging multipurpose/linked trips.
	Engage with public transport providers to maximise use of existing transport infrastructure, e.g.
	Designing our campus to encourage cycling to work, e.g. more cycle storage, dedicated changing areas and showers.

Table 3.1 (Continued)

Activity of College	Creative Recommendations to Address Legislation and Mitigate Impacts of Environmental Risks
Ethical and sustainable procurement	The College could:
	Promote the use of sustainable resources whenever practicable.
	Promote life cycle thinking in the procurement of goods and services.
	Promote the development of central purchasing strategies to ensure efficient and sustainable purchasing practices.
	Embed sustainability into selection criteria, processes, procedures, business cases, budget allowance and early market engagement when purchasing goods and services.
	Manage relationships with suppliers to raise awareness and allow suppliers to respond to additional or different measures in some cases.
The college estate	The College could:
	Ensure that, wherever practicable, all new building works meet the sector standards for sustainability.
	Promote a responsibility culture to improve the internal environment of all College rooms.
	Promote a responsibility culture to improve the external environment of the College.
	Promote the efficient use of all buildings and facilities to meet sector standards for space utilisation.
	Manage the College estate to enhance biodiversity wherever possible.
Education for Sustainable Development	The College could:
	Develop links with local and national groups which encourage education for sustainable development.
	Endeavour to embed education for sustainable development into all curriculum areas.
	Encourage students to participate actively in environmental projects, locally and globally.

etc. and substitution could be achieved through alternative energy sources. In some cases, innovative solutions could fulfil both objectives at the same time. For instance, the College could have reduced the extent of wood waste through recycling (dematerialisation) and utilised the waste to fuel biomass boilers for heating as an alternative zero carbon energy solution (substitution).

3.9.6 Economic, Political and Social Factors Influencing the Evaluation of the Measures

Detailed evaluation of the financial viability and environmental benefits of each measure was required to be undertaken by the College. This approach to business decisions around sustainability is advocated by Ding (2006) who explained that 'Environmental issues and financial considerations should go hand in hand as parts of the evaluation framework of a project.'

Some of the recommendations in Table 3.1 were intended to save money especially when considering life cycle costs. Ayres (2008, p. 289) supported this point and states that '… it may be possible, in principle to cut costs and pollution at the same time in some sectors ... although it is rarely accomplished in practice'. Sant and Carhart (1981) also endorsed this view and suggest evidence through 'least cost' studies that very significant opportunities for savings through sustainability measures have been neglected at the national level. The converse argument is made by Appleton (2006, p. 3) who claimed:

> *An increasing number of corporations and governments have recognized that practices like green buildings, alternative ways of storm water management and certain change to land use practices offer them chances to make additional profits at little downside cost or risk.* (emphasis added)

The College was required to identify projects which should take priority over others and develop overall timescales for instigating our strategies. Factors included complexity, cost resources, funding incentives and the political appetite for change in certain areas. One argument could have been made for priority to be given to those that stand a relatively good chance of yielding a good return on investment. Graham (2003, p. 54) presented a converse argument in that 'sustainability should not be influenced to such an extent by cost alone but by all the environmental and social benefits that generate from it'.

3.9.7 A Sustainable Procurement Process

Using the sustainable procurement processes for the purchase of goods and services the College attempted to implement measures and realise benefits from those initiatives very quickly. The Office for Government and Commerce (OGC) announced in 2015 new commitments for 'greening' public sector procurement to understand and reduce the impacts of its supply chain. It is hardly surprising with an annual public sector spend in excess of £236 billion that government regards procurement as a powerful lever that can reduce emissions and drive innovation. Government also supports the National Procurement Programme which is delivering training on sustainable procurement to public procurers including the FE and HE sectors as well as interest-free loans for set up costs. In addition, there were already procedure, processes and documentation from previous models of good practice as templates that could be utilised for this purpose, which would have precluded the need for a bespoke solution. Financial and resourcing demands would therefore have presented few barriers to changing the procurement strategy of the College in the short term, given the above incentives for change.

3.9.8 A Sustainable Transport Policy

There are other examples of sustainability initiatives such as changes to transport and the introduction of a green travel plan which could take much longer to initiate. Any changes would have had to involve a considerable consultation period and extensive negotiations with partners such as the local authority (as part of their Agenda 21 Sustainable Transport 1 initiative) and public transport providers in the borough. In addition, the need to undertake physical alterations and build new additional facilities on the campus was a consideration. Such measures could have included creating additional changing rooms, bike storage and showers for cyclists could take some time in planning and construction. Whilst this is a valid argument the transport agenda was a major focus for the College. After all, in 2012 domestic transport accounted for 22% of UK greenhouse gas emissions. In addition, the introduction of the Government's Local Transport White Paper in 2010 (HM Government 2011) detailing £560 million enabled local bodies to deliver sustainable transport solutions. This should have provided an incentive to the College to start the process and planning earlier in our programme, to ensure implementation within a reasonable timescale.

3.9.9 The Need for Effective Consultation and Communication

In planning long-term changes to the College's sustainability strategy their success was dependent on undertaking extensive consultation and dialogue with their students and staff as well as third party stakeholders such as their local authority. Innovative ideas were generated to assist the College in developing environmental strategies and sought to discard the former 'identify-and-repair' approach and instead adopt an 'anticipate-and-prevent' approach (Al-Sharrah et al. 2010).

At each stage the College shared and communicated their sustainability strategy as it developed and unfolded. This was aimed at generating maximum support and to be as transparent as possible on the issues and their sustainability objectives. This improved public relations for the College and was intended to enable transparency which could open up avenues of possibilities that would be difficult to achieve without it.

3.9.10 The Need for Commitment, Cultural and Organisational Change

The College decided that success of their sustainability strategy was predicated on securing full commitment from the College's Senior Management Team and Board of Governors. For this reason, all college managers were given responsibility and duties to support the implementation and monitoring of the policy. It was recommended that a new role should be created as Head of Environmental Services, who would be required to report to the Principal, senior management and the Governing Board and provide regular progress reports against sustainability programmes and implement an annual review of all environmental policy.

Finally, in terms of cultural change, educational Institutions, such as the College, did not simply regard themselves as academic institutions that enabled students to obtain qualifications and training. Instead they proclaimed that they offered and created an excellent opportunity to encourage people to adopt more sustainable life styles. In addition, the College believed that the increasing regulation and sustainability agenda affecting organisations like theirs should not be regarded as a negative influence to adversely affect their business operations. Moreover, they professed that they should be encouraged to push the boundaries in the creation of more environment-friendly communities (Architecture Week 2011).

3.9.11 Summary of Case Study

This case study hopefully provides an insight into an organisational sustainability agenda, which in this instance was a College. The example articulated the analysis and evaluation of the implications and impacts from ever increasing regulation alongside the benefits that could be realised from participation which suggested the case for adopting sustainability principles and embedding these within their organisation and everyday activities. The College's corporate social responsibility on ethical grounds alone in promoting the agenda provided many benefits. Furthermore it was intended that the proposed innovative measures would provide the tools and motivation for implementing the necessary sustainability strategies and policies towards promoting a cleaner sustainable environment.

3.10 Examples of Environmental Ethics and Sustainability in Practice: Case Study Two – A UK Northwest University

3.10.1 Introduction to Case Study

This case study relates to achieving sustainability and environmental enhancements through a collaboration and partnering 'toolkit'. The measures undertaken and successful outcomes at the University will be explained and how the development of a carbon management plan assisted the organisation in achieving its environmental target.

The author whilst employed at a university in the Northwest of England worked alongside a sustainability team around improving environmental performance of the organisation. A 'collaboration toolkit' was formulated through the author's research to assist client bodies to adopt a more collaborative model of procurement that integrates the supply chain into the environmental agenda at early stages of projects. Plans were in train for the toolkit to be piloted on a major project as part of the University's master plan and capital programme.

The focus for the toolkit was finding environmental solutions through collaboration with industry and delivering a new sustainability strategy for the University. In this regard the following measures and initiatives were developed and planned to be deployed as part of the case study from many different environmental perspectives including:

- Pursuing a holistic approach across the supply chain to meet the demands of the carbon neutral agenda. All bodies (client end users, design consultants, contractors, subcontractors and other external stakeholders) are being encouraged to collaborate early in design stages on alternative construction methods, systems or solutions to generate more sustainable solutions.
- Reducing embodied carbon in buildings, with academics and industry working collaboratively to formulate pragmatic solutions through workshops.
- Introducing an energy partner to work with the University to assess the ways and means of them reducing its gas, electricity and water consumption.
- Reviewing which renewable technologies are going to give a significant and practical reduction in carbon and are cost effective, respective to site logistics, decanting and appraisal.
- Creating a new sustainable transport policy linked to reducing car travel.

An energy, water and carbon management plan was developed by the University and this is in Appendix A. It gives context and background to the plan and contains an energy and water review. It also incorporates baseline data and key targets for future energy and water usage alongside an action plan and risk assessment. Furthermore, it provides month upon month results from energy audits conducted with data on water and energy usage. In addition to this document, an environmental sustainability strategic plan was prepared by the University which sets out further targets linked to wider sustainability aims and objectives, including procurement, transport, and carbon generation (included in Appendix B). The targets linked with these measures as part of the case study in 2020, three years after implementing the plans were as follows:

- Carbon emissions on target for reduction of 81% by 2030 (from a baseline of 2012)
- Reductions of energy and water consumption by an average of 5% over last three years
- Environmental accreditations achieved in last two years
    1) Green Flag
    2) Eco Campus Platinum Status
- Reduction in waste of 20% over last three years through recycling
- Increase of EV charge points by 30%
- Flexible (Sustainable Procurement) Framework Level 4 achieved
- Robust energy, water and carbon management plan developed consisting of:
    1) Energy audits
    2) Training
    3) Automated systems
    4) Measurement
    5) Switch it off campaign
    6) Water leak surveys
    7) Low flow sanitary fixtures
    8) Rainwater and grey water harvesting
    9) Electric vehicles
    10) LED lighting
    11) Green IT
    12) Cycle to work initiatives
    13) BREEAM Excellent for new builds and refurbishments
- Communication and engagement on sustainability measures and outcomes
    1) Sustainability linked into the curriculum
    2) Active member of the Greater Manchester Green Impact Committee
    3) Active member of Go Green Salford
    4) Green Campus Group at the University

The above successes highlight the measures and consequential beneficial outcomes for the University and how the collaboration toolkit played an integral part in delivery. This will no doubt be of interest to those organisations that are planning for long-term transformation change and raising their sustainability credentials and accreditations such as the Eco Campus Scheme.

3.10.2 Practical Use of the Collaboration Toolkit to Target and Deliver Environmental Sustainability Objectives

The collaboration toolkit provides the practical guidance and the platform for systematic partnering at all design and construction stages with the supply chain. This is intended to provide the conduit for improving outcomes on projects as part of strategies designed around environmental sustainability. Notwithstanding this premise the following considerations need to be integrated into environmental sustainability strategies and policies.

i) The need for an effective Environmental Management System

As 'part and parcel' of the sustainable procurement of construction and engineering projects an Environmental Management System should be put in place as a way for organisations to demonstrate that they have addressed and minimised the environmental impacts of development. In this way it can oversee and manage legal compliance of development whilst striving for continuously improvement on environmental sustainability performance. Furthermore Figure 3.1 shows the environmental measures which should be incorporated into the Environmental Management System as part of the project management process.

In addition to the above measures, having energy, water and carbon management plans in place can greatly improve the success of environmental sustainability projects. The collaboration toolkit can be tailored to ensure that these plan areas developed by the projects team, assisted by the supply chain and then become embedded into construction and engineering programmes.

ii) Waste reduction

On all projects through the collaboration toolkit there should be integrated measures to ensure that waste from construction and engineering activities is minimalised. Every effort

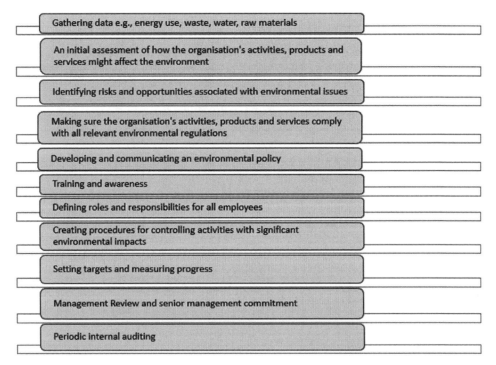

Figure 3.1 Environmental measures which should be incorporated into the environmental management system.

should be made to recycle waste materials and thus reduce the overall amount of arisings disposed off-site. This will serve to reduce the embedded carbon of the build process thus lowering the carbon footprint. Furthermore, measures to improve the green credentials and status of projects will have many advantages for those organisations involved in them, and these revolve around corporate social responsibility, community relations and reputation.

iii) Sustainable and ethical procurement

In a construction and engineering context sustainable procurement philosophies should be deployed as part of the collaboration toolkit when appointing consultants, main contractors, sub-contractors, suppliers and other specialists. In the UK there are various purchasing consortiums, who ensure that 'checks and balances' are conducted on those organisations that are included as part of their framework. Such checks will ascertain the ethical and sustainability credentials of organisations and ensure that only those firms which meet a required minimum standard are included on the framework. In this way, employers can have the comfort factor that companies who are commissioned on projects have already been through a rigorous vetting process to assess their environmental sustainability credentials. Furthermore, organisations should ensure that all tender evaluations required when appointing the supply chain include sustainability weighting as a key quality factor alongside other criteria such as price. Developing and applying ethical and sustainable procurement policies in this way will ensure social and environmental impacts of those employed on construction and engineering projects are appropriately considered.

iv) Sustainable construction and engineering designs and technologies

The collaboration toolkit should be able to assist project teams in making collaborative decisions with their supply chains on the use of environmentally sustainable building technologies in the construction process. This could entail careful and protracted early discussions on the use of renewable technologies. Reducing embedded carbon in construction and engineering projects is a topical issue in the UK and cross discipline dialogue with the supply chain has encouraged more sustainable materials and construction processes to be implemented within projects. Accordingly, sustainability is the core principle of the any project and should embed a high standard of adherence to best practice philosophies.

v) Community engagement

Communication and engagement, encouraged and facilitated through the collaboration toolkit at all construction and engineering stages, are regarded as good practice on most construction and engineering projects. Processes around consultation with the local community into environmental issues and corporate social responsibility to maintain or enhance their well-being, environment and facilities can prove extremely fruitful in most cases. In addition, becoming accredited with schemes such as the Considerate Contractors Scheme in the UK can pay dividends and enhance the reputation of the project team.

Once again, the collaboration toolkit can play an important part on bringing parties together, formally or informally to present innovative ideas and solutions in this regard.

vi) Education around environmental sustainability

Engineering and construction projects should 'promote sustainability and social responsibility' through teaching, learning and research. The collaboration toolkit encourages industry partnerships and therein can encourage teaching and learning. As part of construction and engineering projects using the toolkit, strategies should be focused on the environmental sustainability of all operations and teaching, learning and research around this agenda through education programmes and continuous professional development should be encouraged. In this way construction and engineering projects could be regarded as a 'Living Laboratories' especially when partnering with specialist energy partners.

3.10.3 Successful Outcomes that the Collaboration Toolkit Has Had on Delivering Environmental Sustainability

Although projects, where the collaboration toolkit is being piloted, are at a relatively early stage in terms of delivery, they are proving to have many beneficial outcomes from an environmental sustainability perspective. For the client body these include those benefits shown in Figure 3.2.

3.10.4 Conclusions and Recommendation from the Case Study

The collaboration toolkit has been presented and as previously outlined forms part of action research as part of an on-going case study where it is being piloted on UK

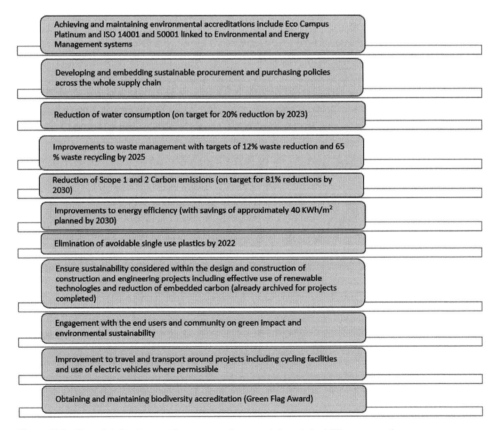

Figure 3.2 Beneficial outcomes from an environmental sustainability perspective.

construction projects. It has so far, since it was introduced in 2018, achieved very successful outcomes for the client organisation in the UK and facilitated and encouraged many different environmental sustainability initiatives and policies to be brought forward. In this way it has assisted the organisation to keep on target with its environmental sustainability objectives. It has created the collaboration and partnering ethos that is so important for innovation and delivery strategies linked to environmentally sustainable construction and engineering projects and processes that could otherwise have been problematic.

The benefits of utilising the toolkit have been clearly articulated in this chapter. Accordingly, it is hoped that the toolkit can provide the right platform for construction and engineering projects in Ajman and therein achieve similar environmentally beneficial outcomes for the emirate. For this reason, it is recommended that the principles and mechanics of the toolkit be deployed when embarking on environmental projects as part of the emirate's aspirations for a more sustainable future.

3.11 Summary

In understanding what sustainability is, there have been many different definitions over previous years. For this reason, sustainability could mean different things to many people. This considers the wide-ranging contexts and environments which construction management operates within. One more prolific definition is that environmental ethics concerns formulation and moral obligations regarding the environment.

The UK Government has committed to delivering significant reductions in greenhouse gas emissions through net zero carbon targets for the future years ahead. In doing so this will strongly influence the way buildings are designed and constructed but also used throughout their life cycle. Accordingly, if the Government pledge and commitments are to be achieved, buildings must become significantly more efficient, not just in the construction process but in terms of energy usage during their operation. For this reason, construction professionals in their respective leadership roles, when procuring projects, should be aware of the growing importance of environmental ethics, as part of their organisations' social responsibilities. Accordingly, there is a strong argument that construction professionals should be increasing political influence in this area, and therein drive the agenda forward. This could take the form of instructing their project teams to adopt environmental and sustainable design measures into their projects as one example.

The role of environmental ethics is to identify the moral obligations in terms of concerns of effects on the atmosphere, climate change, depletion of natural resources, loss of wilderness, dwindling animal and plant biodiversity. Given the increasing concern for environmental issues and the impact that our actions will have on the environments that we live in, the field of environmental ethics will become more important in the future. Political efforts will seek to limit or in some cases reverse the environmental effects of certain delirious actions that have caused harm to the environment. In this regard, it is important for all responsible organisations to uphold a sense of corporate social responsibility. This in turn will have far reaching benefits for their businesses in terms of public perception, marketing, operational efficiency, adherence to ever increasing regulations and maximising future government funding opportunities. However, the global and national agenda on sustainability has in the past had little direct influence on most organisations. Notwithstanding this premise, Government in recent years has contributed to encouraging environmental ethics into the design and construction of projects in the UK. One example of a UK Government initiative in this regard is the introduction by the Department of Trade and Industry (DTI) for Sustainable Progress Reports. This covers principles such as waste, energy efficiency, carbon generation, safety, skills, climate change and costs which have been used in the past for measuring how sustainable projects are. Furthermore, in looking at local sustainability agenda in the UK, and in particular local authority policies, there is strong emphasis on the participation of organisations to encourage reduction of waste, increased recycling, and the adoption of sustainable transport policies. In some cases, county councils have previously had to rely on enforcement of such policies through the planning system (Town and Country Planning Act 1990) for major developments. In addition, it is important and productive for organisations to work more closely with local authorities in partnership, through liaison and consultation, to promote a more robust sustainability strategy for their local area. This would improve the external relations and the public perception of businesses.

Environmental social responsibility is referred to as environmental conservation social responsibility, and curtails to the long-term problem of global warming, increased waste production and contamination on an international scale. Such has been the prominence and exposure of environmental problems, and global climate change that it has become increasingly important for organisations to align themselves in concerted efforts to help the situation. The questions that have been posed for business organisations across the world are whether they have a credible recycling programme in place, and have the programmes and technologies in place for reducing energy consumption. This could take the form of creation of cleaner energy creation through renewable technology such as solar power through photovoltaic technologies and wind power through wind turbines. There are many 'greening initiatives' that companies can adopt which could make a difference and some of which carry funding opportunities. In the UK construction industry, there is much research being conducted in reducing embodied carbon as part of the construction processes. This has called for innovative solutions and challenges to more tradition construction methods and building materials such as concrete and steel which have been commonly associated in being carbon intensive in their production. Notwithstanding the global push for 'carbon friendly' buildings, the promotion and commitment to environmental and sustainability building technologies can sometimes create ethical dilemmas for designers especially. Such dilemmas have revolved around short-term client focus on reducing capital costs at the expense of longer-term outcomes. Accordingly, consultants should be articulating the implications and disadvantages of ignoring the climate emergency and how this could specifically be regarded as irresponsible, leading possibly to reputational damage for organisations. In this way clients could then be persuaded to adopt environmentally sustainable policies and not simply discount them on the grounds of affordability in the short term. This drive to increase client awareness is important as they may not be fully aware of the longer- term implications of not pursuing a policy of environmental sustainability and simply opting for solutions that generate the lowest capital costs in the construction phase.

Two case studies have been articulated and discussed in this chapter relating to environmental sustainability plans and strategies which have been adopted by both a UK college and university. Measures undertaken by these further and higher education establishments have included the introduction of effective environmental management systems, waste reduction initiatives, sustainable and ethical procurement processes, sustainable construction and engineering designs and technologies, community engagement and education around environmental issues. Successful outcomes on delivering environmental sustainability in these case studies have included reduction in energy and water consumption, the award of many environmental accreditations, improvements to travel and transport policies, reduction in waste in carbon generation and waste, decrease in the use of single use plastics, and the introduction of more robust environmentally sustainable procurement policies.

References

Al-Sharrah, G., Elkamel, A., and Almanssoor, A. (2010). Sustainability indicators for decision-making and optimisation in the process industry: the case of the petrochemical industry. *Chemical Engineering Science* 65 (2): 1452–1461.

Appleton, A.F. (2006). Sustainability: a practitioner's reflection. *Technology in Society* 28 (1–2): 3–18.

Architecture Week (2011). *Architectural Global Warming*. Online Available http://www.architectureweek.com/2011/0218/environment_1-1.html (accessed 5th April 2011).

Ayres, R.U. (2008). Sustainability economics: where do we stand? *Ecological Economics* 67 (2): 281–310.

Banerjee, S.B. (2007). Corporate social responsibility, the Good, the Bad and the Ugly. *Journal of International Management* Cheltenham: Edward Elgar Publishing 15 (1): 118–119.

DEFRA (2011). *Embedding sustainability in Government*. (internet). Available http://www.sd.defra.gov.uk/2011/02/embedding-sustainability-in-government. (accessed 28th April 2011).

Ding, K.C. (2006). Sustainable construction-The role of environmental assessment tools. *Journal of Environmental Management* 86 (6): 451–464.

Gore, A. (2006). *An Inconvenient Truth: The Planetary Emergency of Global Warming and What We can Do About it*. New York: Rodale, 17–19.

Graham, P. (2003). *Building Ecology. First Principles for a Sustainable Built Environment*. Oxford: Blackwell Science Ltd.

Guardian (2011). *The guardian home page* (Internet). Available from http://www.guardian.co.uk/politics/2011 (accessed 3rd May 2011).

H M Government (2011). *Carbon plan*. London: DECC Publications.

Harvard Institute for International Development (1992). *Report of the United Nations Conference on Environment and Development*. UNCED Report A/CONF.151/5/Rev.1 13 June 1992.

Hjorth, P. and Bagheri, A. (2006). Navigating towards sustainable development: a system dynamics approach. *Futures* 38 (1): 74–92.

IEA (2011). *A Workshop on Energy Efficiency in Buildings*. Paris. Online Available http://www.iea.org/Textbase/work/2011/cert_slt/announcement.pdf (accessed 15th April 2011).

Roaf, S. (2001). *A Design Guide*. Oxford: Butterworth-Heinemann.

Sant, R.W. and Carhart, S.C. (1981). *8 Great Energy Myths: The Least Cost Energy Strategy.* Pittsburgh PA: Carnegie-Mellon University Press.

Stockport Metropolitan Borough Council (2011). *Local Agenda 21 Sustainable Transport Supplementary Planning Document 2007.* (Internet). Available: http://www.stockport.gov.uk. (accessed 16th April 2011).

Tagawa, S. (2005), Environmental ethics and project management of architecture. *The 2005 World Sustainable Building Conference*, (27–29 September 2005), 4529–4531 Tokyo.

4

Corporate Social Responsibility

Our very lives depend on the ethics of strangers and most of us are strangers to other people.

Bill Moyers

4.1 Introduction

When considering the above quotation, it is important to realise that the communities we live and work within are reliant on companies, and the individuals they employ, doing the right things for others. Ethics linked to the altruistic goodwill of companies and the moral responsibilities they maintain can be very beneficial for them and society at large. Accordingly, this chapter is dedicated to corporate social responsibility (CSR) which has become a highly topical area in recent years. It will discuss what corporate social responsibility actually means in practice and how CSR philosophies can be embedded into organisational cultures, policies and procedures. The benefits that adopting CSR can bring for the company brand, from a reputation and credibility perspective, will be articulated. Conversely, it will examine the reverse scenario where there are breaches of corporate social responsibilities and how this can be extremely damaging for companies. To contextualise this examination, controversial examples of some recent global cases of such breaches will be presented and the financial and reputational fall out that has been inflicted on them.

Different types and models of corporate social responsibility will be discussed from a legal, economic, philanthropic and volunteering perspective, and the interwoven relationship between CSR and economic, social and environmental considerations. The Considerate Constructors Scheme will be examined and how contractors can adopt and deploy different actions and measures on projects, as part of CSR, to better serve their workforce and communities. Ethical statements and corporate codes will also be discussed and examples of these presented. The principles of social value and the different types of intervention that companies can make to its agenda will be explored. In addition, how such interventions can be measured through cost-benefit analysis across projects and programmes will be considered.

Organisational Ethics in the Built Environment, First Edition. Jason Challender.
© 2023 John Wiley & Sons Ltd. Published 2023 by John Wiley & Sons Ltd.

Finally, sustainable and ethical considerations, linked with corporate social responsibility will be examined, alongside examples of how environmental measures can be adopted to reduce carbon generation and waste for the construction industry.

4.2 What Is Corporate Social Responsibility?

According to Fewings (2009) corporate responsibility could be defined as a corporation's obligation to its stakeholders. Stakeholders in this regard are categorised as any individuals or groups that have a stake or interest in a company's business or outputs. This could include suppliers, employees, customers and those communities around the business. Communities may become members of the public in large organisations who have a national or global market.

Depending on the stakeholders, obligations around corporate social responsibility will differ. A customer for instance may be focused on the quality and safety of a business product whereas the needs of employees may revolve around pay, promotion and working conditions. Alternatively, an investor might have a vested interest in company profits, financial resilience. Moreover, a member of the public might have more interest in how the business affects the environment and community they work within, from both a positive and negative perspective. In this way corporate responsibility focuses on optimising the positive impacts on these stakeholders whilst reducing the negative impact.

4.3 The Benefits of Corporate Social Responsibility for Businesses

In today's economy it is widely accepted that businesses must be responsible from a social perspective rather than being purely financially successful and achieving targeted profit levels for their shareholders. Their survival will depend on this commitment and accordingly companies should adopt policies and strategies on important social agenda items. In this way they can build the reputation and credibility of their brand and win the respect and trust of their stakeholders, clients and customers. Accordingly, it is imperative for organisations to embed corporate responsibility into their policies, culture and management approaches. This can allow them to introduce and deliver successful outcomes, not only for their own purposes, but for communities and the public at large. Notwithstanding this premise, the measures by which companies invest in corporate social responsibility initiatives do not come free and in some cases can involve significant sums of money. However, the investment by organisations in undertaking such commitments should be balanced against the reputational benefits and the positive public relations that could be generated. Such benefits should not be underestimated and in some cases, the enhancement to the brands of companies can warrant businesses charging more for their products and services, without seeing any significant knock-on effects of reduced demand. In a construction industry context, along these lines the notion of 'social value' has become more important in recent years.

Conversely, where there are breaches of corporate social responsibilities, this can have extremely negative consequences for organisations, especially those larger businesses that rely on brand and reputation. There have been many controversial global cases that have become exposed in the press in recent years, involving breaches in corporate social responsibility and one example is contained below.

> **Deepwater Horizon oil spill 2010; British Petroleum**
>
> In April 2010 an oil spill which became an industrial and environmental disaster began in the Gulf of Mexico. It was caused by an explosion on an oil rig run by British Petroleum (BP) off the coast of Louisiana and became the largest oil spill in the history of the petroleum industry with an estimated five million barrels discharged into the sea. The devastating effects include it becoming the largest environmental disaster in American history and the loss of 11 lives. There was a massive response to stop the spill spreading to water courses, beaches and wetlands. Unfortunately, however, the damage to the environment was inevitable owing to the scale of the spill which had a significant effect on wildlife and marine habitats.
>
> The clean-up extended for many years and this had for many years a detrimental effect on tourism and fishing in the area with the effects spreading for many hundreds of miles in all directions. A long-drawn-out investigation of the cause of the explosion ensued. It was found by US Government report that the cause was related to defective cement on the well and blame was attributed to both BP and their rig operator Transocean in 2011. The blame largely related to cost-cutting decisions which compromised safety systems and procedures designed to prevent this type of event occurring. It found that there was a systematic culture of breaching established industrial safety practices and without significant improvement measures the event could happened again elsewhere. A major legal case was brought in the United States against BP and this culminated in BP pleading guilty to 11 counts of manslaughter as well as lying to Congress. The court found also that BP would have to be subjected to four years of intense tracking and monitoring of their ethics and safety practices, with a clear intention of preventing any future similar breaches. Furthermore, BP was banned from any new contracts by the US environmental Protection Industry and in 2015 was forced to pay nearly $17 billion in fines, which represented the largest corporate settlement in US history.
>
> That was not the end of the story for BP, and in the period up till 2018 the company had expended more than $65 billion in clean-up costs. Notwithstanding the financial impact this had on BP and the resultant drop in the value of the company, owing to a massive fall in their share price, the most significant cost was to their reputation. With reports coming out of the US, for many years showing the environmental damage this caused a reputational and trust issue for BP and could have taken them out of business altogether. With press coverage on news stations showing oil spills on beaches and figures releases showing the numbers of marine life and birds killed in the disaster and irreparable damage to wildlife habitats their international reputation was severely tarnished.

4.4 Social Value as Part of Corporate Social Responsibility

Social value has been defied by the Sustainable Procurement Task Force definition as:

> *A process whereby organisations meet their needs for goods, services and utilities in a way that achieves value for money on a whole life basis in terms of generating benefits not only to the organisation, but also to society and the economy, whilst minimising damage to the environment.* (emphasis added)

4.4.1 Sustainable Procurement Task Force

Increasingly, organisations are contemplating their management activities from a holistic perspective, taking into perspective the wider social, economic and environmental effects of their actions. In this regard 'social value' is an umbrella term for these wider effects and businesses and institutions should make a conscious effort to ensure that they have positive outcomes. These benefits are regarded as contributing to increasing and enhancing social value and therein benefitting the well-being of people, communities and the general public at large. Accordingly, the public sector particularly should consider social value in all their policies and strategies to maximise the advantages to the communities they represent. Private sector organisations should also incorporate corporate responsibility, social value and sustainability in their management approaches which should allow them to address social, economic and environmental issues in their local community.

In terms of measuring social value, this is becoming increasingly standardised. In the UK the National Social Value Measurement Framework or National Themes, Outcome and Measures (TOMS) has been created by the Social Value Portal for this purpose. One of the main advantages of a standardised approach to measurement is to create a tool that is based on a consistent framework across the board in terms of methodological approach. In this way it can provide the important link between social value strategy and delivery and aligns themes and outcomes to appropriate measures. Such measures can be translated into financial terms drawing on a broad range of data sources including the Unit Cost Database and the Office of National Statistics. This allows a comparison to be made on the cost and benefits of a project across different sectors and organisations. However, according to the Social Value Portal (2020), these financial terms should not solely be used to assess the 'social value added' owing to the subjective nature and difficulties in quantifying of outcomes such as well-being.

In the UK construction industry, large main contractors especially now place social value at the heart of everything they do and employ community liaison officers to manage their social value delivery programmes. Community benefits are treated in this way as core deliverables for these contractors alongside the physical process of constructing buildings. Such deliverables are integrated into their overall project planning from the outset. There are many different case studies and examples of contractors deploying clear strategies linked to the delivery of quantifiable and clear social value outputs and outcomes in alignment with local government or community aspirations. In the past these have included employment and career opportunities for local people which have extended to apprenticeship and internships. In the latter case main contractors have worked with local universities and colleges to provide such placements within their respective organisations. These can provide local people with the skills, experience and knowledge to progress careers in the construction industry and make a large impact in areas where the numbers of those not in employment, education or training (NEET) are high.

Notwithstanding the above premise, it is important for main contractors to understand local needs before embarking on social value initiatives. After all, it would not be a wise proposition to invest effort, energy and resources into projects for the community which local people do not want. For this reason, it is important for main contractors to undertake a consultation process with the local community which would normally include ward councillors, local business owners, elected officials, residents, colleges, universities and

4.4 Social Value as Part of Corporate Social Responsibility

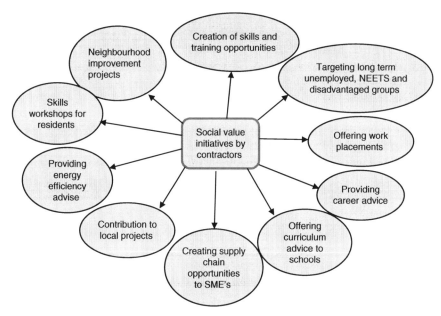

Figure 4.1 Examples of the type of social value initiatives provided by contractors.

community leaders. Taking this approach should create the 'buy in' from these stakeholders and therein make the process simpler and more effective in its delivery. Defining what form the social value projects and initiatives will take for a particular area is critical in this regard. Such projects and initiatives depend on the area, the local community needs alongside the duration, value and nature of the construction project. If main contractors are developing construction projects, they can also be expected to deliver local economic growth perhaps using local supply chains to safeguard that money is put back into the local economy. Social value could also take the form of initiatives around community wellbeing, and this could relate to improvements in health and reducing crime. Other examples of social value initiatives are illustrated in Figure 4.1.

It is normal practice for these social value initiatives to be encapsulated into a social value charter which is published on the website of the contracting organisation. One example of a contractor's social value charter is that of Morgan Sindall who are a large national UK main contractor. Their ethical commitment to social value is articulated in their social value charter below.

Social Value Charter of Morgan Sindall Contractors

We are delighted to have been instrumental in the development of Building Social Value. Morgan Sindall pioneered the development of the building social value checklist, instigating the reaction and testing of it. Recognising the positive benefits to local communities that our construction activities can bring, we trialled it on our infrastructure and construction projects. The benefits that a third part check by our approved consultants brings is evident in helping our construction clients demonstrate compliance with the Social Value. Whether it is providing work opportunities for local people,

(Continued)

> **(Continued)**
> undertaking educational engagement activities, using locally sourced products or materials, reusing and recycling resources, all of these activities help to enrich the local communities we serve, and we are delighted to have played a part in Building Social Value (Morgan Sindall 2019).

4.5 Different Types and Models of Corporate Social Responsibility

There are different types of corporate social responsibility identified and these relate to legal, ethical, economic, environmental, philanthropy, volunteerism and diversity and inclusivity requirements in the workplace. Carroll (1991) referred to 'The Pyramid of Corporate Social Responsibility' which many organisations have adopted and widely utilised to articulate the main duties that organisations owe to their stakeholders. These types of social responsibility are described and discussed below.

> **Legal social responsibility**
> All organisations must ensure that their working practices are fully legal at all stages. Strict conformity to regulations and legal requirements is essential for all businesses to protect all stakeholders including their clients and consumers. It is important to present accurate information especially to clients and investors and be truthful in reports and forecasts. Any breach of regulations or illegal activity can result in criminal proceedings which could jeopardise the future of those organisations.

> **Economic**
> It may be surprising to learn that one of the key responsibilities for businesses, according to Woiceshyn (2011), is for them to be profitable. This is largely linked to providing owners and shareholders with a return on their investment, creating employment in communities and contributing to the economic prosperity of the local area. Become economically responsible could involve a degree of efficiency measures designed to find the optimum level to operate their businesses. It also could relate to investing profits into research and development and innovation to increase future sales, and therein increase the financial performance of the organisation further. Ethically it is important for companies to be transparent in their accounting procedures and owing to the reliance of investors to make base their decisions on accurate data. Notwithstanding this premise, it is sometimes tempting for companies to hide less than attractive financial accounts on their balance sheets. In recent years legislation has attempted to curve this practice owing to the devastating effects that loopholes can have on individuals if things fail. This has happened many times in the financial services industry where innocent victims, possibly employees or shareholders, have been disadvantaged by unscrupulous large businesses destined to make a profit at all costs.

Philanthropy

The biggest global organisations are closely aligned with philanthropy, sponsorship and fund raising. One example is the founder and former Chief Executive of Microsoft Bill Gates who went to create The Bill and Melinda Foundation to offer poor communities from developing countries around the world the technologies they require to prosper. This was aimed at benefitting the younger generation and equipping them with the means for success and innovation. Philanthropic efforts are not mutually exclusive to large conglomerates and small-to-medium-sized companies can benefit from the philosophies behind fundraising and supporting good causes. An example could be a local clothes retailer who provides free football kits to a local team or a restaurateur who offers their venue free of charge for a community event. There are many good examples of such offering winning the respect and trust of the public and this can have benefits not just for the recipients but for the philanthropic organisations. This emanates from the good marketing that such altruistic interventions make in their local communities. In such cases, the public can see those companies in a positive way which can enhance the value of their brand.

Supporting volunteer efforts

Charities and local communities frequently rely on assistance from local business leaders. This could take the form of those business becoming involved in a community project. An example could be related to addressing a local problem such as litter collection, homelessness, or improving local amenities in the area. Other examples could take the form of company directors joining the board of governors of their local school, college or community sports centre. Having the businesses volunteering in this way and working directly with community groups in this way can create collaborative relationships and build trust. Clearly with smaller businesses time and resources need to be carefully considered in any proposition of this kind to enable them to play a part and make a meaningful contribution.

The aforementioned corporate social responsibility areas, according to Fewings (2009), can be brought together in a business ethics model which can be underpinned by transparent processes, governance and controls. This is illustrated in Figure 4.2 and the areas grouped in to three main themes namely social, economic and environmental. Each theme is supported by robust governance and regulation. For instance, the economic responsibilities are verified by corporate governance in the form of company reports and shareholder communication, and the social responsibilities predicated on employee equality and diversity procedures and community engagement. Furthermore, environmental responsibilities can be closely managed and tracked through environmental and energy audits, e.g. carbon management reports. Policies need to be instigated and managed by organisations to ensure that targets associated with each are of corporate social responsibility are tracked and achieved through established key performance indicators (KPIs). These KPIs will provide the metrics to measure success in these areas and determine whether improvements measures are required where targets are not being met. Such improvement measures may

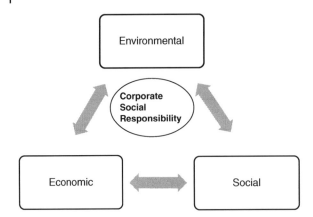

Figure 4.2 The corporate social responsibility model (adapted from Fewings 2009).

entail changing systems and procedures and in some cases policies and strategies themselves. They could simply involve raising awareness of compliance with each area. Training and education for staff as part of continued professional development in this regard coupled with introducing incentives for compliance.

Some would argue that organisations themselves are not the right bodies to be advocating contributions and responsibilities to society at large as they are potentially conflicted with business-related priorities, linked to increasing share value and profits. Sceptical views have emerged that corporations cannot be trusted and relied upon to look after the welfare of people and their environment and the needs of society generally. Accordingly, altruistic offerings for social benefits may become very low on their priority list. However, in the construction industry, contributions to the society and social investments and contributions are dealt with through the planning process. In this regard it is normal as part of gaining planning approval for a Section 106 agreement to be executed. This agreement commits the developer to undertaking works and providing a capital contribution as part of a social value offering. It could include providing sports facilities for local communities or contributing to local infrastructure projects. The scale of the contribution will normally be linked to the size and value of the development to be proportion and fair in this regard.

4.6 The Considerate Constructor Scheme

In the context of the UK construction industry one example of an intervention to uphold the principles of the corporate responsibility model could include contractors committing to the Considerate Constructors Scheme. This is a national initiative set up to raise standards in the construction industry. It is a voluntary Scheme for contractors to look after the well-being of the workforce and general public and take on board environmental issues. It is predicated on an eight-point code of practice based on being considerate to individuals and the communities we live and work within. The codes relate to issues that include cleanliness, good neighbourly practices, the environment, being respectful, responsible, safe and accountable. A Code of Considerate Practice under the Scheme seeks to improve the image of the construction industry by striving to promote and achieve best practice. It outlines the Scheme's expectations and describes those areas that are considered fundamental for registration. The Code is structured into five parts, with each one containing a series of bullet points. Each section of the Code contains an aspirational supporting statement and four bullet points which represent the basic expectations of registration with the Scheme and this is illustrated in Figure 4.3.

Considerate Constructors Scheme
Code of Considerate Practice

Organisations registered with the Considerate Constructors Scheme make a commitment to conform to the Code of Considerate Practice. Through consideration to community engagement, the environment and workforce wellbeing they strive to improve the impact of the construction industry and leave a positive legacy through the implementation of best practice. The Code of Considerate Practice defines the fundamental expectations for registration with the Scheme. The Code applies equally to all organisations registered with the Scheme, regardless of size, type or location.

Respect the Community

Constructors must manage their impact on their neighbours and the public to support a positive experience, by:

- Ensuring courteous and respectful language and appropriate behaviour in and around the construction activity.
- Providing a safer environment, preventing unnecessary disturbance, and reducing nuisance for the community from their activities.
- Proactively maintaining effective engagement with the community to deliver meaningful positive impacts.

Care for the Environment

Constructors must minimise their impact and enhance the natural environment, by:

- Prioritising environmental issues to protect the natural environment and minimising negative impacts.
- Optimising the use of resources, including minimising carbon throughout the value chain.
- Engaging with the community to improve the local environment in a meaningful way.

Value their Workforce

Constructors must create a supportive, inclusive, and healthy workplace, by:

- Actively encouraging and supporting an inclusive and diverse workplace.
- Proactively supporting safe working, mental and physical wellbeing at work.
- Providing workplaces that are, well maintained, clean and secure from physical and biological hazards.

For more information, call 0800 783 1423
www.ccscheme.org.uk

Figure 4.3 Code of considerate practice. Adapted from Considerate Constructors Scheme https://www.ccscheme.org.uk/ last accessed February 13, 2023.

Under the Scheme scores are allocated against each code and registration normally means sites are visited and monitored regularly. Construction companies registering under the Scheme would normally make regular contact with the communities around them and make them aware well in advance of any construction activities likely to cause them any disruption. Clearly mitigation of any such disruption is also a prerequisite for those contractors to take on board at every stage of projects. Furthermore, it is normal for construction companies to issue residents and companies in proximity to their construction sites, newsletters and other communications. These are designed to keep these companies and individual appraised on all issues likely to be of interest to the local community. Such issues could comprise the communication of timescales for implementing projects and any events that the community may be invited such as an opening ceremony. From an environmental ethics perspective there could be 'dashboard' communications on site boards detaining such variables as the amount of waste recycled, energy used and safety records on site. These could be useful for demonstrating commitments to sustainability measures alongside managing a safe site and reducing risks of injuries and/or fatalities.

The Considerate Constructor Scheme is a not-for-profit independent organisation. Construction sites, companies and suppliers voluntarily register with the Scheme and agree to abide by the Code of Considerate Practice, designed to encourage best practice beyond statutory requirements. The Scheme should include and focus on any area of construction activity that may have a direct or indirect impact on the image of the industry as a whole. The main areas of concern fall into the three categories: the general public, the workforce and the environment and therein aligned with the Corporate Social Responsibility Model. The Scheme is designed as a conduit to allow contractors to address the ethical means by which they can work with local communities and reduce disruption to neighbourhoods where their developments are taking place. Scores are allocated against the individual categories. Good scores can be regarded as ethically positive and this can be used by contractors for marketing purpose to bolster their reputation, nationally and locally. The Scheme is predicated on a Site Registration Monitors Checklist, and this is contained in Appendix C. The Checklist supports the Code of Considerate Practice and asks a number of questions to establish what level a site is performing to. It is made up of five sections with ten questions in each section. Closed questions in the Code are aligned to the points of the Code and allow the Monitor to establish whether the site or company has achieved compliance with the Scheme. In addition, open questions directly support the bold questions by asking 'what' is being done and 'how' things are being addressed to meet and exceed the required standards. The Code of Considerate Practice Checklist's bold questions establish if a site is meeting these requirements, and its non-bold questions then ask 'how' these requirements are being met or exceeded, as shown in Figure 4.4.

Each question in the Checklist is supported by a number of 'prompts' and these indicate specific items the Monitor may consider when establishing whether the site has met the expectations of that Checklist question. It is important to note that this list of prompts is not exhaustive and inevitably other items will need to be considered. Some of these items will apply on all sites while others and therefore generic, but some may refer only to certain sites will not. The Monitor will look at the size, type and location of the site, and use judgement to decide what is expected. The prompts can also be used by site managers to understand the

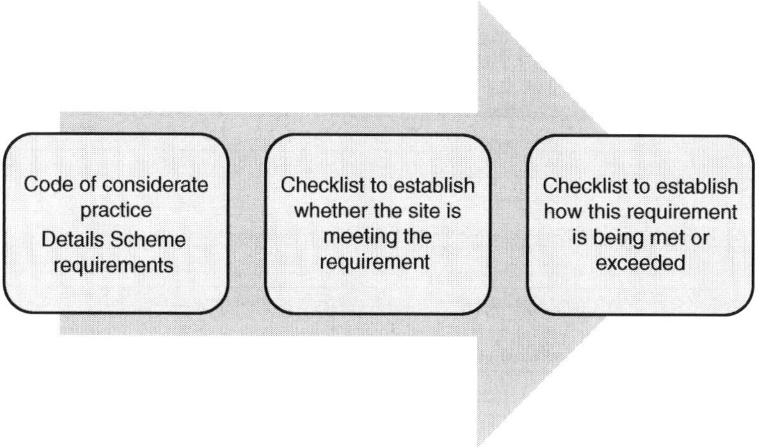

```
┌─────────────────┐   ┌─────────────────┐   ┌─────────────────┐
│ Code of         │   │ Checklist to    │   │ Checklist to    │
│ considerate     │──▶│ establish       │──▶│ establish how   │
│ practice        │   │ whether the     │   │ this requirement│
│ Details Scheme  │   │ site is meeting │   │ is being met or │
│ requirements    │   │ the requirement │   │ exceeded        │
└─────────────────┘   └─────────────────┘   └─────────────────┘
```

Figure 4.4 Code of considerate practice checklist questions.

questions but, as above, the list is not exhaustive and simply addressing each prompt will not necessarily provide a high score. It is expected that contactors will use their initiative to decide what they can and should do to achieve the expectations of the Scheme. Overall scoring against the checklist will fall under one of the five categories in Table 4.1.

Table 4.1 Categories for action and possible measures to address as part of considerate constructors scheme.

Categories for Action	Possible Measures That Address Action
Care about appearance	Constructors should ensure sites appear professional and well managed • Ensuring that the external appearance of sites enhances the image of the industry • Being organised, clean and tidy • Enhancing the appearance of facilities, stored materials, vehicles and plant • Raising the image of the workforce by their appearance
Respect the community	Constructors should give utmost consideration to their impact on neighbours and the public • Informing, respecting and showing courtesy to those affected by the work • Minimising the impact of deliveries, parking and work on the public highway • Contributing to and supporting the local community and economy • Working to create a positive and enduring impression and promoting the Code
Protect the environment	Constructors should protect the environment • Identifying, manging and promoting environmental issues • Seeking sustainable solutions, and minimising waste, the carbon footprint and resources • Minimising the impact of vibration, and air, light and noise pollution • Protecting the ecology, the landscape, wildlife, vegetation and water courses

(Continued)

Table 4.1 (Continued)

Categories for Action	Possible Measures That Address Action
Secure everyone's safety	Constructors should attain the highest levels of safety performance • Having systems that care for the safety of the public, visitors and the workforce • Minimising security risks to neighbours • Having initiatives for continuous safety improvements • Embedding attitudes and behaviours that enhance safety performance
Value their workforce	Constructors should provide a supportive and caring working environment • Providing a workplace where everyone is respected, treated fairly, encouraged and supported • Identifying personal development needs and promoting training • Caring for the health and well-being of the workforce • Providing and maintaining high standards of welfare

4.7 Corporate Codes and Ethical Statements

It is common, especially for large organisations, to market their companies and their brands on corporate codes and ethical statements as part of their corporate social responsibility. In this regard corporate codes and ethical statements should be relevant to the nature of the business and articulate an organisations commitments and stance on a particular issue. By way of an example Marie Curie is a UK organisation committed to caring for families living with a terminal illness. It has clear ethical statements on care and research, fundraising, investments, the environment and purchasing and this is contained below.

Example of an ethical statement: Marie Curie

Care and research
This policy lays out Marie Curie's commitment to delivering the best care to patients, their carers and families.

We strictly comply with all appropriate UK law and Care Commission standards as a minimum and have been commended on a number of activities for our outstanding levels of service and ethical approach.

All Marie Curie staff involved in research and the provision of care have a responsibility to be aware and have a thorough understanding of the ethical issues referred to in this policy.

Fundraising
Marie Curie is committed to its charitable aims and fundraising in order to provide care and research. We concur with the fundraising regulators' promise to be open, honest, fair and legal.

This policy seeks to cover the ethical issues and social responsibility within fundraising. All Marie Curie staff involved in fundraising have a responsibility to be aware and have a thorough understanding of the ethical issues referred to in the policy.

(Continued)
Investments
The objectives of the investment policy are to maximise the total investment return from the funds invested, whether through income or capital growth within a medium risk profile.
The trustees, in delegating their investment management, require their investment managers to pay attention to the standard investment criteria, namely the suitability of the class of investment and the need for diversification in so far as appropriate to the circumstances of the charity. Any restrictions on the type of investments or markets in which the manager may invest on the charity's behalf will be given in writing.
Environment
The policy lays out Marie Curie's commitment to continuous improvement in our actions to decrease our environmental impact.
We comply with relevant environmental legislation as a minimum level of performance and strive to better our practices to minimise waste, energy and carbon footprints whilst achieving the charity's service objectives and ensuring patient care is not adversely impacted.
Purchasing
This policy seeks to cover the ethical issues and social responsibility within supply chains when managing business supplier relationships and the purchasing of goods and services from suppliers.
Upholding the principles of this policy with a shared responsibility between Marie Curie and its suppliers and the policy will be adopted at senior level within the charity. Marie Curie will expect its suppliers to assign and accept similar responsibility.
All Marie Curie staff involved in the procurement of goods and services have a responsibility to be aware and have a thorough understanding of the ethical issues referred to in its policy.

Many organisations are taking the opportunity to publish their ethical statements to promote the reduction in greenhouse gas, carbon production into the atmosphere and as a way to address the climate emergency. The University of Salford are one public sector organisation keen to raise public awareness of their environmental plans and this is articulated in their environmental ethical statement below.

University of Salford Ethical Statement on environmental and sustainability measures
The University of Salford is committed to managing the sustainable design, construction, refurbishment and post completion occupancy of its buildings as part of their life cycle. This is intended to reduce its environmental impact, enhance the well-being of staff and student users of the building, minimise operating costs and comply with all relevant sustainable building legislation.
The University of Salford through its Environmental Sustainability and Energy and Water Policy Statements is committed to addressing the environmental implications of all its Estates and Facilities operations. The University aims to ensure that the adverse environmental impacts of its actions are minimised, and opportunities for delivering

(Continued)

> **(Continued)**
>
> environmental improvements are maximised. The University recognises the value of designing and constructing sustainable buildings.
>
> The University recognises that its sustainability is part of a high-performance building design process and as such sustainability will be embedded in the process of design developments and throughout the life cycle of its operation components which will be reviewed continuously. Clear sustainability requirements for each project will be adopted and communicated to all people involved in the design process including key consultants and contractors. A member of the University Environmental Sustainability Team will be consulted on all construction activities to advise on sustainability risks and opportunities.
>
> Our aim is to demonstrate continual improvement in our environmental performance delivered through an Environmental Management System, certified to ISO 14001:2015, and an Energy Management System, certified to ISO 50001:2011. The University is proud to have achieved certification to EcoCampus Platinum, ISO 14001 and ISO 50001. The ISO standards (for environment and energy respectively) are evidence that our Environmental and Energy Management System meets international environmental standards; we are aware of and compliant with environmental obligations and continually improving environmental and energy performance. The EcoCampus Scheme has been developed specifically for the higher and further education sector and the platinum award is the highest level and equivalent to ISO 14001.

4.8 Sustainable and Ethical Construction Linked to Corporate Social Responsibility

Data from the Office of National Statistics found that the UK construction industry provides built environment assets that are worth approximately 8% of the UK's Gross Domestic Product. Furthermore, it is responsible for an energy usage of 32% of the total domestic energy consumption, when considering construction processes and the life cycle of those facilities. This is quite an alarming reminder of the importance of the built environment in reducing its carbon footprint, and to seek ways and means to reduce non-renewable energy consumption by promoting more sustainable and renewable sources. The built environment is also associated with other environmental issues in term of waste generation and the significant amount of carbon that is generated from what some would describe as antiquated building processes. In the latter case, the UK Government in a concerted effort to promote more sustainable and ethical construction methods earmarked £72m in 2018 as part of its Industrial Strategy; Transforming Construction to address this dilemma.

Industry has been tasked to work with research and development partners such as universities to spearhead cleaner, more energy efficient construction techniques, possibly linked to modular construction technologies. This is regarded as a positive steppingstone to 'breaking the mould' and moving from traditional building methods to more appropriate and less environmentally costly forms. This approach has been heralded as a much overdue initiative according to Farmer (2016). The Farmer Review of the UK Construction Labour Model, commonly known as 'Modernise or Die' was commissioned by the British Government. It addresses the lack of technological advancement in construction and the lack of skilled workers required to deliver the Governments infrastructure and housebuilding targets. Notwithstanding this positive initiative, the UK Government has in the past created carbon reduction targets and failed to deliver on these. In 2009 its Department for Trade and Industry produced Sustainable Construction Reports which forecasted target for zero carbon construction for all new housing by 2016 and all other new build by 2020. These targets have not been met albeit a concerted effort to reduce carbon have been achieved in the period. New targets have now been set by the UK government's committee on climate change for net zero carbon emissions by 2050. This is in contrast to Denmark where Copenhagen is on track to become carbon neutral by 2025 having adopted a climate plan in 2009. For the UK to achieve its 2050 target, it will be necessary to not only reassess the building methods used in the construction industry but also consider how buildings are designed and managed. Criticism has emerged of some building being designed and commissioned with more capacity than they will ever require, and this is leading not only higher construction costs but increased environmental impacts. This inefficiency could be driven by overdesigning buildings as a factor of safety in complying with technical requirements and current design codes. Mechanical and electrical designers have been criticised for 'over engineering' new building services. In this regard, they have been accused of opting for the safe position of designing buildings to be hermetically sealed from their outside environment and incorporating large air conditioning installation to control humidity, temperature and clean air. Similar environmental performance may simply be achieved through opening windows allowing natural ventilation at a fraction of the cost of artificial means but sometimes harder to prove in design calculations.

According to the Association of Consultancy and Engineering there are four key stages that the UK construction industry must make to make the initial push towards carbon neutrality. These are articulated in Table 4.2.

Table 4.2 The four key stages towards carbon neutrality.

Stage One	Take on board the positive examples set by others. Copenhagen airport has recently been certified as CO_2 neutral based on climate compensation and has targets to become a zero-carbon airport by 2030.
Stage Two	Utilise the Construction Sector Deal to focus on new sustainable and environmental performance standards. These could be brought about by strategic innovations and new sustainable building performance standards for the UK construction industry. Such standards should be focused significantly reducing or eliminating waste and reducing carbon emissions.
Stage Three	The UK government should introduce clear and concise strategies and policies to accelerate change. Although there has been much written in the past on revolutionising the UK, with many government reports and commitments to reducing environmental damage. However, the construction industry has been slow to follow and there is a risk of construction organisations becoming embroiled in self-interest at the cost of modernising. This risk should be avoided as it will hamper much needed reforms in the industry.
Stage Four	Despite many government reports over a number of years (Latham 1994, Egan 1998, 2002) the construction industry is still transfixed on procurement based on lowest price tenders rather than best value. Those tenderers which could offer added value in terms of making the improvement to buildings to lower carbon generation are simply being overlooked in a number of cases. Accordingly, the industry needs to establish a robust framework by which environmental performance become a valid and established measure of success for project outcomes.

One excellent example of an organisation that is upholding its corporate and social responsibility is the University of Salford which has published an Environmental and Sustainability Strategy document, and this was referred to in Chapter 3.

4.9 Summary

Corporate social responsibility (CSR) has become a highly topical area in recent years and can be defined as a corporation's obligation to its stakeholders including suppliers, employees, customers and those communities around the business. CSR focuses on optimising the positive impacts on these stakeholders whilst reducing the negative impacts. On this basis, there is a growing requirement that businesses must be responsible from a social perspective rather than being purely focused on financial success and achieving targeted profit levels for their shareholders. Accordingly, for this reason, in today's economy it is imperative for organisations to embed corporate responsibility into their policies, culture and management approaches, and therein bringing reputational benefits from enhancement of their brands and positive public relations. Conversely, where there are breaches of corporate social responsibilities, these can have extremely negative consequences for organisations, especially those larger businesses that rely heavily on brand and reputation. There have been many controversial global cases which have contextualised this potential dilemma. Recent examples include the Deepwater Horizon oil spill in 2010 which became the worse environmental disaster in American history with the loss of 11 lives and which caused untold reputational and financial damage for British Petroleum.

Social value can be affiliated with corporate social responsibility and is a process whereby organisations meet their needs for goods, services and utilities in a way that achieves value for money on a whole life basis; therein generating benefits and positive outcomes. These benefits are regarded as contributing to increasing and enhancing social value and therein benefitting the well-being of people, communities and the general public at large. Organisations should also incorporate social value initiatives and sustainability in their management approaches which should allow them to address social, economic and environmental issues in their local community. The Social Value Portal provides a standardised approach to measurement of the benefits against costs. Notwithstanding this premise, it is important for main contractors to understand local needs before embarking on social value initiatives. Accordingly, it is important for main contractors to undertake a consultation process with the local community to ascertain needs and aspirations and therein gain their 'buy in'. Community consultees in this regard would normally include ward councillors, local business owners, elected officials, residents, colleges, universities and group leaders.

There are different types of corporate social responsibility, and these relate to legal, ethical, economic, environmental, philanthropy, volunteering and diversity and inclusivity requirements in the workplace. These can be brought together in a business ethics model which can be underpinned by transparent processes, governance and controls. Policies need to be instigated and managed by organisations to ensure that targets associated with each corporate social responsibility area are tracked and achieved through established key

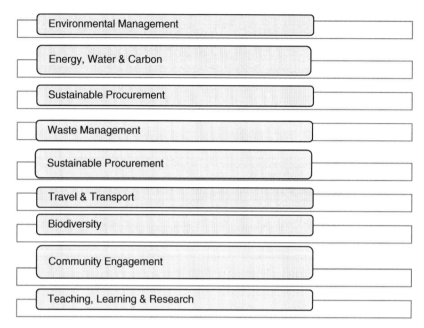

Figure 4.5 Main area of environmental and sustainability strategy; University of Salford.

performance indicators and properly reported on. Figure 4.5 shows the main area of environmental and sustainability strategy of the University of Salford to demonstrate their corporate social responsibility in this regard.

In the context of the UK construction industry one example of an intervention to uphold the principles of the corporate responsibility model could include the Considerate Constructors Scheme which is a not-for-profit independent organisation. Construction sites, companies and suppliers voluntarily register with the Scheme and agree to abide by the Code of Considerate Practice, designed to encourage best practice beyond statutory requirements. It is predicated on an eight-point code of practice which includes cleanliness, good neighbourly practices, the environment, alongside being respectful, responsible, safe and accountable. It seeks to improve the image of the construction industry by striving to promote and achieve best practice. Each section of the Code contains an aspirational supporting statement and four bullet points which represent the basic expectations of registration within the Scheme and designed to encourage best practice beyond statutory requirements.

It is especially important for large organisations, to market their companies and brands on corporate codes and ethical statements as part of their corporate social responsibility and this frequently includes commitments to environmental sustainability. For many years the construction industry has been associated with environmental issues related to excessive carbon and waste generation from what some would describe as antiquated and extremely wasteful building processes. The industry has been tasked to work with research and development partners such as universities to spearhead cleaner, more energy efficient construction techniques, possibly linked to modular construction technologies to improve environmental standards as part of its corporate social responsibility.

References

Carroll, A.B. (1991). The pyramid of corporate social responsibility: toward the moral management of organizational stakeholders. *Business Horizons* 34: 39–48.

Egan, J. (1998). *Rethinking Construction. The Report of the Construction Task Force*. London: DETR.TSO. 18–20.

Egan, J. (2002). *Accelerating Change. Rethinking Construction*. Strategic Forum for Construction. London.

Farmer, M. (2016). *Modernise or Die: The Farmer Review of the UK Construction Labour Market*. London: Construction Leadership Council.

Fewings, P. (2009). *Ethics for the Built Environment*. London: Routledge.

Latham, M. (1994). *Constructing the Team*. London: The Stationery Office.

Social Value Portal (2020). *A Tool for Measuring, Managing and Reporting Social Value*. Available at https://socialvalueportal.com (accessed 1st December 2020).

Woiceshyn, J.A. (2011). Model for ethical decision making in business: reasoning, intuition, and rational moral principles. *Journal of Business Ethics* 104: 311–323.

5

Relationship between Ethics and Other Factors

> *To have a right to do a thing is not always the same as to be right in doing it.*
> K. Chesterton

5.1 Introduction

This chapter will consider the relationship between ethics and other factors including risk, quality, reputation, trust, leadership, and motivation. It will start by looking at the links between ethic and risk and will explain how risks have been traditionally managed in the construction industry by utilising risk assessments and introducing mitigation measures. The ethical questions and dilemmas which arise when considering the probability of risks occurring and the potentially devastating effects they can have will be discussed. Hypothetical and real-life practical examples will be given where risks were taken due to the unlikelihood that they would occur and the catastrophic consequences that ensued. The dilemma for the construction industry whereby known risks are simply passed down the line by clients to main contractors and then by main contractors to subcontractors, who may not have the capacity, competency, and resources to manage such risks, will be examined and the potential pitfalls of such approaches. Other links between risk and ethics will be identified and discussed including scenarios where one party purposely does not give another adequate warning of an avoidable issue and is consequently disadvantaged by such action. Furthermore, risks associated with breaches of ethical standards in commercial dealings and certification of building works, sometimes brought about by the imbalance of power between parties, will be explored.

   The relationship between ethics and quality will then be discussed, especially in the context of the construction industry within developing countries. Various examples and scenarios will be presented wherein unethical behaviour, dereliction of duty and 'cutting corners' of one party has severely compromised construction quality in some cases, resulting in catastrophic failures of buildings and structures and loss of life. The potential ways and means of approaching this problem will be debated through the introduction of ethical codes of conduct and more robust ethical leadership models, especially in developing nations.

The relationship between ethics and reputation of organisations will then be examined. In this sense the chapter will explain where organisations have experienced reputational damage, emanating from acts of fraud and corruption that they had been party to and the negative publicity leading to loss of trust by the public that has ensued. Conversely from a positive perspective the chapter will articulate cases where organisations have developed strong reputations built on the trust and confidence of their customers and the general public through responsible and ethical practices and behaviours.

Ethics linked to trust and collaboration will then be analysed. This will consider the strong partnership relationships that companies can build if they are trusted by other organisations and individuals and the benefits that this can bring to achieving successful project outcomes. Thereafter, leadership traits linked to ethics will be discussed. This will identify the practices which are deemed essential for aspiring leaders including courage, conviction and assertiveness when faced with difficult choices and decisions. It will also highlight the importance of leader's behaviours and actions being grounded in fairness, integrity, and a sense of responsibility. Finally, the relationship between ethics and motivation will be explored in this chapter of the book. This will articulate the benefits that can be gained from pursuing construction management initiatives on projects, especially around job satisfaction and personal development, to motivate staff and therein lead to more successful outcomes.

5.2 The Relationship between Risk and Ethics

According to Fewings (2009) the acceptability of taking risk is based on likelihood of risk materialising. Taking on board this premise, the construction industry is to a large extent no different from other industries in how it manages risk. The degree of risk on a particular project is dependent on several different factors and generally the larger and more sophisticated the proposed building is the more risk is carried by the project. Risk assessments are normally compiled at the conceptual stages and regularly reviewed throughout the project life. In normal scenarios it would be normal practice to address each risk in terms of the likelihood it will occur and its potential impact. Risk assessments should also consider potential mitigation measures to either reduce the probability of the risk materialising or lessening its impact. The ethical debate comes into the project planning stages in considering the probability of the risks emerging and if so their overall effects on the project. For example, if the probability of a risk materialising is low and its effects not that significant, organisations may choose to accept a particular risk, especially where mitigation measures are expensive and potentially disruptive. One could debate whether this is ethically the right thing to do when a potential problem area has been identified, especially when it could lead to devastating outcomes including loss of life. The answer may very much depend on the context of the risk and the circumstances surrounding it. What is probably an easier question to answer could relate to a situation where a risk is known and very unlikely to emerge from a probability perspective, but potentially catastrophic if it occurs. Such devastating effects could be to the organisation itself, its clients, or the public at large. A hypothetical example of this scenario is contained below.

5.2 The Relationship between Risk and Ethics | 65

> **Example of an unlikely risk which caused devastating effects**
>
> A main contractor is constructing a building which involves major excavation and earthworks to create a two-storey basement as part of the build. A structural engineer is employed to design the temporary works necessary to support the excavation during the earthworks. The engineer gives the main contractor two different options for the temporary works.
>
> The first option (Option 1) is the cheapest option and incorporates a scaffolding type of strut arrangement which is easy to erect and dismantle. The structural engineer points out this solution would in all probability be structurally sound but there might be consequential damage in abnormal climatic conditions. The second option (Option 2) involves a much more expensive solution consisting of deep sheet piling capable of withstanding every type of weather and abnormality.
>
> A risk assessment carried out by the main contractor confirms that Option 1 in all probability would be sufficient but still carries a 1% risk of damage in severe weather conditions. The main contractor opts for Option 1 on his basis as it presents a more cost-effective solution.
>
> Four weeks into the project there is record breaking inclement weather which involves strong winds, heavy rainfall over a prolonged period and major flooding of the site. The ground being supported by the temporary works gives way under the weight of the flooded site, the scaffolding collapses and the site excavation caves in. At the time there are ground works in the excavation who become buried in the ensuing catastrophic collapse. Four men lose their lives and there is widespread condemnation of the contractor's actions. The main contractor is subjected to claims that they have taken short cuts for financial benefits which has led to a severe and flagrant breach in health and safety legislation. A public enquiry ensues couple with a criminal case against the main contractor for the statutory breaches and counts of manslaughter. The directors of the main contractor are convicted and receive custodial prison sentences. The main contracting organisation with a reputation in taters, goes into liquidation under the pressure and falls in work orders.

The above example hopefully demonstrates that even a small unlikely risk can materialise and one which should have really been properly considered during the decision-making process on the final temporary works solution. The flagrant disregard for the safest option on economic grounds has caused untold damage to the contracting company and a loss of life. If a more ethical position would have been taken, risks of this kind could not have simply been ruled out on cost and the likelihood of the risk occurring, especially where the potential disastrous consequences are known.

There have been many similar cases to the above example in other sectors and where disastrous consequences have ensued. In 2020 Whirlpool recalled over half a million washing machines sold in the UK between 2014 and 2018 under the brands of Hotpoint and Indesit. These were deemed unsafe following a succession of fires. The court of public opinion was quick to condemn them for insufficient due diligence in their technical research and development of new products. This was the second time in two years that Whirlpool had been required to recall potentially unsafe washing machines. This no doubt caused them untold reputational damage and financial damage for replacement machines and decreases in sales resulting from customer confidence.

It is not always easy for companies to identify every problem but nevertheless their reputations are fragile, and they are only as good as the people who work for them. Even those companies that have ethical strategies, processes and procedures cannot always guarantee that some of their staff will simply ignore them! Companies should have both robust ethical standards and contingencies to deal with unexpected events to respond proactively to these. In this regard, most public sector organisations have disaster recovery strategies in place which covers a multitude of potentially catastrophic events. One could argue that this is both a responsible and ethical position to take given the potential aftermath of disasters if no contingencies are in place.

According to Uff (2003) there are many dilemmas and problems around ethical issues associated with the issue of risk. In the past risk has sometimes been 'passed down the line' with clients passing most risks to their contracted main contractor and those main contractors passing it on to their subcontractors. In most cases these subcontractors, which tend to be smaller companies with lesser resources than their main contractor counterparts, are not able to correctly manage such risks and projects have got into difficulties as a result. Such placing of risk on those organisations who are sometimes clearly do not have the necessary resources, expertise, or experience to manage those risks could be regarded as unethical in this sense. Ideally risk workshops at early project stages should through discussion and collaboration with the wider supply chain should identify which parties are best able to manage risks identified on projects. An example of this could be the argument, when undertaking refurbishment works, that building owners or occupiers are best able to manage certain risks around the possible contamination of a building they own or occupy. This argument is predicated on the notion that building owners/occupiers have greater and more in-depth knowledge of their own buildings than the main contractors they engage to refurbish their premises. In this regard, responsible building owners should have conducted surveys on their premises for contaminants such as asbestos and already should have registers and management plans in place. Another example could relate to main contractors passing on financial risks relating to their 'domestic subcontractors'. Domestic subcontractors are those who are known, selected, and appointed solely by main contractors as opposed to nominated or named subcontractors where clients have an involvement in the selection and appointment process. In this case, it could be regarded as unfair for clients to take on this responsibility when they were not involved in the selection or appointment of these domestic subcontractors.

In some cases, contractors and subcontractors are not given any choice in negotiating or agreeing standard contractor terms provided to them by their employers. This can sometimes put a relatively large amount of risk on them, which could be regarded as unreasonable and therein unethical. Even when an organisation might be the right type of skill sets and experience to take such risks, it could also be seen as unethical if they are deemed not to have adequate financial resources for taking on board such risks. An example of such a scenario could be where a small fire protection consultancy has been requested to take all the associated risks in the event of a building fire. Accordingly, construction professionals and their appointed legal representatives owe a reciprocal ethical duty of care to avoid imposing risk on another party which cannot be economically or practically borne.

Furthermore, Uff (2003) suggested that construction professionals owe an ethical duty to prevent cheating and unconscionable conduct, in a similar way that they owe duties to avoid corruption and bribery under criminal law.

During the construction stages of projects there may be ethical considerations associated with one party not giving another adequate warning of an avoidable issue. This raises the question of a 'duty to warn' where there might not be a contractual duty to do so, but more an ethical and moral obligation to do so (Uff 2003). In the past the legal system has found such a duty in the law of tort. An example could be where a construction professional is aware of a possible impeding danger associated with a project, which could result in a subsequent injury or fatality. This may explain why in recent years, with tightening regulations around health and safety on construction sites, main contactors have adopted rigorous process and procedures related to reporting near misses.

Other ethical issues could apply to those construction professionals such as architects and project managers who hold the balance of power between clients and contractors in terms of certifying various aspects of construction management. An example could be where project managers are tasked to agree or not agree to an extension of time for a main contractor, whilst they are employed by their clients, who may suffer a financial impact as a result. In some cases, project managers have taken a less than arbitrary position in favour of their paymasters which could be classes as unethical and unfair practices. The significance of the certifying process has led to disputes and adjudication proceedings in extreme cases. For these reasons, Uff (2003) suggested that all construction professionals should be placed under an ethical duty to ensure that their projects are managed in a fair manner, avoiding the temptation to act in adversarial ways and upholding the interests of the other parties. Accordingly, some would argue that a multi-disciplinary ethic should be created and applied to all construction professionals.

Ethical issues can also emanate after construction projects have been completed. An important part of any construction project normally comes post completion, and this relates to agreeing final accounts. This can sometimes not purely relate to contracted work valuations but transgress into claims and counterclaims where variations or 'critical incidents' have occurred. An example could be where a contractor submits financial claims for a critical incident that occurred many weeks or months before, and for which their clients were unaware of. This can lead to potential disputes where financial claims are significant and considered unfair by clients who may not have the increased budget to settle such claims. To overcome these type of scenarios, most partnering building contracts such as the New Engineering Contract (NEC 3) have 'use it or lose it' clauses that stipulate that they must inform their clients of any critical event as they arise, or alternative forfeit any associated claims.

5.3 The Relationship between Ethics and Quality

According to Abdul-Rahman et al. (2010) the issue of poor quality of construction projects, linked with questionable professional ethics has been widely publicised, especially in those developing countries around the world. With the move over recent years globally to focus on quality-related outcomes rather than primarily cost and time, this has highlighted the need for construction professionals to focus more intensely on ethical aspects. This argument is supported by Hamzah et al. (2010) who advocated that construction quality is strongly influenced by ethical practice and behaviours. The correlation between ethics and quality could be possibly explained from the position that they are both built upon the promise of doing the right thing. In this sense, good ethical conduct can create the right conditions for trust to build and to optimise value for money, improve competitiveness and increase client satisfaction levels. Moreover, low ethical standards can have the reverse affect and lead to low-quality outcomes.

The following practical examples around issues relating to the relationship between quality and ethics, highlight the dilemmas that exist and the devastating aftermaths that can ensue from dereliction of duties on projects.

> **Case Study 1 An unethical practice affecting a catastrophic and tragic building failure: Kansas City Walkways Hotel**
>
> An example of a scenario where quality standards were compromised by poor ethical practices was played out in the case study of the Hyatt Regency Kansas City Walkways Hotel project in 1973. This project incorporated a raised walkway spanning an atrium within the hotel. Intent on saving costs on the project, the walkway was compromised through 'value engineering' measures which weakened the steel design and ultimately the structural integrity of the walkway structure.
>
> One year following completion, the walkway, where people were dancing, collapsed into the crowded atrium causing one hundred fatalities and more than two hundred people injured. Investigations revealed that the original design, albeit compliant with the design codes at the time, was responsible for the collapse and not designed to cater for the numbers that were loading the walkway on the evening in question. The consulting structural engineers were found to be negligent, and it was revealed there was too much emphasis and responsibility placed on the steel work subcontractor. The structural engineers were subsequently charged with misconduct, negligence, and incompetence and subsequently lost their license to practice engineering as a direct result of the case.

> **Case Study 2 An unethical practice affecting the design of a new hospital: Johor Bahru Hospital**
>
> A newly opened hospital in Johor Bahru, capital of the Malaysian state of Johor, was found to have a life-threatening mould and fungal attack on its wall, floors, and equipment. This problem emanated from compromises which were made to specification of the hospital leading to a poor overall design. Owing to a poor specification for mechanical installations, condensation was allowed to accumulate on surface areas. Furthermore, the omission of properly designed air conditioning filtration led to the introduction of mould and fungus which could have caused the spread of diseases to patients. It was deemed that the hospital project team were responsible for a serious dereliction of ethical duties in the hospital design.

(Continued)

> **(Continued)**
>
> Fortunately, there were no fatalities in this case but nevertheless illustrates the threat and associated risks from unethical behaviour which compromised the health and safety of the hospital community and vulnerable patients. The reputations and track record of the contractor and project team were severely tarnished in this case which resulted in financial burden of expensive remedial works and reduced future work orders.

5.4 Potential Impact of Professional Ethics on Construction Quality

Research carried out by Besterfield et al. (2003) concluded that the quality of service is dependent on ethical behaviour. Furthermore, their studies confirmed that ethics and quality have an aligned premise built up on a desire to do 'right things right'. In this way they found that this to be a proven way to improve competitiveness by reducing costs whilst improving client satisfaction. A similar study was carried out by Abdul-Rahman et al. (2010) where most respondents (93%) agreed that acts of unethical behaviour or practices contribute to quality-related problems in the construction industry. In addition, most of the same respondents (72%) agreed that unethical conducts could be the predominant cause of poor-quality outcomes in the construction industry. Similar findings which revealed that a majority of its respondents (63%) agreed that unethical acts are commonplace within the construction industry in developing countries. More disturbing data from the same study also found that 84% of respondents had witnessed, experienced, or encountered business dealings that they perceived to be unethical. The same respondents felt that such a shift to unethical conducts of one form or another could lead to poor-quality service delivery and project outcomes, alongside monetary loss for their clients. In the latter case they estimated that as much as 10% of global construction output in 2004 was lost to some form of corruption.

Analysis from the study by Abdul-Rahman et al. (2010) determined that quality and unethical behaviours and practice are positively correlated. More than 90% of respondents articulated that quality-related problems in construction can emanate from unethical conduct. This correlation result gives strength to the prognosis that quality and ethics are not only related but can both have a strong impact on the quality of construction outputs. In view of the findings from Abdul-Rahman et al. (2010), the research also offered ways to improve professionalism in the construction industry. In this regard, the research recommended that greater levels of professional practice and behaviour could be generated by establishing standard codes of ethics for the construction industry and by leaders serving as role models.

5.5 Reputation and Ethics

The reputation of organisations and individuals can be heavily affected by how well or badly they maintain ethical standards and adhere to codes of conduct within their professional practice. There have been many examples of organisations which had experienced reputational damage, emanating from acts of fraud and corruption that they had been party to. Such cases will be discussed throughout the remainder of the book and have caused negative publicity for these companies leading to loss of trust by the public.

The converse situation can arise where organisations commit to ethical practices and have policies and procedures geared around upholding moral values and professional codes of conduct. In such cases they have gained positive publicity and their reputations have been enhanced in this way. Those organisations which have 'gone the extra mile' in committing to staff development and improving job satisfaction for their employees have found that this can be very rewarding for them and raised their ranking on schemes such as the Best Companies. The Best Companies benchmark and accreditation scheme represents a significant achievement and shows that employers are taking workplace engagement seriously. It is a model based on ethical leadership predicated on employee interaction and an unconditional and unequivocal commitment to the workforce. In a similar way, organisations that have invested heavily in commitments around sustainability and environmental measures have experienced positive public support which in some cases has led to increased work orders.

5.6   Ethics and Trust

Trust and professional ethics should be regarded as a collaborative necessity which construction professionals should be encouraging and closely managing on their projects. It is crucial to consider the importance of professional ethics for building trust, adopting collaborative policies, and developing good working relationships, as there is a reciprocal correlation between all these factors. There are many interdependencies between these groups where the behaviour, successes, or failures of one party could affect another in a positive or negative way (Challender 2019). It is for this reason that trust built upon partnership and collaboration working relationships has a strong reciprocal correlation with professional ethics.

Furthermore, there is an argument that building client trust is very important for consultants and contractors, especially in the early stages of relationship building. Many construction professionals may report that, owing to the fragility of trust, a loss of client confidence can result in loss of credibility and permanently damage relationships. Such is the importance and significance of the relationships between professional ethics and trust; this area has been covered separately under Chapter 8.

5.7   Ethics and Leadership

There have been many different academic theories that have forged various links, correlations and relationships between ethics and leadership. When leadership is particularly good, certainly within a construction context, this can gain the confidence from other members of the project team and give them the assurance that ethical actions, behaviours and decision-making are being practiced. Applied in practice it is essential for aspiring leaders to have courage, conviction and assertiveness when faced with difficult choices and decisions and their behaviours and actions should always be subject to fairness, integrity and a sense of responsibility. For this reason, it is essential for aspiring leaders to understand what makes a good leader, to perform their roles and motivate others for achieving successful outcomes. Benefits which can derive from good ethical leadership include developing the workforce and improving the service to their clients and supply chains whilst maintaining the reputation of their respective organisations. This can make staff feel valued, listened to and engaged in their businesses. It can also lead to a feeling of being proud of the organisations they work for. This positivity can have ethical considerations in them being more prepared to abide by professional codes of conduct and not put their employer's reputation in jeopardy. In this sense training and development are regarded as being the pillars which underpin an ethical, successful, and motivated workforce and excel a company to achieve successful outcomes and increased growth. In a UK organisational context, the Best Companies benchmark and accreditation scheme represents a significant achievement and shows that employers are taking workplace engagement seriously. It is a model based on ethical leadership predicated on employee interaction and an unconditional and unequivocal commitment to the workforce.

Imperial College Business School (2020) advocated those responsible leaders should focus on others rather than solely themselves, consider the longer-term ramifications and implications in their decision-making and thought processes and follow a moral compass in their management approaches. Accordingly, it is important when considering the traits of ethical leadership for people in authority to lead by example and make decisions that make those around them feel safe, valued, and listened to. Conversely, the relationship between poor leadership and ethical adherence can result frequently in project failure especially in the areas of project initiation and planning, governance, people and technical competent. For these reasons, leaders should ensure that measures are all in place which include adopting the right project controls, communication strategies, ethical codes of conduct and collaborative approaches to work practices. Such is the importance and significance of the relationships between professional ethics and leadership; this area has been covered separately under Chapter 9.

5.8 Ethics and Motivation

Leaders should be aware of the advantages in keeping project teams motivated by encouraging a culture of staff development through training and education, especially around ethical issues, and compliance with professional values. In cases where there are no financial incentives or where pay for undertaking a particular role in the construction industry is extremely low, this could be deemed unethical in not paying the going rate and possibly illegal if below the minimum wage in the UK. This can not only compromise motivation levels on behalf of workers, leading to poor productivity but also cause reputational damage for companies and allegations of potentially exploiting staff.

There are significant benefits that can be gained from pursuing construction management initiatives on projects, especially around job satisfaction, to motivate staff and therein lead to more successful outcomes. Such is the importance and significance of the relationships between professional ethics and motivation; this subject area has been covered separately under Chapter 10.

5.9 Summary

This chapter has identified and discussed the relationship between ethics and several other factors including risk, quality, reputation, trust, leadership, and motivation.

When considering the relationship between ethics and risk, the acceptability of taking risk is based on likelihood of a particular risk materialising. The ethical debate in a construction context comes in the project planning stages when considering the probability of the risks emerging and their overall effects on projects. For example, if the probability of a risk materialising is low and its effects not that significant, organisations may choose to accept a particular risk, especially where mitigation measures are expensive and potentially disruptive. One could debate whether this is ethically the right thing to do when a potential problem area has been identified, especially when it could lead to devastating outcomes including loss of life. Several examples were provided in this chapter to identify catastrophes that have occurred in the past to reinforce this point. To add further complexities around risk decision-making, even those companies that have ethical strategies, processes and procedures cannot always guarantee that some of their staff will simply ignore the risks. Accordingly, companies should have both robust ethical standards and contingencies to deal with unexpected events to respond proactively to these. In this regard, most public sector organisations have disaster recovery strategies in place which covers a multitude of potentially catastrophic events.

In the past risk has sometimes been 'passed down the line' with clients passing most risks to their contracted main contractor and those main contractors passing it on to their subcontractors. Such placing of risk on those organisations who are sometimes clearly do not have the necessary resources, expertise, or experience to manage those risks could be regarded as unethical in this sense. Ideally risk workshops at early project stages should through discussion and collaboration with the wider supply chain should identify which parties are best able to manage risks identified on projects and this is considered the most ethical approach. During the construction stages of projects there may be ethical considerations associated with one party not giving another adequate warning of an avoidable issue. This raises the question of a 'duty to warn' where there might not be a contractual duty to do so, but more an ethical and moral obligation to do so. Other ethical issues could apply to those construction professionals such as architects and project managers who hold the balance of power between clients and contractors in terms of certifying various aspects of construction management. The significance of the certifying process has led to disputes and adjudication proceedings in extreme cases. Ethical issues can also emanate after construction projects have been completed and can involve risks around financial claims for events and variations which can invariably lead to disputes and litigation. An example could be where a contractor submits financial claims for a critical incident that occurred many weeks or months before, and for which their clients were unaware of.

When considering the relationship between ethics and quality, the issue of poor quality of construction projects, linked with questionable professional ethics has been widely publicised. This is especially in those developing countries around the world construction where quality is strongly influenced by ethical practice and behaviours. The correlation between ethics and quality could be possibly explained from the position that they are both built upon the promise of doing the right thing. In this sense, good ethical conduct can create the right conditions for trust to build and to optimise value for money, improve competitiveness and increase client satisfaction levels. Conversely, low ethical standards can have the reverse affect and lead to low-quality outcomes. Two real life examples to reinforce this dilemma have been given in this chapter.

Research carried out by Besterfield et al. (2003) focused on the potential impact of professional ethics on construction quality, concluded that the quality of service is dependent on ethical behaviour. Furthermore, their studies confirmed that ethics and quality have an aligned premise built up on a desire to do 'right things right'. Analysis from the study by Abdul-Rahman et al. (2010) determined that quality and unethical behaviours and practice are positively correlated. From this study most respondents (93%) agreed that acts of unethical behaviour or practices contribute to quality-related problems in the construction industry. In addition, findings showed that a shift to unethical conducts of one form or another could lead to poor-quality service delivery and project outcomes, alongside monetary loss for their clients. This correlation result gives strength to the prognosis that quality and ethics are not only related but can both have a strong impact on the quality of construction outputs.

The reputation of organisations and individuals can be heavily affected by how well or badly they maintain ethical standards and adhere to codes of conduct within their professional practice. Reputational damage, emanating from acts of fraud and corruption that they had been party to, can cause them negative publicity leading to loss of trust by the public. The converse situation can arise where organisations commit to ethical practices and have policies and procedures geared around upholding moral values and professional codes of conduct. In such cases organisations have gained positive publicity and enhanced their reputations. The Best Companies benchmark and accreditation scheme represents a significant achievement and shows that employers are taking workplace engagement seriously. It is a model based on ethical leadership predicated on employee interaction and an unconditional and unequivocal commitment to the workforce.

Trust and professional ethics should be regarded as a collaborative necessity which construction professionals should be encouraging and closely managing on their projects. According to Challender 2019, trust built upon partnership and collaboration working relationships has a strong reciprocal correlation with professional ethics. In addition, there have been many different academic theories that have forged various links, correlations and relationships between ethics and leadership. When leadership is particularly good, certainly within a construction context, this can gain the confidence from other members of the project team and give them the assurance that ethical actions, behaviours, and decision-making are being practiced. Leadership decisions and their behaviours and actions should always be subject to fairness, integrity, and a sense of responsibility. Benefits which can derive from good ethical leadership include developing the workforce and improving the service to their clients and supply chains whilst maintaining the reputation of their respective organisations. Conversely, the relationship between poor leadership and ethical adherence can result frequently in project failure especially in the areas of project initiation and planning, governance, people and technical competent. Furthermore, leaders should be aware of the advantages in keeping project teams motivated by encouraging a culture of staff development through training and education, especially around ethical issues, and compliance with professional values.

References

Abdul-Rahman, H., Wang, C., and Yap, X.W. (2010). How professional ethics impact construction quality: perception and evidence in a fast-developing economy. *Scientific Research and Essays* 5 (23): 3742–3749.

Besterfield, D.H., Michna, C.B., Besterfield, G.H., and Sacre, M.B. (2003). *Total Quality Management*, 3e, USA: Prentice Hall.

Challender, J. (2019). *Building Collaborative Trust in Construction Procurement Strategies*. Oxon: Wiley.

Fewings, P. (2009). *Ethics for the Built Environment*. London: Routledge.

Hamzah, A.R., Chen, W., and Wen Yap, X. (2010). How professional ethics impact construction quality: perception and evidence in a fast-developing economy. *Scientific Research and Essays* 5 (23): 3742–3749.

Imperial College Business School (2020). *How to think like a responsible leader.* Available at https://www.imperial.ac.uk/business-school/ib-knowledge/strategy-leadership/how-think-responsible-leader (accessed 2nd December 2020).

Uff, J.P. (2003). Duties at the legal fringe: ethics in construction law. *Society of Construction Law* October 2010. Retrieved from www.scl.org.uk.

6

Ethical Human Resource Considerations

We're trying to move away from looking at ethics as scandal to ethics as strategy. We're lifting the hood and seeing the concrete strategies companies are using and this might help us to see what other companies can do.

Peter Singer

6.1 Introduction

Leaders of businesses across the globe recognise that ethical human resource considerations are becoming increasingly important for their organisations. This chapter has been dedicated to this area and covers aspects including inclusivity, equality and diversity. It suggests best practices to enable organisations to uphold policies and strategies in this regard and articulates the benefits that this will have for their staff and their businesses. The issues of discrimination and the consequences for organisations in breaching regulations around human resource management are also articulated. For this reason, a toolkit for change and reflection on how the construction industry can improve diversity and inclusion has been created to help organisations to help deter breaches and bad practices. The chapter also discussed the ways and means that organisations can improve the culture of their workplaces and suggests initiatives and measures that can be adopted in this regard. Finally, the chapter deals with the requirements of the Modern Slavery Act 2015 and also requirements as part of the General Data Protection Regulations (GDPR).

6.2 Inclusivity, Equality and Diversity

Inclusivity, equality and diversity in the workplace are not only beneficial but, in most countries, governed by law with severe penalties for breaches in employment law. Policies, procedures and instruments should be carefully monitored by human resource departments to ensure that these are applied to all employees, both existing and new, and at every level with organisations. Most larger companies operate Equal Opportunities Policies linked to offering staff of all genders, race, religion and disability the same employment rights and do not differentiate on the grounds of any of these variables. In the UK and many other countries there are act of parliament to prevent discrimination in such cases and these include the Disability Discrimination Act (1995) and Equality Act (2010).

Management policies and protocols should also be in place for dealing with complaints and potential breaches and violations. This can avoid any costly legal action brought by employees who feel they have been treated unfairly and unjustly and can create a culture of positivity in the workplace. Having this culture and recognising a diverse and varied workforce of employees should build the foundations for high morale and motivation alongside future prosperity.

Some high-profile cases in the global media should emphasise to us that no organisations are immune from claims of sexual harassment, prejudices, racism or any other form of treatment to a minority. Damage to corporate brands from such cases has in the past reminded us how potentially fragile organisations and individuals can be to allegations of wrongdoing. This clearly can have very damaging consequences.

6.2.1 Maintaining Cohesive, Inclusive and Safe Working Environments

The statutory requirements of the Equality Act 2010 play a major part in maintaining diversity and equality in the workplace. Notwithstanding this assertion, the creation of a truly inclusive organisation is the creation and sustaining of respect and trust between disparate groups and individuals. This involves raising awareness of the different perspectives of people and is fundamental for the development and maintenance of good relations as building blocks for good relations. If good relations exist between different communities, this will facilitate and encourage the recruitment of a wide range of construction-related personnel across the industry. Accordingly, it is important for organisations to consider how an agenda for inclusion is managed. This does not imply that every person has to think and act in similar ways and certainly stereotyping needs to be firmly avoided. It is more about respecting differences within the context of the wider construction society and listening to others across the whole supply chain.

6.2.2 Allowing Freedom of Speech and Open Dialogue

Freedom of speech is protected amongst employees in the UK. It is not so much related to endorsing a particular viewpoint, commonly known as Groupthink, and suppressing what people should believe and think. Conversely, it is more about asking people what they think like they do and asking them to consider other people's rights, beliefs and perspectives. Having an effective and pragmatic framework of operation allows for open debate in safe environments.

Construction personnel are emerging into an increasing globalised world, with the UK made up of many different nationalities, mostly the European Union. Organisations should create opportunities for employees to acquire intercultural competencies in which they are able to negotiate, faith and cultural boundaries, with respect and tolerance. This can be a challenge for some migrant workers and other minorities at times and important in creating a sense of belonging that is shared by the whole construction industry. Promoting the development of open forums where debates and new idea can be facilitated is a key role for construction organisations in this regard. They should effectively manage diverse viewpoints and orchestrating respectful debates on a whole range of issues for the benefit us all. Notwithstanding the above, freedom of speech does not mean anyone can say anything they want to. It is the context and how we express our views that are important, so personal and organisational boundaries need to be put in place.

6.2.3 Ethics around Equality and Diversity and Inclusion

Equality relates to ensuring that everyone has an equal opportunity and is not discriminated against or treated any differently as a result of their background or characteristics. Diversity revolves around how we think, look and behave. It takes account of the differences between people and groups of people and places a positive value on those differences. Differences in this regard could relate to personal characteristics, socio-economic status and location.

Inclusion goes further than just recruiting and retaining a diverse workforce. It is about the creation of an environment where differences are regarded as an advantage, where differences and perspectives are shared, leading to better decision-making and working cultures. An inclusive environment is one in which everyone feels that the contributions they make, large or small are valued. Furthermore, it creates the means by which individuals can reach their full potential irrespective of who they are and what their circumstances may be.

While UK legislation around certain characteristics as gender, disability and age sets down minimum standards, an effective diversity and inclusion strategy goes beyond legal compliance. It seeks to add value to an organisation, contributing to employee engagement and well-being. Most medium to large-scale organisations in the construction industry have many different policies, strategies and initiatives linked to embracing equality, diversity and inclusion and not these are underpinned by training workshops and continuous professional development. These have in some cases improved the workforce and had the added benefit of generating good public relations and image for the organisation. Those organisations that deploy such moral and ethical policies may see increased staff retention and increased work order resulting from such reputational benefits. Open communications routes are fundamental to making progress on diversity and inclusion and these are illustrated in Figure 6.1. These communication routes create dialogue with people based on their narratives and experiences in terms of disability, faith and beliefs, sexual orientation or transgender identity.

It is important for mangers in construction industry to demonstrate empathy and understanding of peoples experiences, whilst acknowledging that we have all have different perceptions and realities in any given situation. The concept of diversity involves an understanding that everybody is different and unique and so recognising such differences and treating them with the respect they deserve. In this regard all construction employees should be treated with kindness, courtesy and politeness. They should be encouraged to speak to their colleages to express ideas and opinions, which is what drives innovation in construction management, especially where people's ideas improve or change working practices. Furthermore, individuals should listen to what others say before expressing their own views and opinions. This will build trust and establish rapport.

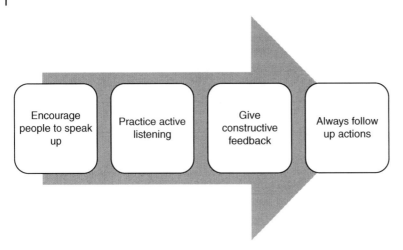

Figure 6.1 Open communications routes for inclusion.

Construction personnel may be diverse in nature in terms of religion, nationality, gender, ethnicity, age disability, sexuality, background, education, social class and many other respects. It is important to embrace these differences and acknowledge the diversity of the workforce. In respecting their beifefs, goals, experiences and situations and developing strategies that respond to this diversity is the most importmat step in making construction personel feel that they can succeed.

6.2.4 Disability Equality

Construction personel with disabilities should be accomodated wherever possible within the construction industry. Clearly there may be boundaries to the extent to which certain disabilities can be accommodated on construction sites owing to obvious health and safety issues such as blindness and those confined to wheel chairs. However, organisations should think creatively about the opportunities that can be made available for the more severely disabled within their overall business. Notwithtsanding this assertion, construction personel with disabilities often require additional support and depending on the severity of their disabilities, this may not always be feasible to accommodate their needs. However, there are many different degrees of disability and it is important not to stereotype the needs and situations of all disabled people into one or two categories. The more ethical organisations will have policies, procedures and processes in place for ensuring that disbled personel are not given less favourable treatment to those colleagues who are able bodied. Normally this approach at an organisational level would be part of an organisation's Equal Opportunities Policy, which would make commitments to undertaking reasonable ajustments to workplaces and their operations to ensure compliance.

6.2.5 The Equality Act 2010

It should be made clear that equality and diversity is not just an ethical consideration. In the UK, The Equality Act 2010 is there to protect every employee from unlawful discrimination and this is supported by other related legislation such as the Public Sector Equality Duty. The Equality Act 2010 created nine 'protected characteristics' that cover age, disability, sex, religion and belief, race, sexual orientation, pregnancy and maternity, marriage and civil partnership and gender reassignment. The Act makes it unlawful to discriminate on any of these protected characteristics. For instance, an example could be direct discrimination around pregnancy and maternity leave. In this case the female employees would not need to demonstrate that she has been treated less favouarbly in comparison to someone who is not pregnant, but more so she has been treated unfavourably because she is pregnant or on maternity leave. In the latter case this could involve the employee being reassigned a lower grade position wihin the organisation directly as a result of being on maternity leave. Some organisations have for ethical and legal compliance reasons undertake an Equality Assessment (also know as an Equality Analysis) whereby they consider the effects of policies and decision have on different groups protected from discrimination by The Equality Act 2010. It relies on utilising information on equality and the results of engagement with groups that are protected and others to comprehend the potential and actual effects of functions, decisions and policies. It can assist in identifying those practical steps necessary to address any discrimination or negative effects, to foster good working relationships and advance equality.

All staff should ideally be inducted and receive regular continuous professional development (CPD) to ensure that they are familiar with the key requirements of discrimination law in this regard and therein understand the associated risks to their organisation and themselves for breaches. Such breaches could result in disciplinary actions for employeess for acts which could be deemed as misconduct and possibly even dissmisal for acts of gross misconduct. Breaches can have widespread implications also for the organisation, leading to fines, reputational damage and reduction in staff engagement.

6.2.6 Liability for Discrimination

Employers are legally responsible for acts of discrimination, victimisation and harrassment carried out by their employees in the course of their employment. It is not dependent on what the employer knew or did not know about the act of discrimination. Construction organisations have an overiding responsibility to protect their employee from discrimination. Individual employees can be held responsible for acts of unlawful discrimination provided their organisations can prove that they have provided their employees with a resonable level of ability to carry out their duties and an awareness of their ethical and legal responsibilities. It is also a requirement of organisations to demonstrate that they have taken all reasonable steps to prevent employees or agents from acting unlawfully, to avoid legal responsibility.

6.2.7 Types of Discrimination

There are many different types of discrimination and these include direct, indirect, associative and perceptive discrimination.

Direct discrimination would normally apply where an individual treats another individual less favourably than they otherwise would do, primarily as a result of their protected characteristic groups; age, sex, religion or belief, sexual orientation, disability or gender reassignment. Indirect discrimination would normally occur when a provision, practice or criterion is neutral on the face of it, but its impact particlarly disadvantages others with a protected characteristic. The exception would be where the individual applying the provision can provide justification as a proportionate means of achieving a legitimate aim. Associative discrimination is related to direct descrimination based on a persons association or friendship with another person blonging to a protected characteristic. Finally, perceptive discrimination would refer to direct discrimination where a person believes than another has a protected characteristic, when in reality they do not. Accordingly, this type of discrimination does not just impact on minority groups but anybody who is treated differently because of perceived or actual personal characteristics.

6.2.8 Best Practice for Construction Organisation around Embedding Equality in the Workplace

Organisations have an ethical, moral and legal responsibility to address equality issues in the workplace with policies, procedures and processes in place to promote best practice. They should minimise or remove disadvantages suffered by individuals who share a relevant protected characteristic and devise and implement processes and procedures that are consistent, accessible, transparent and fair. In addition, organisations should take steps to satisfy the requirements of people who have protected characteristics. One example of this could be to undertake reasonable adjustments for those persons who have physical disabilities and making construction workplaces safer and more accessible accordingly. Construction companies should also encourage their employees to share a relevant protected characteristic and not feel reticent to reveal personal characteristics. This will raise awareness of some of the requirements that this group of people will benefit from initiatives and measures to make working practices more appropriate.

6.2.9 Victimisation, Harassment and Bullying

Victimisation involves penalising someone for exercising their rights under equality legislation, or for making an internal complaint or supporting someone else in doing so. Harassment normally involves penalising someone for exercising their rights under equality legislation or for making an internal complaint or for supporting someone else in doing so. Bullying can be defined as 'coercion or intimidation which serves to undermine the competence, effectiveness, confidence and integrity of another'. In most cases bullying within an organisational context involves the abuse of power, position or knowledge by one person to unfairly criticise or humiliate another, causing the individual being bullied to feel vulnerable, damaged and lacking in confidence.

The most ethical of companies have robust policies, procedures and management approaches linked to reducing and eliminating victimisation, harassment and bullying in the workplace. Clearly these issues are not only ethical issues and covered by employment law and employers have a duty of care to their staff to protect them and pursuing policies not linked to a 'one size fits all' approach. This is particularly the case in the construction industry which is one of the largest sectors for employment and composed of many different individuals and cultures. In this context managers need to be also mindful that when recruiting staff, processes allow their organisation to employ the best person for the job, reap the benefits of a diverse workforce, and avoid allegations of unlawful discrimination. It is ethical unacceptable and a breach of The Equality Act 2010 to treat an applicant less favourably because of a protected characteristic at any point in the recruitment and selection process. Taking this into account, employers should avoid making assumptions about the type of person who will be able to perform the job. One example of this may be to deter any phrasing of job advertising which may imply a particular suitable age group or gender. Furthermore, particular attention should be focused on not including unjustifiable requirements which purposefully or inadvertently discriminate against disabled people. Additionally, reasonable adjustments should be made by employers and made clear at the recruitment stages, to encourage non-able-bodied persons to apply and be considered for vacancies. Accordingly, this can attract a more diverse workforce. In conducting fair and ethical processes construction employers should also avoid unconscious bias.

Every individual is unique and albeit individuals have many things in common, they are also different in many ways. It would be unethical to treat individuals differently on the basis of their background, personality, accent, work style, language or culture. Sometime the first step to overcoming the influence of bias is to recognise that it exists. After all, arguably everyone has biases, and these can affect our ability to make good decisions. It is therefore important to identify them and put them aside to support the development of all staff. Stereotyping groups of people should also be avoided. An example of this could be related to foreign trade's persons not possessing the same level of expertise and experience as the domestic workforce. Similarly, prejudices based on opinions and attitudes formed beforehand without sufficient information or from misinformation should be avoided. Such prejudices could unfavourably affect the thinking about a particular person or groups of people and make them feel less positive about them.

In summary, strategies linked to employers seeking to adopt an inclusive diversity strategy should look to the following measures and interventions:

> Enable reporting of inappropriate behaviours to support protection from discrimination, harassment and bullying at work
> Assess what the diversity impacts are and how to mitigate those aspects
> Build inclusive and regular communications processes
> Create a working environment and culture based on dignity and respect for everyone
> Include diversity aspects and impacts in all planning aspects of businesses
> Ensure development activities and learning opportunities are available to all staff
> Undertake fair and transparent recruitment
> Establish what the diversity impacts are and how to mitigate those aspects

6.3 Ethical Issues around Culture

Managers in construction-related organisations, like other sectors, need to contribute to developing and maintaining a culture which makes diversity and inclusion a reality. Some of the policies and initiatives that this could entail in practice could include the following:

Employees may encounter individuals who behave in an unacceptable and inappropriate manner or know other individuals who are victims of such treatment. The effects of this can be very damaging and should be taken seriously. Accordingly, it is vital for construction professionals to know how to approach such situations if it impacts them or someone around them. Individuals should be encouraged to speak up without fear if they believe something to be wrong. It should be safe and easy for people to raise sensitive matters and concerns without reprisals from other people within the organisation. Sometimes whistle-blowing processes assist in this regard, to enable staff to raise issues anonymously and this will be covered later in the chapter. Unacceptable behaviours should be challenged at early stages. If managers do not challenge bad behaviours, they could be construed as condoning such behaviours and it becomes integrated within the culture of businesses. Every employee has a responsibility to challenge unacceptable and inappropriate behaviours or actions and report them to the relevant person within the organisation and therein it is not solely the responsibility of the person who is directly affected. Construction professionals should be able to determine what the diversity impacts are and how to mitigate those aspects in their respective organisations. Diversity can sometimes lead to situations, partly because there are potential 'diversity tensions' within the wider community and thus by extension in the workplace. Failing to confront these difficulties effectively can have a profoundly negative effect on experiencing a supportive and positive working culture. The construction organisation's policies are intended to protect and support all employees, but it is people that determine workplace cultures.

The construction industry, certainly in the UK, continues on a journey towards increasing its diversity and being inclusive, but equality of opportunity and outcomes cannot just be assumed. The construction sector has made progress through asking fundamental question about Equality, Diversity and Inclusion. Some of these questions could be:

> What is the makeup of the construction sector in terms of different groups of individuals?
> Who is progressing and succeeding in the construction sector; who is not and why?
> What areas in construction are over- or under-represented by particular groups and why?

The important things for construction organisations to consider are what they are doing about the answers they discover. Some recurrent diversity issues have included:

> Female employees being under-represented in the UK construction sector when compared with their male counter parts
> Gender pay gap
> Differential attainment for different groups of construction employees
> Career progression and representation in senior posts
> Mental well-being

When reflecting on the drivers for change within the construction industry, we need to consider that cultural change and transformation are driven by the passion of the employees who work in the construction sector. In addition, there should be a rigorous approach to reviewing and evaluating progress in driving such changes through a systematic process. Apart from the legal duties construction organisations face which drive equality (plus the ethical case for inclusion), the construction sector employees expect the highest standards of equality and inclusion. This makes a powerful business case in terms of attracting different groups of employees.

6.4 Initiatives and Accreditations Available for Construction Organisations to Drive Positive Change for Diversity and Inclusion

There is a scheme in the UK known as Disability Confident which has been set up to help organisations recruit and retain disabled people, and individuals with health conditions. It supports employers to make the most of the talents disabled people have and can bring to the workforce. It is a powerful signal to potential and actual employees that the construction industry is committing to challenging outdated attitudes towards disability. This scheme promotes a better understanding of disability and seeks to remove barriers to disabled persons (and those with long-term health conditions) in order that they can have the opportunity to fulfil their potential, and career aspirations. Stonewall is a campaigning charity that actively works to promote and deliver equality and inclusivity for lesbian, gay, bisexual and transsexual (LGBT) people. Their Diversity Champions Programme provides a framework to create a workplace ensuring that LGBT staff work in an inclusive and accepting environment. Many construction companies also participate in the Stonewall Workplace Equality Index, a benchmarking tool that evaluates institutional initiatives, achievement and progress. It assists companies in being able to achieve a better understanding of the challenges in creating an inclusive work environment for LGBT employees.

Time to Change is another initiative and is a social movement that works to change the way people think and act about mental health problems, focusing on improving behaviours and attitudes. The global construction industry has traditionally perceived as exhibiting and possessing a macho culture and accordingly the notion of employees coming forward and disclosing mental health problems has been limited in the past. Notwithstanding this dilemma, a substantial number of construction companies have now signed up to be pledge employers with a commitment to change how we think and act with regard to mental health.

6.5 A Toolkit for Change: Reflection on How the Construction Industry Can Improve Diversity and Inclusion

The construction industry should reflect on ways and means by which it can learn from previous lessons and focus on many examples of actions that can be undertaken to improve best practice. Specifically focusing on groups with protected characteristics these actions might compose of the following sub-categories.

6.5.1 Disability

The full spectrum of companies that work in the construction industry should be sending a positive message about disability. Workplaces, including building sites wherever possible, should be accessible and audited. There is absolutely no reason as to why most of those who have disabilities should not be able to work in most working environment, through reasonable adjustments being made. Organisations should communicate their accessibility policies to potential employees and advise on the reasonable adjustments that they can make to facilitating work opportunities for non-able-bodied individuals.

6.5.2 Gender

Organisations need to ensure the means by which both females and males are actively encouraged to apply for all roles that become available, and how do they proactively address occupational stereotyping? This is in the context of the UK construction industry being traditionally under-represented by women employees when compared to their male counterparts. The question of how organisations tackle stereotyping with robust policies and procedures will therefore be important for organisations.

6.5.3 Pregnancy, Maternity and Carer Responsibilities

Construction organisations should deploy policies for those members of staff who are pregnant or on maternity leave. They should ask themselves whether such policies are suitable and offer the right level of support for employees who are pregnant or need time away from the organisation for maternity leave. There are moral and ethical questions about offering the correct level of support to staff with carer responsibilities, whether this relates to childcare or care responsibilities for adults. In the latter case this could relate to offering flexible working, possibly in the form of creating part-time positions or offering flexi-hours for staff. Those construction companies that offer such support to employees, anecdotally attract more under-represented females into the construction industry and are perceived as better employers than those who offer less support. Accordingly, all companies involved in the construction industry including clients, main contractors, subcontractors and suppliers should be asking themselves what provisions and schemes they have in place and consider what further measures they could introduce. Furthermore, they should be considering how their organisations have investigated the impact of these responsibilities on their employees.

6.5.4 Religion, Belief and Race

The whole construction industry supply chain should be evaluating how it ensures an environment which respects all. In this regard, organisations should be considering initiative and proactive measures that promote and generate good relations between different groups. This is especially important for the UK construction industry which has seen a sharp rise in recent years of foreign labour and management staff especially migrants from the European Union. In such cases, organisations should be adopting employment practices that are welcoming for all by creating facilities and services suitable for the entire community. Companies should be able to demonstrate their inclusivity in relation to the organisational culture and shown that all groups have an equal opportunity to achieve and progress their careers. To be able to demonstrate this pledge construction firms should confirm that their policies, procedures and approaches respect and reflect the needs of all.

In addition, there should be no disparity when considering different backgrounds and nationalities of all employees.

6.5.5 Age

Construction organisations should ensure that when recruiting for staff there should be nothing to suggest that they are targeting a particular age group. They should ensure that they are diverse in their employment profile, and actively seek to encourage employment from all age groups. They should consider the workings of their organisation and be able to confirm that the workplaces and employment prospects are suitable for employees of all ages. Career prospects therein should be based on ability rather than chronological age.

6.6 Ethical Supply Chain Management

In the procurement of a new building there is clearly a heavy reliance on many different components, elements, equipment, materials and fixtures and fittings. The ethical dilemma surrounding such a large supply of these component parts is where the parts originate from and are these supplies ethically sourced. In recent years, the supply chain has extended from domestic companies to many different and dispersed companies across the world. The motivation to go outside of the domestic supply chain is normally linked to procuring those goods more efficiently and this normally means at a much lower cost, e.g. mechanical and electrical components from China.

In this global arena it is not always apparent whether those international suppliers are adopting principles linked to ethical procurement of labour. There have in other sectors such as retail been reports that many large stores are using companies in developing countries for their supply of goods which are exploiting staff on exceptionally low wages, coupled with terrible working conditions. In the UK, Primark was exposed as one major chain who were selling clothing manufactured through 'sweatshop' child labour in Bangladesh, which caused reputational damage for them. Investigations revealed that there was an obvious and planned attempt on their behalf to ignore regulatory standards linked to the modern slavery act, linked to financial gain. Clearly there are ethical issues and questions which arise from using cheap labour from oversees. Organisation may sometime justify their decisions to source products from developing countries in that this offers customers better value are more range of products. An ethical question nevertheless remains, and this relates to possible them putting financial benefits ahead of ethical procurement.

The other ethical dilemma for the global construction industry relates to the use of migrant labour on construction sites. Migrant labour brought about by an increase in recent years of Eastern Europeans and Polish migrants in the UK has provided an abundance of available labour for construction projects, especially in the Southeast Regions of England. In most cases, this skilled and semi-skilled labour from foreign workers is cheaper than employing local labour. In recent years it is estimated than one in ten workers on UK building sites could be migrants, and many of these numbers could relate to illegal workers. The shortage of labour in the UK construction industry has fuelled a propensity for construction companies to employ this group of employees.

The vast majority of large construction companies operate fair pay regimes linked to staff remuneration and conduct diligent checks on the capacity of migrants to work legally in the country. There are, however, other smaller and less reputable contractors who would not operate such fair-trade policies and would not think twice about exploiting migrant labour on pay rates well below the norm for increased profits. The other ethical dilemma could be that some migrants are not fluent in the national language which can lead to problems in communicating with health and safety repercussions. Other associated problems could arise from migrant workers not being properly inducted and trained before being set to work on construction sites. Clearly this can have safety implication for them and their fellow workers on site, given that the construction industry has still one of the highest death rates throughout the world, when compared to other sectors. Ethics around health and safety will be discussed further in Chapter 7.

6.7 Modern Slavery

6.7.1 What Is Modern Slavery?

The UK government strategy on tackling modern slavery defines it as 'slavery, servitude, forced and compulsory labour and human trafficking'. It is an increasing problem across the world with an estimated 13,000 slaves working in the UK, many of these working in the construction industry. It is frequently the scenario that traffickers will coerce and force people into a situation where they receive inhumane treatment and abuse. In the construction industry most of the large-scale constructors have human relations policies for their organisations and the companies that make up their supply chains to avoid cases where individuals will be exposed to modern slavery practices. However, in the global construction industry the majority of organisations are small to medium-sized businesses. With these companies it is sometimes more often the case that employees will be treated unfairly and their conditions of contracts at odds with national policies on such things as wages and working conditions. In the UK this has been exacerbated in recent years with the advent of cheap migrant labour from Eastern European countries such as Poland and Lithuania.

6.7.2 The Modern Slavery Act 2015

In the UK, under the Modern Slavery Act 2015, all businesses with over £36m turnover are required to publish an annual statement demonstrating that they have taken adequate steps to combat modern slavery. This includes risk assessments to ensure that the supply chain is free of slavery regardless of the complexity of this undertaking in practice. The Act lays down that it is the responsibility of all staff to be vigilant for red flags which could highlight possible issues. These include workers who cannot talk freely or whose movements are restricted. Failure to comply could involve employees losing their jobs and on an organisational level, negative public relations and fines. The requirements of the Act do not solely extend to large organisations and smaller organisations also have a responsibility to know what to look for and report concerns to. It is therefore important that if there are any concerns over modern slavery in the workplace that it is reported, and the police are contacted immediately.

6.7.3 Examples of Modern Slavery

Modern slavery can find its way into almost any construction organisation's supply chain. Please see below a hypothetical example of how a construction company might find themselves involved with illegal activities, inadvertently or otherwise in this regard.

Hypothetical example of modern slavery within UK construction company

A large UK construction company decided to source a new, more local steel frame installers to form part of their supply chain framework. Several regional steel frame installers submitted bids for the new contract, and the construction company agreed a favourable agreement with one of them, predicated on a very competitive tender price. The construction company visited the steel frame installers regional offices and did not witness anything that would raise any concerns. However, they did not ask to see their statement on modern slavery or request any further information on ethical supply chain management. Content and satisfied that they did not need to propagate any further investigates any further, the construction company signed the contract with the steel frame installers for the provision and erection of all new steel frames on their projects. The construction company and the steel frame installers had both worked successfully together on two previous projects before, which may have influenced their decision to award the contract to a supplier they had experience of and were familiar with.

Subsequent to the award of contract it emerged in the press that the steel frame installers are sourcing their steel from China. Unbeknown to the steel frame installers their steel supplier is employing low paid migrant labour from neighbouring Tibet and under unacceptable working conditions. Some of the employees are children as young as twelve and others contracted by day release from local prisons. The fallout from the scandal causes reputational damage to both the steel frame installers but the more nationally recognised large UK main contractor. In the latter case the main contractor is asked what due diligence and scrutiny they conducted in selecting the steel frame installer as part of their supply chain framework. Their neglect in not asking to see their subcontractor's statement on modern slavery is deemed to be both negligence and a flagrant breach of due process on their part. Articles are published in national newspapers and on the internet and in a concerted effort to reduce and mitigate any further reputational damage they are forced into giving a full and unequivocal public apology.

The above hypothetical example can be compared with real-life cases where companies have inadvertently and indirectly engaged modern slavery in their procurement processes, through their dealings with disreputable and unethical companies on their supply chains.

Notwithstanding this dilemma, there is a paradigm that it can sometimes be very difficult for organisations and institutions to know where to start when it comes to assessing their supply chain for potential slavery violations. This notion is predicated on the basis that modern slavery is often well-hidden and can present itself in many different guises; there is no single red flag when it comes to indicating whether modern slavery is present.

6.7.4 Methodologies for Best Practice in Identifying Modern Slavery

In trying to identify whether modern slavery is present within an organisation, the first useful step is to begin with a risk assessment of current suppliers. In this pursuit, organisations should divide their suppliers up into high, medium and low risk, based upon their geographical location and the nature of their products and services. If companies identify high or medium risk suppliers within their supply chain, the best practice step for them would be to complete an on-site audit. A good audit seeks to ask suppliers sensitive questions and apply appropriate scepticism to the responses given to uncover potential abuses. The audit must include a thorough check of the information received from the suppliers against other sources of information derived from employees' interview, management reviews, documents and records. The purpose of worker interviews should be to corroborate or refute what the managers of the organisation have advised, either face-to-face or in their reports. It is important to conduct these interviews in a safe location where the interviewees can speak freely without fear of reprisal and without manager oversight. When it comes to selecting who to interview, it is best practice to choose a fair representation of the workforce and to ensure that the interview is conducted in the native language of the worker to avoid communications difficulties. The interviewees' identity should be kept anonymous. When selecting managers for interview it is a good idea to choose from all levels of management within the fields of human resources, finance, procurement and those within organisations that deal to ethical compliance and corporate social responsibility. During the interviews the aim is to gain an understanding the organisations recruitment process, management of wages, working conditions, workers hours, pay and benefits. It is also important to gain an insight into employee training and development programmes alongside termination and grievance procedures. If the results of these interviews can be validated against worker interview and their documents and records, then it can normally be deemed that there are no instances of modern slavery.

It is normal to review documentation, policies and procedures at the same time as the management interviews. The typical documentation is included in Figure 6.2.

Once organisations have completed their investigations, the audit report should be shared with their suppliers and used as the basis for a 'Correct Action Plan' (CAP). The CAP can also be used as evidence for addressing concerns about suppliers, for reporting abuses to government agencies and for information purposes to stakeholders. The contents of the report should include both good practices and non-compliance practices based upon the evidence they have collected during the audit. It is important for organisations to follow up their reports with a CAP so they can clearly define the corrective actions necessary to be in compliance. A good CAP should include the following information in Figure 6.3.

If cases of modern slavery are encountered during investigations and the suppliers are located in the UK, then the police should be contacted immediately. If modern slavery is suspected to have occurred abroad, the first step should be to outline this in the audit report

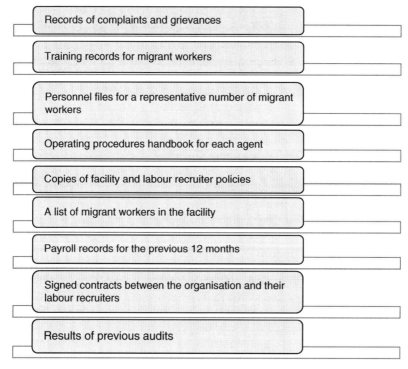

Figure 6.2 Typical list of documentation, policies and procedures required to be created and maintained as part of compliance with Modern Slavery legislation.

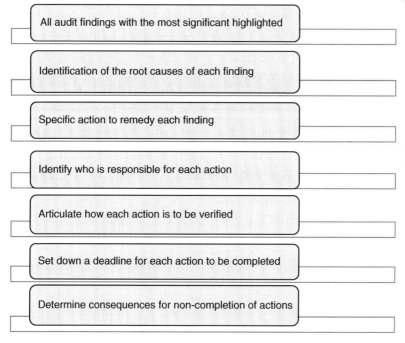

Figure 6.3 Correct action plan.

and CAP. If the abuses are severe, then it may be more appropriate to contact local law enforcement and local government. In other instances, it could be more appropriate to contact local industry bodies and trade unions to attempt to remedy the situation. In either case the organisation should consider which approach is the safest outcome for the potential victims. If the local response is inadequate and the supplier is not taking the issue seriously, then ultimately as an ethical organisation, the company should seek to terminate their relationship with the supplier. These actions should then be included in the next statement report.

6.7.5 Modern Slavery Statements

A commercial organisation is required to publish an annual modern slavery statement if all the criteria below apply, and these are shown in Figure 6.4.

In this regard they may wish to seek legal advice to decide if their organisation needs to produce an annual statement. In normal circumstances, it would be the directors of companies or in the case of public sector organisations, the governing bodies who are responsible for publishing modern slavery statements. Notwithstanding this, in order to ensure that due diligence and risk assessments in the supply chain are practiced effectively throughout the organisation, it would be the responsibility of anyone involved procurement and the hiring of new suppliers to perform an audit. Once the statement is completed it must be published on the website of the organisation or institution.

There can be penalties for failure of organisations to provide and publish annual statements linked to modern slavery and these could include heavy fines and negative public relations. In addition, the Secretary of State may seek an injunction through the High Court requiring compliance. If a court order has been issues, and the company still fails to comply then they will be deemed to be in contempt of court and may be liable for an unlimited fine. For this reason, it is important to have in place the correct training, policies and

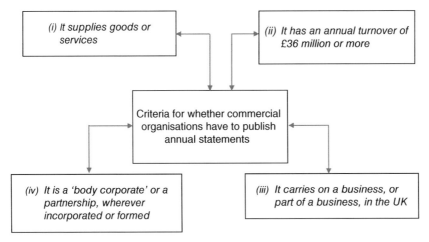

Figure 6.4 The criteria for whether commercial organisations are required to publish an annual modern slavery statement.

procedure to ensure you have the information to confidently submit a statement. Failure to complete a statement could also include cases where organisations have not published the statement on their websites.

6.8 Data Protection

6.8.1 Professional Ethics Related to Data Protection

Professionals who work in the construction industry, as in other sectors, need to be aware of ethical and regulatory requirements related to the General Data Protection Regulations (GDPR). The GDPR is the European Union (EU) regulations covering the personal data of individuals and are designed to protect data in an era of mass digital use. The GDPR is in force across the EU and supersedes previous data protection laws and is applies alongside member state regulations such as the Data Protection Act 2018. It imposes much tougher restrictions on how personal data is used and applies globally to anyone who uses the personal data of European Union (EU) citizens. Almost all organisations use some kind of personal data, even if it is only their employees. Accordingly, the GDPR applies to nearly all businesses and that instils stringent regulations upon them for personal data in all forms including hard copies as well as digital. Personal data, whatever the form, must be stored securely and should only be accessed or shared with those individuals and organisations that need access to the data for the reasons it was originally collected.

The GDPR applies to organisations anywhere in the world that use or stores the personal data of anyone in an EU country, when offering goods or services or monitoring their behaviour. It also applies to organisations that process any personal data where the controller or processor is established in the EU. The GDPR enhances the rights of individuals over their personal data, including their rights be provided with information when their personal data is collected about how it will be used. It also outlines the rights of people to access their data and check that it is accurate. Individuals have the ability to request that their personal data is corrected if it is inaccurate or in certain instances that their personal data be deleted or object to its use is not aligned with the purpose for which it was collected.

There is both an ethical and legal obligation set down by the GDPR to ensure the security of data. Accordingly, security should be considered at the outset when designing data systems. In addition, it is important to minimise the collection of personal data to only that necessary for the purpose of processing, and to encrypt and anonymise data wherever possible. There are several points of compliance as part of the GDPR which allows organisations to collect and process personal data. Organisations must document and evidence how they are protecting data and using it fairly, ethically, lawfully and transparently.

6.8.2 Breaches of Data Security

Data security is not simply about cyber security and protecting information and documentation from phishing or external attempt to hack into systems. Human error represents the most common form of data breaches in the UK and this frequently occurs through emails being sent to the wrong person. In such cases there are strict rules for what to do in these cases. Any breaches of this kind are required to be reported to the organisation. For instance, if a construction site manager inadvertently sends private and sensitive data to the wrong person, then this needs to be formally reported within the company to the relevant person which could include the Human Relation (HR) Director. At this stage, it will be for the organisation to review the nature of the breach and take whatever appropriate and action it deems necessary. The organisation is required by law to report the data breach to the national regulator which for the UK is the Information Commissioner's Office (ICO).

Organisations should employ a Data Controller who should be responsible for the security and use of personal data. They should ensure the following measures with regard to data:

> Collect data only for legitimate reasons that are made clear
> Ensure that any data collected is up to date and accurate
> Provide the means by which to edit and update information held
> Avoid data that is no longer required
> Prevent data being used for any other purpose than what it was originally intended without the express permission of those individuals for which it was obtained

6.8.3 Data Subject Rights

Data subjects have a right to request that a Data Controller supply them with a copy of their personal data. This includes correspondence to and from an individual. Individuals have the right to obtain a restriction of processing in certain circumstances. Data Processors must maintain only accurate information about individuals. Individuals have the rights to object to processing in certain limited circumstances and to delete their personal data in certain limited circumstances or transfer it to another Data Processor. Furthermore, there is also a right for people to claim compensation for damages caused by infringement of the GDPR from organisations that breach the rules.

Certain types of data are considered more sensitive than others and are referred to as special categories. Accordingly, there are ethical implications to collating data for these categories and these include the following:

> Physical or mental health
> Political opinions
> Trade union membership
> Sexual orientation
> Religious or philosophical beliefs
> Political opinions
> Racial or ethnic origin
> Restrictions and consequences

Under the GDPR there are strict penalties for breaches to the rules on how data is processed and used. These normally include large fines which are normally dependant on the severity of the breaches. The range of severity would extend from not keeping proper records to flagrant and purposeful violations of data protection policies and procedures.

As a checklist of key steps that organisations should take to ensure data protection compliance the following list of requirements should be identified and addressed:

> Ensure key people in the organisation are aware of the requirements of GDPR
> Collate a register of all the personal data that is held
> Review the privacy notices and if consent is required
> State data subjects' rights
> Identify lawful basis for data processing
> Review processes for handling breaches
> Decide how to implement privacy by design into processes and procedures
> Assign the Data Protection Officer role

6.9 Summary

Inclusivity, equality and diversity in the workplace is not only beneficial but, in most countries, governed by law with severe penalties for breaches in employment law. The statutory requirements of the Equality Act 2010 play a major part in maintaining diversity and equality in the workplace. Policies, procedures and instruments should be carefully monitored by human resource departments to ensure that these are applied to all employees, both existing and new, and at every level with organisations. Management policies and protocols should also be in place for dealing with complaints and potential breaches and violations.

Equality relates to ensuring that everyone has an equal opportunity and is not discriminated against or treated any differently as a result of their background or characteristics. Diversity revolves around how we think, look and behave. Inclusion is about the creation of an environment where differences are regarded as an advantage, where differences and perspectives are shared, leading to better decision-making and working cultures. Most medium- to large-scale organisations in the construction industry have many different policies, strategies and initiatives linked to embracing equality, diversity and inclusion and some of these are underpinned by training workshops and continuous professional development. Open communications routes are fundamental to making progress on diversity and inclusion communication routes and creating dialogue with people based on their narratives and experiences in terms of disability, faith and beliefs, sexual orientation or transgender identity. Construction personnel may be diverse in nature in terms of religion, nationality, gender, ethnicity, age disability, sexuality, background, education, social class and many other respects. It is important to embrace these differences and acknowledge the diversity of the workforce.

The Equality Act 2010 created nine 'protected characteristics' that cover age, disability, sex, religion and belief, race, sexual orientation, pregnancy and maternity, marriage and civil partnership and gender reassignment. The Act makes it unlawful to discriminate on any of these protected characteristics. There are many different types of discrimination and these include direct, indirect, associative and perceptive discrimination and construction organisations have an overiding responsibility to protect their employee from discrimination. However, individual employees can be held responsible for acts of unlawful discrimination provided their organisations can prove that they have provided their employees with a resonable level of ability to carry out their duties and an awareness of their ethical and legal responsibilities.

Organisations have an ethical, moral and legal responsibility to address equality issues in the workplace with policies, procedures and processes in place to promote best practice. They should minimise or remove disadvantages suffered by individuals who share a relevant protected characteristic and devise and implement processes and procedures that are consistent, accessible, transparent and fair. The most ethical of companies have robust policies, procedures and management approaches linked to reducing and eliminating victimisation, harassment and bullying in the workplace. Such measures should seek to ensure that it is safe and easy for people to raise sensitive matters and concerns without reprisals from other people within the organisation and sometimes whistle-blowing processes assist in this regard. In addition, managers need to be also mindful that when recruiting staff, processes allow their respective organisations to employ the best person for the job, reap the benefits of a diverse workforce, and avoid allegations of unlawful discrimination.

Ethical supply chain management is becoming an ever-increasing area for construction organisations especially those where international suppliers are adopting principles linked to ethical procurement of labour. Notwithstanding this premise, modern slavery is an increasing problem across the world with an estimated 13,000 slaves working in the UK, many of these working in the construction industry. Under the Modern Slavery Act 2015, all businesses with over £36m turnover are required to publish an annual statement demonstrating that they have taken adequate steps to combat modern slavery. In addition, the Act lays down requirements to identify whether modern slavery is present within an organisation, and this entails a risk assessment of current suppliers. If companies identify high or medium risk suppliers within their supply chain, they should complete an on-site audit to uncover potential abuses. Having completed investigations, the audit report should be shared with their suppliers and used as the basis for a 'Correct Action Plan' (CAP).

Professionals who work in the construction industry, as in other sectors, need to be aware of ethical and regulatory requirements related to the General Data Protection Regulations (GDPR). The GDPR is the European Union (EU) regulations covering the personal data of individuals and are designed to protect data in an era of mass digital use. The GDPR applies to organisations anywhere in the world that use or store the personal data of anyone in an EU country, when offering goods or services or monitoring their behaviour. There are several points of compliance as part of the GDPR which allows organisations to collect and process personal data. Organisations must document and evidence how they are protecting data and using it fairly, ethically, lawfully and transparently. Under the GDPR there are strict penalties for breaches to the rules on how data is processed and used. These normally include large fines which are normally dependant on the severity of the breaches.

7

Ethics around Construction Health and Safety

An incident is just the tip of the iceberg, a sign of a much larger problem below the surface.
Don Brown

7.1 Introduction

With reference to the above quotation, normally when there is a serious breach of health and safety, it is usually synonymous with management failings of some kind of another. Such failings if left unchecked could lead to further breaches, possibly resulting in serious accidents or fatalities. This chapter will introduce the legal and regulatory requirements in the UK which are designed to protect the health and safety of employees and members of the public. It will examine the role of the Health and Safety Executive in policing and managing the regulations in this regard and the potential consequences for breaches. Furthermore, it will articulate the measures and initiatives that the construction industry has adopted over recent years to reduce accidents and fatalities on site. Near miss reporting and improved supply chain management will be discussed in this regard and examples given to explain how the measures can be implemented in practice.

The Construction Skills Certification Scheme in the UK will be discussed and the importance of having method statements and risk assessments in place for any given work operation. In addition, the significance of the Institution of Occupational Safety (IOSH) will be examined and the role of this body in enforcement of health and safety standards in the construction industry. Codes of conduct for members of IOSH will be articulated alongside the measures that the institution have introduced to raise health and safety awareness and management practices in recent years. Notwithstanding such measures, there are still some health and safety dilemmas which are prevalent in the construction industry, and which create major challenges for it. Such dilemmas associated with completing projects on time and within budget where shortcuts on health and safety will be examined in this regard.

The Construction Design and Management Regulations (CDM) 2015 will be articulated and the responsibilities and duties that it imposes on clients, designers, CDM coordinators and principal contractors explained. Finally, the purpose and importance of legal requirements around documentation and recording of health and safety for control, compliance, competency will be discussed.

7.2 Governance and Management of Health and Safety in the UK

All employers are ethically required to provide a safe and healthy workplace for their employees. Health and safety professionals (HSP) may often be employed either full-time or part-time to achieve this. According to Fewings (2009) the health and safety professionals should be viewed as non-partisan providers of safety and health. In the UK the government agency that oversees all aspects relating to health and safety is the Health and Safety Executive (HSE). The construction industry especially in the UK has been instrumental around improving health and safety on construction sites over many years. Despite this, numbers of injuries and fatalities in the construction industry remain higher than most other sectors. Reported accidents and deaths in Health and Safety Executive (HSE) statistics have year on year shown that the construction industry represents one of the most dangerous industries in the United Kingdom and in 2019 alone, the UK suffered 30 construction workers deaths. Whilst this can be conceived as tragic, it also represents the lowest number of fatalities in the UK on record and consequently does suggest a trend for improvement. This figure compiled by RIDDOR (Reporting of Injuries Diseases and Dangerous Occurrences Regulations) data equates to 1.31 deaths per 100,000 workers, which places the UK as one of the safest places to work globally. Nevertheless, the rate of deaths is still regarded as being too high when compared with other industries such as manufacturing and the petrochemical industry. Accordingly, there continues to be a concerted effort to improve health and safety management on construction sites with more rigorous procedures for ensuring that risk assessments and method statements are strictly adhered to for all works. Working to a zero-accident regime is the aim but regarded by some as being unrealistic. Nevertheless, the human costs caused by health and safety failings in the UK construction industry, coupled with over £1billion worth of damages a year have predicated the need for continued investment in health and safety to further reduce dangerous incidents. In this pursuit there has been a particular focus on certain roles and types of construction work which naturally present more apparent risks than others.

Measures to reduce the number of accidents have in the past been brought in by Government through regulations and legislation. In addition, some of the larger main contractors have also chosen to exceed the regulations and enhance safety measures on their construction sites. Such an undertaking has been predicated on ethical grounds for the added welfare of their workforce and members of the public. One example of an initiative designed to support an enhanced commitment to health and safety management is the introduction of systems to report near misses on construction sites. This has proved instrumental in reducing accidents through pro-active measures to address hazards before they bring about accidents. An example of near miss reporting is contained below.

Example of near miss reporting

A national building contractor has decided to enhance their commitment and pledge to health and safety by introducing a near miss reporting procedure on their construction sites. This initiative goes further than simply complying with current legislation and represents a concerted and proactive response to addressing hazards before they become potentially harmful for construction staff.

Every member of construction staff including all subcontractors and visors entering onto site are required to be inducted in the process and procedures around the near miss policy. The induction explains what is required of each individual in certain circumstances and the responsibilities that they have for formal reporting of any hazards that they become aware of or any near miss incidents. Furthermore, they are required to sign a declaration that they have fully understood the site rules and regulations around the policy and will strictly adhere to them at all times.

In one particular case, a ground worker installing drainage witnesses a hammer falling to the ground in proximity to where operatives have been previously working. Fortunately, there was no workman in the area at the time and consequently no one injured. Notwithstanding this premise, the ground worker adhering to site rules formally reports the incident to the construction project manager, with overall responsible for the site. An investigation is carried out to ascertain the circumstances of the incident and how it occurred. It is established that a steel erector working on scaffolding at height had placed a hammer on the scaffold board and inadvertently knocked it off when manoeuvring to a different location. It is established that the particular area of scaffolding did not have any edge protection preventing objects falling from height. Furthermore, the steel erector's method statement and risk assessment outlined that hand tools will at all times be properly secured when not being used. Clearly this undertaking was not adhered to in this instance.

It was concluded that a serious incident and possible fatality could have resulted from this incident and that there were two factors which led up to the event. The missing scaffolding edge protection boards would have prevented the hammer falling off the scaffold and the failure of the steel erector to comply with his own method statement and risk assessment in securing tools when working at height should not have occurred. Both factors were addressed; the steel erector received a written warning and was temporarily suspended from the site pending whilst undertaking out additional training requirements. The scaffold edge boards were replaced immediately, and additional safety measures introduced to ensure further checks and inspections of scaffolding before commissioning.

Author's Verdict

The incident did not result in an accident of any kind and consequently it could have been simply ignored as an occupational hazard of working on a construction site. Conversely, the near miss procedures instigated in this instance ensured that it was fully reported and investigated, and improvement measures brought in to address any future hazards of this nature. This is crucial to preventing such an incident occurring again and in which case could have led to a more serious outcome.

The example above may not represent a single occurrence as it is not unusual for an average construction site to have many different types of near miss reported in a given week. This should be regarded as a positive rather than negative as reporting near misses is fundamentally important to being able to follow up on them with essential improvement measures. Such measures have included more stringent induction requirements and training for all construction workers entering on to construction sites. In addition, some main contractors have mandatory special safety accreditation requirements for all their supply chain. To obtain such accreditation normally involves training of those staff that are required to work on sites, and this is a pre-requisite for the award of subcontractor work orders. One of the most common accreditation schemes in the UK of this kind is the CSCS (Construction Skills Certification Scheme) which involves individuals completing an online test in order to qualify as accredited members. Staff would then be issued with a card to verify their membership which they would be required to display on site. Notwithstanding these measures, it is especially difficult for smaller contractors to invest heavily in a robust health and safety regime but nevertheless they should still comply with the statutory requirements and invest in training for their workforce. Accordingly, it is probably not surprising that the in the UK there are more deaths on smaller contractor's sites than those of larger contractors.

Most construction organisations take adequate precautions in methods of work and having the right personal protective equipment (PPE) to limit the risks to an absolute minimum. Risk assessment and method statements are normally prepared for all work operations and mitigation measures formulated to address such risks. However, some construction companies are still failing to maintain acceptable health and safety standards, and this could be endangering the staff that they employ. According to the HSE most construction industry deaths are attributable from falls from height which accounts for approximately half of all cases. This has led to trade association and especially access equipment manufacturers to spearhead an increased number of campaigns and educational events geared to safety training, equipment and awareness.

It is not just accidents in the construction phase which are important but also during the occupation, maintenance or refurbishment stages. For this reason, it is important for risk assessment and method statements (commonly referred to as RAMS) to be in place for work operations. When accidents are reported through RIDDOR 1995 the incident may be investigated externally especially when there is a presumed major breach of health and safety regulations at stake. When external investigations are carried out in the UK, they are normally conducted by the HSE who look for negligence and potential breaches of the Health and Safety at Work Regulations. There is also a need to examine the accident from an internal perspective and consider what the organisation could have done differently. Such analysis is important to reflect into changes and improvement in their overall procedures to avoid such accidents in the future.

Accidents on construction sites are sometimes blamed on risks associated with non-coordinated or regulated working operations. This could involve several trades working within a particular area of a building and as a consequence affecting the safety of each other in the process. This is particularly prevalent when carrying out particularly hazardous work stream such as welding or cutting using mechanical equipment. Other accidents are considered to have occurred as a result of a failure to adhere to method statements that have been prepared and could involve construction workers not wearing the required level of personal protective equipment. Most cases of accidents have emanated from a dereliction of duty in not following by a prescribed set of requirement and site rules, for possible reasons of wanting to take shortcuts and speed the construction process. The health and safety of the workforce should not simply look at construction-related accidents as there are wider effects on health including allergies, back pain, asthma, and vibration white finger and hearing loss to name just a few.

7.3 The Institution of Occupational Safety (IOSH)

In the UK the Institution of Occupational Safety (IOSH) is one of the main professional bodies that deals with health and safety and has been granted a Royal Charter to act in the interests of the general public. The institution has a Code of Conduct which provides a strong ethical foundation for its members working towards its vision of 'a world of work which is safe'. The purpose of the code is to help its members demonstrate their value and reliability through adherence to high standards. The Code of Conduct in Table 7.1 applies principally to the activities of members as health and safety professionals. Members of

Table 7.1 Code of conduct IOSH.

Code of Conduct
1 Integrity
Members are required to:
1.1 Be honest
1.2 Abide by the law
1.3 Notify the Institution in writing promptly if they have been
(a) Convicted of a criminal offence by a Court
(b) Charged with a criminal offence contrary to health and safety legislation
(c) Adjudged bankrupt or have entered into a formal arrangement with their creditors
1.4 Avoid conflicts of interest wherever possible and if one arises promptly take appropriate steps to manage it
2 Competence
Members are required to:
2.1 Ensure they are competent to undertake proposed work
2.2 Ensure persons working under their authority or supervision are competent to carry out the tasks assigned to them
2.3 Undertake appropriate continuing professional development and record it in the manner prescribed by the Institution
2.4 Ensure that they make clients, employers and others who may be affected by their activities aware of their levels of competence

(Continued)

Table 7.1 (Continued)

Code of Conduct
3 Respect
Members are required to:
3.1 Cooperate fully with IOSH and abide by its Charter, Bye-laws and Regulations
3.2 Respect the rights and privacy of other people and organisations
3.3 Cooperate with employers or clients in fulfilling their legal duties under the terms of their employment or consultancy contracts
3.4 Have due regard for the effect their professional activities may have on others
3.5 Ensure that their professional and business activities are reasonable
3.6 Respond promptly and appropriately to disputes and complaints
4 Service
Members are required to
4.1 Ensure that the terms of appointment and scope of work are clearly recorded in writing
4.2 Ensure that where necessary they have adequate professional indemnity insurance when providing consultancy services
4.3 Carry out their professional work in a timely manner
4.4 Ensure that professional advice is accurate, proportionate and communicated in an appropriate format
4.5 Have due regard for levels of service and customer care reasonably expected of them
4.6 Inform any person overruling or neglecting professional advice of the potential adverse consequences and keep a written record of the date, time and nature of this action

IOSH owe a primary loyalty to those at risk and should seek to ensure professional independence in the execution of their duties.

7.4 Ethical Dilemmas around Construction Health and Safety

Fewings (2009) argued that risk is likely to be a factor for construction despite all the necessary mitigation measures to reduce the risk occurring or reducing the impact of the risk. This largely relates to the nature of the work in what effectively is an environment where hazards such as working from height are commonplace. Furthermore, the fact that buildings are in a state between start and completion poses an additional risk and thus calls for additional safety measures. Unfortunately, in the past commercial considerations have temped and influenced the degree of health and safety measures in construction. There are always the added pressures to complete construction projects to time and cost constraints and have buildings ready for occupation on a given date. When delays occur and time is at a premium it frequently imposes pressures on main contractors to flood construction sites with labour and extend working days. There is an ethical dilemma in this respect whether this places additional unreasonable pressures on the workforce to deliver projects and phraseology such as 'acceleration of the work' is commonplace in these scenarios. In these cases, the emphasis sometimes switches from a controlled method of safe working to a method of working which speeds up the construction process without all the necessary safety measures in place.

Where accidents have occurred, compensation claims and fines have ensued. Other dilemmas for risk and commerciality have been raised in the past of whether it is economically feasible to eliminate or minimise all risks. Contractors have traditionally looked at risk from the perspective of probability that the risk will occur and the severity of the event if the risk occurred. In addition, by reducing or possibly eliminating one risk, might lead to new risks emerging. Financial assessments in risk avoidance or mitigation are normally considered at tender stages and contractors will decide on how they will manage risks but still be able to submit a competitive price to win the job. In the past, contractors have gained a competitive financial advantage in not including the required degree of safety measures on a project to reduce their costs and increase their profit margins. For this reason, it is important for clients and their design team consultants to evaluate tenders not just on price but on the overall quality of the bids taking into account safety measures included. This notion should then ensure that compromises and shortcuts in the construction process are not occurring. Setting criteria at the tender stage for how bids will be evaluated and scored is therefore an important part of the tender process. It is not uncommon to have weighed scores for criteria based on an overall 70% quality and 30% price, to deter such unethical tendering practices.

7.5 The Construction Design and Management Regulations (CDM) 2015

In addition to health and safety in the construction phases of projects, the UK Government has introduced the Construction Design and Management Regulations (CDM) 2015 which superseded the CDM Regulations 2007. The regulations describe the law that applies to the whole construction process on all construction projects from conception through to completion. They also articulate what each duty holder must or should do to comply with the law to ensure projects are carried out in a way that secures health and safety. Whatever role of an individual in construction, the CDM Regulations aims to improve health and safety in the construction industry by helping to:

> Sensibly plan the work so the risks involved are managed from start to finish
> Have the right people for the right job at the right time
> Cooperate and coordinate work with others
> Have the right information about the risks and how they are being managed
> Communicate this information effectively to those who need to know
> Consult and engage with workers about the risks and how they are being managed

These regulations impose duties and responsibilities for all roles in construction including clients and their professional consultants for designing buildings to a standard that does not create specific risks during the operational phase or 'life cycle' of buildings. A summary of duties in this regard is contained in Table 7.2.

Table 7.2 Summary of duties under the Construction Design and Management regulations (CDM) 2015.

CDM Duty Holders	Main Duties and Responsibilities
Commercial clients Individuals or organisations for whom a construction project is carried out that is done as part of a business	Make suitable arrangements for managing a project including ensuring that: • Other duty holders are apported as appropriate • Sufficient time and resources are allocated Make sure: • Relevant information is prepared and provided to other duty holders • The principal designer and principal contractor carry out their duties • Welfare facilities are provided
Domestic clients People who have construction work carried out on their own home (or the home of a family member) that is not done as part of a business	Through in scope of CDM 2015, their client duties are normally transferred to: • The contractor for single contractor projects • The principal contractor for projects with more than one contractor However, the domestic client can instead choose to have a written agreement with the principal designer to carry out their duties
Designers Individuals or organisations, who as part of a nosiness, prepare designs for a building, product or system relating to construction work	When preparing or modifying designs, eliminate, reduce or control foreseeable risks that may arise during: • Construction • The maintenance and use of a building once it is built Provide information to other members of the project team to help them fulfil their duties
Principal designers Designers appointed by the client in projects involving more than one contractor. They can be an organisation or an individual with sufficient knowledge, experience and ability to carry out the role	Plan, manage, monitor and coordinate health and safety in the pre-construction phase of a project. This includes: • Identifying, eliminating or controlling foreseeable risks • Ensuring designers carry out their duties Prepare and provide relevant information to other duty holders Liaise with the principal contractor to help in the planning, management, monitoring and coordination of the construction phase
Principal contractors Contractors appointed by the client to coordinate the construction phase of a project where it involves more than one contractor	Plan, manage, monitor and coordinate health and safety in the construction phase of a project. This includes: • Liaising with the client and principal designer • Preparing the construction phase plan • Organising cooperation between contractors and coordinating their work Make sure: • Suitable site inductions are provided • Reasonable steps are taken to prevent unauthorised access • Workers are consulted and engaged in securing their health and safety • Welfare facilities are provided

Table 7.2 (Continued)

CDM Duty Holders	Main Duties and Responsibilities
Contractors Those who carry out the actual construction work. Contractors can be an individual or a company	Plan, manage and monitor construction work under their control so it is carried out without risks to health and safety For projects involving more than one contractor, coordinate their activities with others in the project team. In particular, comply with directions given to them by the principal contractor For single contractor projects, prepare a construction phase plan
Workers Those working for or under the control of contractors on a construction site	• Workers must: • Be consulted about matters which affect their health, safety and welfare • Take care of their own health and safety, and of others who might be affected by their actions • Report any event they see which is likely to endanger either their own or others' health and safety • Cooperate with their employer, fellow workers, contractors and other duty holders

One example of a duty under the CDM Regulations could be for the project team to design a building that can be cleaned at high level through installation of access gantries rather than rely on platform lift, scaffolding or even ladders to access those high level and inaccessible components. In the past many lives have been lost to accidents during maintenance processes where poor building designs have created hazards for accessing building elements. One could surmise that interventions brought about by changes in legislation to lessen accidents, during both construction and thereafter in the building operational stage, are ethically justified in making environments safer and reducing injuries and fatalities.

When considering health and safety on construction sites, it is important to understand the responsibilities of the wider project team members. Clients in this regard have responsibilities for safeguarding that projects have the most appropriate resources to deliver them at the construction phase but also allow them to be safely operated into the future. There are moral and ethical duties on them to ensure that teams who they appoint are competent and experienced to carry out their work and resist the urge to select inferior companies based on lowest cost. The Construction Design Management Regulations (CDM) 2015 places such duties of care on clients, and they are legally obliged to employ a health and safety coordinator. Under the regulations they are responsible for providing information related to contamination, hazardous materials and site conditions. They are also under a duty to ensure that there is a comprehensive and robust health and safety plan prior to commencement of the works, and to play a vital role in developing the plan into a final health and safety file on completion.

The designers of building also have responsibilities under the Construction Design Management Regulations. Their duty under the regulations is to design buildings which are safe to construct and operate safely. In this regard they need to carefully consider forms of construction that do not carry with them obvious risks. Equally they need to have foresight in the operational phase and consider how buildings can be safely occupied and maintained. Simple factors as ensuring there are safe access measures designed into the building that allow maintenance and cleaning to be carried out without risk to human life need to be considered. If there is an unavoidable risk, they have a duty to inform their clients of these and offer innovative ways on measures to mitigate such risks. Fire safety is clearly an area that needs to be closely factored into building designs and especially important for tall buildings. For more complicated buildings and those that have large atriums which can spread fire, they need to work with specialist consultants to develop fire engineering solutions where building regulations may not be the most appropriate standard of compliance.

Under the Construction Design Management Regulation (CDM) 2007, the health and safety coordinator is appointed by the client and has the responsibility for assessing the suitability of the main contractor's health and safety plan and to jointly develop the health and safety file with them for completion at the end of the project. The principal contractor under the CDM Regulations is normally the main contractor and is responsible for creating and developing the construction safety plan. They have a duty to ensuring safe operations on site and that the workforces, mostly made up of subcontractors, comply with site rules and methods of working that are safe. When breaches of rules take place, they must take reasonable measures to ensure that sanctions are made to ensure future compliance. In some cases, subcontractors have been forced to leave sites after repeated breaches of safety standards. In formulating measures for safe project delivery, the role the principal contractor extends to sequencing and coordination of the works to avoid risks and hazards associated with different trades being undertaken at the same time. Particular consideration should be given to particularly hazardous works such as working from height, heavy lifting, hot works and works involving potentially dangerous vapours to safeguard that these follow strict method statements, formulated from risk assessments.

Principal contractors need to ensure that the entire workforce is properly inducted before they enter on to site and know the respective arrangements for accessing and working in their designated part of the building. They need to enforce that the minimum appropriate level of personal protective equipment (PPE) is worn by the workforce and specialist PPE in some cases, e.g. visors when welding and ear defenders when using load power tools. In addition, normally they would insist that all operatives have been trained and accredited to the Construction Skills Certification Scheme (CSCS) scheme and required to display a card to confirm they have valid membership of the scheme. Their responsibility extends to creating the safe means for plant and vehicles to enter on the construction sites and ensure that there is a safe mode of operatives moving around the site. This is particularly important with the COVID-19 virus where sites have had to introduce one-way systems to ensure social distancing measures and reduce densities of people on site to avoid risks of virus transmission.

Safety measures imposed by principal contractors may need to be carefully tailored and enhanced when working within existing building or a 'live site' where members of the public are potentially vulnerable and need to be protected. A good example of this is the construction of a building on a university campus where students, staff and visitors need to be isolated from construction sites, through carefully designed hoardings and working practices. Having scenarios where construction works interferes or compromises the safety of those occupying adjacent sites should be avoided at all costs. Another example is when refurbishment works are carried out on buildings that remain partially occupied. In this scenario, safe methods of access for the works should be devised so that construction operations are separated from occupants and therein do not present risks. It is also important to consider the disturbance factor in such cases, where disruption to normal working operation within a building should be agreed with the building users to avoid cases of noise, vibration and dust affecting business as normal. Refurbishment works create particular added risks when dealing with potential issues of asbestos and other hazards within existing building. Clients and the health and safety coordinators that they employ should ensure that there is an appropriately detailed register and record drawings identifying all hazards within existing building, for the respective safety measures or removal works to be factored into the project.

7.6 Procedures and Documentation around Construction Health and Safety

There are many different documents that professionals should source from their construction partners prior to works commencing on site. These are largely related to health and safety, compliance and competency and designed and implemented by construction professionals to assess contractors' suitability prior to them being included on tender lists. Construction professionals normally adopt schemes such as the CSCS scheme and their questionnaires relate to prequalifying criteria for such affiliated accreditation. It is essential for construction professionals to ensure that their contracting partners are checked to have the right compliance measures in place, prior to commencing work. Consequently a 'Monitoring Checklist' for construction professionals or their project managers should be in place to ensure that all documentation is in place in this regard. The checklist should serve construction professionals and allow them to verify that their contractors have compliance measures in place and covers all the necessary insurances, risk assessments, method statements, permits to work, COSH assessments, completion of site and health and safety inductions, contacts and awareness of procedures for emergencies. This became more important than ever when dealing with the Coronavirus global pandemic where contractors were required to produce detailed documentation to ensure the health and safety of their workforce on site. Real-life examples of such documentation are contained in the appendices and include the following:

- COVID-19 Project Review of Site Operating Procedures (Appendix A)
- Supply Chain Communication Coronavirus Business Continuity Planning (Appendix B)
- COVID-19 Management System (Appendix C)

In addition to the above, it is important to issue permits to work and to make contractors aware of known hazards. This should include asbestos removal, confined space entry, excavations, higher risk electrical work, hot work, work on fire alarm systems and emergency lighting, work on fire alarm systems and emergency lighting, roof access and working at height. Furthermore, it is good ethical practice for construction professionals to compile their own health and safety contractors' handbooks, which is normally tailored to their particular organisations and estates. These should incorporate important information on what contractors need to know about construction professionals' sites and buildings and safety precaution that must be strictly adhered to.

7.7 Summary

All employers are ethically required to provide a safe and healthy workplace for their employees. In the UK the government agency that oversees all aspects relating to health and safety is the Health and Safety Executive (HSE). The UK construction industry has a high death rate, albeit less than most other countries on a global scale, when compared with other industries such as manufacturing. Measures to reduce the number of accidents have in the past been brought in by Government through regulations and legislation. Notwithstanding these regulatory measures, some of the larger main contractors in the UK have chosen to exceed the regulations and enhance safety measures on their construction sites. Introducing near miss reporting with essential improvement measures is one example of an initiative in this regard.

It is important for main contractors to have mandatory special safety accreditation requirements for all their supply chain. The most common accreditation schemes in the UK of this kind is the CSCS (Construction Skills Certification Scheme). Despite this scheme, some construction companies are still failing to maintain acceptable health and safety standards. This dilemma has led to trade association spearheading an increased number of campaigns and educational events geared to safety training, equipment and awareness. When considering health and safety management it is important to not just focus on the construction phase but also consider the occupation, maintenance or refurbishment stages of buildings. In this regard when works are planned it is imperative to have risk assessments and method statements (commonly referred to as RAMS) to be in place for the task in hand.

Past commercial considerations have temped and influenced the degree of health and safety measures in construction. There are always the added pressures to complete construction projects to time and cost constraints and have buildings ready for occupation on a given date. Most cases of accidents have emanated from a dereliction of duty in not following a prescribed set of requirement and site rules, for possible reasons of wanting to take shortcuts and seeking to speed up construction processes. For this reason, it is important for clients and their design team consultants to evaluate tenders not just on price but on the overall quality of the bids taking into account safety measures included.

The Institution of Occupational Safety (IOSH) has a Code of Conduct which provides a strong ethical foundation for its members. It applies principally to the activities of members as health and safety professionals. In addition, the Construction Design and Management Regulations (CDM) 2015 imposes duties and responsibilities for all roles in construction including clients and their professional consultants for designing buildings to a standard legislation to lessen accidents, during construction and thereafter in the building operational stage. Under the CDM regulations health and safety coordinators are appointed by the client. They have the responsibility for assessing the suitability of the main contractor's health and safety plan and to jointly develop the health and safety file with the principal contractor and client for completion at the end of the project. Principal contractors need to ensure that the entire workforce is properly inducted before they enter on to site and know the respective arrangements for accessing and working in their designated part of the building. Safety measures imposed by principal contractors in this regard may need to be carefully tailored and enhanced when working within existing building or a 'live site' where members of the public are potentially vulnerable and need to be protected.

Finally, there are many different documents that professionals should source from their construction partners prior to works commencing on site and throughout the duration of the construction projects. These include prequalification questionnaires, monitoring checklists and development of health and safety files. They should be designed to assess contractor's suitability and competency for projects in the first instance and to ensure that projects are completed with the required level of health and safety compliance.

Reference

Fewings, P. (2009). *Ethics for the Built Environment*. Oxon: Taylor and Francis.

8

Ethical Considerations around Trust and Collaborative Working

In relation to issues concerning ethics and trust; individual people possess different propensities to trust; some are trusting, some not so trusting and some distrustful.

Walker, 2009

8.1 Introduction

When considering professional ethics, one cannot fail to consider the importance of building trust, adopting collaborative policies, and developing good working relationships. Although there are many other definitions, trust in this sense can be defined as the willingness of parties to become vulnerable to others. This is especially applicable in the context of construction management, as the sector relies heavily on interaction between many different groups of individuals and disciplines. In addition, there are many interdependencies between these groups where the behaviour, successes or failures of one party could affective another in a positive or negative way (Challender et al. 2019). It is for this reason that trust built upon partnership and collaboration working relationships has a strong reciprocal correlation with professional ethics. To extenuate this notion, consider a scenario where one party suspects the other of unfair or unprofessional practices. This unethical behaviour can lead to another party being unable or unwilling to work and engage fully with them for fear of exposure to said unethical practices. In this scenario 'guards are very much up' and there is a propensity to avoid sharing information, reluctance to collaborate but most of all to enter into trusting relationships. The opposite could be enacted where positive and trusting relationships emerge from working practices predicated on open, considerate and ethical behaviours and practices.

In consideration of the above, this chapter focuses on trust and collaboration as a way and means of encouraging improved ethical practices and behaviours and improving business relationships. It will discuss the duties of trust in the context of professional ethics and how trust can be gained and lost through the actions and behaviours of individuals. In this sense the fragility and robustness of trust will be examined in the constantly changing and dynamic environment of the construction industry. In addition, the importance and interdependence of trust for professional ethics and collaboration will articulated alongside the associated barriers and deficiencies that present challenges for the UK construction

Organisational Ethics in the Built Environment, First Edition. Jason Challender.
© 2023 John Wiley & Sons Ltd. Published 2023 by John Wiley & Sons Ltd.

industry. Finally, the factors associated with partnering principles which influence successful project outcomes will be explained in the context of building trusting collaborative relationships between individuals and organisations.

8.2 Professional Ethics and Duties of Trust

Construction practitioners, like professionals in other disciples such as medicine, law and accountancy, have a duty of care to their clients. Such duties are regulated by institutional code of conducts and regulation. It is worth considering at this stage that professionals also by their nature and status have an ethical responsibility to maintain the trust of their clients, associates, employees and the general public. This trust comes from their standing in society and the trust that is vested in them as professionals. The trust of the professional is predicated on their qualifications, experience and institutional membership. It has been articulated that professional are identified as trustworthy, because they act in their clients' interests, and they are experts in their respective fields. Consider the hypothetical example below.

> **Hypothetical example of a duty of trust**
>
> A client approaches an architect with regard to a project to build them a new specialist facility. The facility consists of a new prison facility with purpose-designed security measures. The architect demonstrates to the client that they have a successful track record of experience in designing prisons and has the right level of expertise and understanding around some of the specialist requirements. This expertise has been built up over many years and lessons learnt from many different projects in the prison sector. The architect gains the trust of the client, as a result of their track record and previous achievements.
>
> The architect gives particulars of previous projects and evidence to support why they would be particularly suited to the client's project. Furthermore, they articulate how they procured successful outcomes for past projects and how they would approach the new project and apply similar methodologies. The trust the clients have in the architects has not been developed over time through a previous relationship, or not generated by the kindness and consideration of the architect. Moreover, the trust has been built in a relatively short period of time by the perceived knowledge and expertise of the architect.

In the above example the trust that is generated from knowledge, experience and qualifications is known as 'Cognitive -based' trust (Challender et al. 2019). This is related to knowledge, understanding and interaction. For this reason, it could be potentially important to collaborative working and partnering with the aim of reducing conflicts and underpinning successful project outcomes. It is also linked to reputation, track record and financial status (Wong et al. 2008). A similar view was provided by Parayitam and Dooley (2009) who described it as 'the perceived trustworthiness of individuals based on their competence'. In the medical profession, the same type of trust could come from a consultation with a general practitioner or to a higher level, a specialist medical practitioner. This trust is predicated on the expertise and knowledge of a professional that is perceived by others to have spent many years gaining their qualifications and experience in a particular field.

What can also be extrapolated from the example is that the client is trusting the architect to deliver the project successfully and this construction professional owes a duty of care to the client to represent their best interests. However, this trust could be lost where the architect does not undertake a professional service or simply lacks the knowledge, skills and experience that a particular project requires. For this reason, it is therefore critical for construction professionals to consider a potential project and reflect carefully on whether they have the right credentials to manage it. Furthermore, if the architect that the client approaches to build the prison has little or no previous experience of similar specialist projects then they should declare this, as it clearly represents a risk for the project. Furthermore, they should carefully consider whether their skills and knowledge will be sufficient to deliver the project successfully. Given they have little or no previous experience in such a specialist project it would be probably unethical for them to allow the client to employ them. It is normally the case that when construction professionals accept commissions for projects, they really should not be engaged in that problems can come about. This practice is unethical and could lead to claims of professional negligence when things go wrong. In a court of law, a professional who is defending a lawsuit against them would be in a precarious position if it was deemed that they should have not taken on the project in the first place through lack of specialist expertise. Notwithstanding this premise, it may be acceptable from a professional ethics perspective to accept the appointment subject to bringing in additional specialist design consultants that have the transferable specialist knowledge from previous prison projects. In this way they could demonstrate that, although they do not have the right specialist experience and knowledge themselves, they have addressed the problem by bringing in those who can bridge this deficiency gap.

It is important for consultant professionals at the initial stages, ideally before they are commissioned, that they discuss with their clients the scope of services required of them to deliver the project and support the client's organisation. It is essential for professionals to then clearly outline in detail, the precise extent of tasks and services which they have included for in their fee quotation or tender proposals. One of the potential sources of conflict and loss of trust between clients and their design consultants is related to additional fees for services that clients thought were initially included. Furthermore, it is important for consultants not to feel compromised and pressurised to carry out an additional scope of services without recompense. To avoid both these scenarios a detailed brief and scope of services should be carefully prepared and agreed between the parties to avoid any potential problems at later stages. The other potential source of a loss of trust could come from a situation where the expected costs of a project are exceeded. Clients, in situations where they have not instructed any additional services or works (commonly known as 'scope creep'), may feel that projects haven't been managed well or that value for money has been achieved.

8.3 The Fragility and Robustness of Trust in a Constantly Changing and Dynamic Environment

> It takes 20 years to build a reputation and five minutes to ruin it.
> Warren Buffett, 1982

Perhaps this statement demonstrates the fragility of trust as something that is potentially temporary in nature and needs to be carefully managed and maintained. There are widespread opinions that business environments exist within a constant state of flux, largely brought about through constantly changing government policies. Such challenging and dynamic climates could be compromising the continuity required to maintain partnerships between organisations (Huxham and Vangen, 2000). In this way, McKnight et al. (1998) explained that levels of trust over time will change as they respond to changing conditions of a positive or negative nature and differentiated trust as 'fragile' or 'robust'. In the former case, the status of trust could be classified as unstable or at best easily influenced, with the possible negative outcomes of a downward spiral under extreme circumstances.

Applying this theory to practice, one could argue that fragile trust is synonymous within the UK construction industry, where adversarial behaviours still exist, and mistrust is embedded in some quarters. However, in circumstances where trust is regarded as robust, then arguably it is less influenced by changing events and conditions and does not change to the same degree (McKnight et al. 1998). An example could potentially be applied to strategic partnering agreements and framework agreements, predicated on repeat contracts and established relationships. In such cases it may be argued that working relations are easier to maintain, especially in difficult circumstances. Notwithstanding this, expectations of both fragile and robust trust when broken can have emotional consequences, with parties feeling violated, and can signal those relationships have become damaged. Such violations can frequently be caused by unethical practices and behaviours and where one party has been damaged or disadvantaged by such actions. Furthermore, such events can change the dynamics of trust between parties and in extreme cases can lead to its complete collapse.

In less extreme cases, however, specific behaviours may need to be changed in an effort to repair and prevent further damage to relationships (Jones and George 1998). Applying this theory into practice, perhaps justifies the claim that more is required to train project teams to deal with these types of situations as they arise. This view can be corroborated by recent findings from Strahorn et al. (2014), who reported that in UK construction management 'trust repair skills appear to be rare', especially following disputes. Furthermore, there is an argument that building client trust is very important for consultants and contractors, especially in the early stages of relationship building. Many construction professionals may report that, owing to the fragility of trust, a loss of client confidence can result in loss of credibility and permanently damage relationships. A construction project manager's perspective on the issue is given below.

A construction project manager's view on the fragility of trust

'One thing that I have learned over my career in the construction industry is that the trust you gain from individuals is fragile. You can lose money on a job and recover from that, but if you lose the trust of people, especially clients, you also lose their confidence and the relationship or even the partnership is over. You lose credibility once and that can be it. Trust is about taking the time to consider the other parties. One example could be to avoid setting meetings up for first thing on a Monday morning and last thing Friday evening, if others have vast distances to travel. Small considerations and interventions such as this will demonstrate benevolence that may make a big difference.'

Table 8.1 Examples of potential breaches of trust and their effects on contracting relationships.

Potential Breach of Trust	Possible Effects on Contracting Relationships
One party purposefully uses sensitive or privileged information to serve their own needs and give themselves commercial advantage over their partners.	The disadvantaged party will feel violated and reticent to disclose any further sensitive or confidential information in the future. In extreme cases they could seek retribution measures to 'settle a score' leading to a complete loss of mutual trust breakdown of the relationship.
Having relied on the assurances of a partner to deliver against a certain target or milestone on a project, their expectations are not met.	Disappointment and frustration from false assurances could lead to resentment from the party who have been reliant on them. This may adversely affect reliance on any future assurances given to them. If partners cannot rely on others this could limit the extent of trust, they place on them, resulting in low levels of conditional trust.
One party unwittingly or deliberately releases sensitive information which causes damage to the reputation of their partner.	Reputational damage at the very least can cause embarrassment and frustration to the affected party. In extreme cases, where the damage is perceived to be malicious, there is a strong probability that relationships will be permanently scarred and beyond repair.

In understanding the characteristics of different types of trust and their fragility, it is also necessary to consider the effects of breaches. Violations of impersonal trust, for instance, could be regarded as deviance and include non-adherence to procedures and breaches of confidentiality, whereas violations of personal trust could be seen as a betrayal (Elangovan and Shapri, 1998). In the latter case, the catalyst for such violations could be linked to pursuit of individual self-interests rather than organisational goals, and deceitful practices may be deployed to this end. Applying this argument within a construction context, breaches of trust, particularly in partnering arrangements, can be extremely damaging to relationships and there is a risk of reversion in such cases to traditional adversarial practices or legal recourse (Walker 2009). Perhaps what this demonstrates, is the fragile and dynamic nature of trust in construction contracting. Some examples of potential breaches of trust and their effects on contracting relationships within such scenarios is detailed in Table 8.1.

8.4 The Importance of Trust for Collaboration and Professional Ethics

> *Without trust we don't truly collaborate; we merely coordinate or, at best, cooperate. It is trust that transforms a group of people into a team.*
>
> Stephen MR Covey (2010) (empasis added)

McDermott et al. (2005), using content analysis from interviews with many different construction professionals, found trust to be the most important factor for long-standing ethical relationships to succeed. Notwithstanding this, there is an argument that when companies enter into highly complex, uncertain and potentially risky projects as relative strangers, it is not unsurprising that frequent conflicts and disputes arise, resulting in unsuccessful projects (Chan et al. 2004). It has been reported through many different sources that to potentially lessen the risks of such conflicts and disputes, the use of partnering approaches can increase cooperation and develop more sustainable relationships built on trust (Diallo and Thuillier, 2005).

The extent and degree of trust has also been shown to underpin relationships of key stakeholders which in turn can lead to project success (Pinto et al. 2009). The emergence of these arguments appears to affirm the findings of Challender et al. (2019) who identified trust as being the most important factor to risk minimisation on projects. Notwithstanding these findings, it should be stressed, however, that trust and adherence to professional ethics *per se* do not guarantee success on projects, as the mechanisms to initiate and maintain trusting relationships are extremely complicated (Bresnen and Marshall, 2000). One example, for instance, could be where organisations are 'engineered' to work together as one combined entity under partnering arrangements. In this context such variables as social interaction, power, relationships, identities, expectations and commitments could, if not managed correctly, result in anxieties, conflicts and tensions between organisations (Cicmil and Marshall 2005).

One case study as an example of a partnering contract that lacked trust and was not successful, is detailed below:

An example of a partnering contract that lacked trust

The project involved a new build further education construction and engineering teaching facility with a value of approximately £29m. The contract was awarded, by the college client, through a competitively tendered OJEU process and owing to financial constraints had to be completed within a demanding and highly challenging budget. The main selection criteria were based on the tender price of the successful bid, albeit quality of submission in terms of the contractors' experience, track record and understanding of the proposed project were also taken into account. A partnering contract, namely the New Engineering Contract 3 (NEC3), was entered into between the client and the main contractor. Despite this, however, there was very little dialogue, understanding or commitment from either party on how they would work collaboratively throughout the partnering contract. Instead, there was much negotiation of how each party could pass risk on to the other in terms of unforeseen events and abnormal costs. An example of this included unforeseen ground conditions where the client expressly passed on the risks to the main contractor in an attempt to avoid costly additional ground works. Such provisions should have been included within the brief that each tendering contractor submitted their respective bids on, but unfortunately this was omitted from the documents. Some would argue that this type of approach is not collaborative working and only constitutes partnering in name alone. Furthermore, there was an inherent reluctance for both parties to work together and share potential pain from any risks which materialised or share gain from buildability, value engineering and innovation leading to reduce build costs.

This was one of many issues that compromised project team relationships at the early negotiation stages of the contract. Other problems and conflicts related to perceptions of unfairness of contract terms. For example, the appointed main contractor considered the extent of liquidated and ascertained damages for contract delays leading to late completion to be disproportionate and excessive, despite being specified in the original OJEU notice.

(Continued)

> (Continued)
>
> Perhaps unsurprisingly behaviours between the two contracting parties became adversarial at an early stage of construction. Collaboration workshops were non-existent, and monthly on-site construction progress meetings rapidly transcended into blaming the other party for problems that emerged from design and construction difficulties. Relationships between the client team and the main contractor became worse as construction works progressed, and an environment of mistrust unfortunately became the norm. Attitudes and behaviours displayed by the whole project team were reflective of the adversarial nature of such relationships. The project was delivered three months behind schedule owing to disagreements over the cost of potential variations and additional works and a whole series of technical problems, which could have been resolved earlier had the project team been working in an integrated and joined up way. This delay caused the College major operational difficulties in not being able to open the new teaching and learning facility to students at the start of the new curriculum year in September. Furthermore, there were post-contract claims on both sides with the client claiming for liquidated and delay damages for the disruption brought on by the delay and the contractor claiming for prolongation costs (preliminaries, profits and overheads) for being on site three months longer than expected. One could be misled in believing that this was a partnering contract that went badly wrong, when it is probably more accurate to conclude that this was a partnering contract in name alone. The fact that a partnering form of contract was used was no guarantee that collaborative working approaches and partnering philosophies would be deployed. In fact, in this particular case, the approach was much more akin to a highly commercial and adversarial tradition contractual procurement route, built on a win-lose strategy on either side. As things turned out, with damage to the reputation of both organisations and a major dispute on completion, it turned into a lose-lose outcome for the client and contractor.

The above case study, hopefully, demonstrates that one cannot simply create a partnering relationship based on the type of contract that is entered into. It probably also highlights that the OJEU, NEC form of partnering contract and early contractor involvement do not always sit well together. In this example, there was little, if any, collaborative working and no sensible apportionment of risk. The success of projects requires much more than this if they are to be truly collaborative. Bresnen and Marshall (2000) supported this claim and concluded that

> *Collaborative working and partnerships practices engineered without understanding of the underlying factors, shared beliefs, attitudes and values, and simply imposed by senior management will have only limited success. They should involve embedding partnering principles and mechanisms for building trust at an early stage and throughout the whole design and construction processes.* (emphasis added)

The previous case study and many others like it underpin the argument from construction professionals that trust is a very important aspect for integrating project teams. The common consensus indicates that a trust regime allows projects to move forward more effectively within an environment where problems can be solved easily, and barriers are broken down. Some construction professionals, supporting the theory of Thurairajah (2006), believed that where trust is lost it can make working relationships untenable leading to a 'downward cycle of trust'.

The case study also reinforces the premise and opinions that trust is not something that can be engineered instantaneously through contractual and procurement routes alone but needs to be built and earned over time. Where trust is developed in this way, clearly there is a tendency for working relationships to become more reliable and consistent between parties. It provides the conditions where each party can rely more on the other for getting what they expect from partnering agreements. However, some would suggest that these benefits are only applicable if trust is 'full' and unequivocal to make an important contribution to collaborative working arrangements. In situations where it is not the case and partial trust only exists then such benefits will not be realised. These findings would appear broadly to underpin theories of trust from perspectives of 'deep interdependence' (Shepperd and Sherman, 1998) and 'unconditional trust' as explained by Jones and George (1998). Deep interdependence is characterised by relationships which have shared meaning, values, products and goals and organisations with sense of strategic alignment and common membership. A scenario to illustrate where this could be applicable to a construction project is detailed below.

Practical hypothetical scenario of unconditional trust and deep interdependence

A client is working closely with a contractor to deliver a highly challenging project. This may entail an extremely demanding budget and programme or simply relate to a high degree of technical or logistical complexity on the project. In such a scenario there may be win-win or conversely at the other extreme a lose-lose for both the client and contractor in terms of realising successful project outcomes. The extent of such levels of success or failure may vary and be reflected in pain-share/gain-share mechanisms or provisions built into the contract terms and conditions. Where projects finish on time, under budget and to the client's satisfaction, clients and contractors may share cost savings or a separate bonus provision made payable to the contractor. Conversely, where projects are not delivered to meet clients demands including the budget, then contractors may have to meet a proportion of overspend or suffer delay damages claim for late completion of the works.

Such provisions to cater for successful and non-successful outcomes could motivate and incentivise both parties to work more collaboratively then they may otherwise do in an alternative scenario. They would be heavily reliant on each other to share relevant information and ensure that decision-making is done in a timely manner to ensure progress is maintained by both parties at all times. In such relationships both organisations would be completely integrated and aligned in terms of ensuring that the other is working at an optimum level, given the high stakes involved and the mutual desire for project success. They are aware that any failure of the other parties' ability to perform will ultimately adversely affect their performance and thus lead to a lose-lose outcome for the project. In this way they are sharing the risks and are highly motivated for project success to share gains brought about. Furthermore, there is trust for each other as a consequence of the environment of teamwork and collaboration that has developed between them. Owing to the mutual aims and the objectives for

(Continued)

> **(Continued)**
>
> the project, such trust could be deemed as unconditional with no limits to such things as sharing potentially sensitive or confidential information. Reciprocation of trust and further development of working relationships can prosper in this scenario leading to an 'upward cycle of trust'. Conversely, the opposite affect could be experienced where conditionality on trust restricts such things as information sharing. Such a scenario may be severely damaging and could lead to a 'downward cycle of trust' occurring. One example of this could be where clients will work alongside contractors up to a point to value engineer a project, to meet levels of affordability but not trust them enough to the point where they share their budgets.

The next interesting issue to explore is whether the extent, and conditional or unconditional nature of trust, depends upon who you are and who that potential trust is to be bestowed upon. For instance, some construction professionals would indicate that trust is influenced by the position and role of people within project teams. Trust between clients and their design team consultants, for instance, could be regarded as being critical and fundamental whereas their relationships with contractors, potentially less important. A possible explanation for this could be that clients employ consultants to represent them 'as an extension of themselves' whereas they employ contractors to undertake a distinct tangible piece of work under prescriptive and well-defined contractual conditions. This could suggest that clients do not naturally always want to trust contractors possibly under the misguided perception by some in the industry that their interests are still regarded as being diametrically at odds with each other. Perhaps this explains why Pinto et al. (2009) opined that behaviours and factors which may encourage trust for clients may not be the same for contractors and vice versa.

Notwithstanding the above, the notion of trust has attracted its critics from some construction professionals who may not advocate trust as a critical factor for generating successful partnering outcomes. Whilst some would regard trusting, collaborative working relationships are desirable in a business environment, they may consider they will not guarantee profit margins and survival in very difficult and competitive economic climates. Views have emerged in the past that partnering could disadvantage contractors in some cases, especially when dealing with more powerful clients, preventing them from receiving fair and reasonable commercial returns from projects. Furthermore, cases have been described where clients had made significant changes but still expected these to be absorbed into the original price in 'the spirit of the collaborative arrangement'. Perhaps unfair relationships can make one partner feel that the other is taking advantage, which could undermine trust. These views support the theory of Das and Teng (1998) which stressed the importance of maintaining fairness in partnering relationships. It would suggest that if returns could be guaranteed for contractors in terms of reasonable profits, then these perceptions may change accordingly.

Interestingly arguments have emerged amongst construction professionals that partnering arrangements could still exist with little or no trust present. In lieu of trust, there would need to be legal frameworks, documented processes and procedures adequately set down to manage each stage of projects. Such provisions would serve to protect organisational and individual interests. However, this is regarded as a minority view and there are still strong views expressed that trust is something that is probably more important in collaborative working generally rather than formal partnering arrangements *per se*. Collaborative working has long been felt to be more reliant on trust in building informal personal and working relationships and thus promoting teamwork, integration, and collegian between individuals.

This represents a potential differentiation of how participants perceive formal partnering and 'informal' collaborative working in its widest sense. Perhaps it could justify why some construction professionals believe that 'formal' partnering arrangements do not necessarily need to be in place to create an environment of collaboration or what Gadde and Dubois (2010) described as the 'relationship atmosphere'. Some have explained that if trust exists between parties, then traditional contracts could become equally as collaborative as formal partnering agreements. This demonstrates that they view collaborative working in much broader terms than simply partnering, with collaboration potentially being practiced under any form of contract or procurement arrangement. It may also suggest that trust is equally as important for any type of building contract, collaborative or non-collaborative. The key issue here, however, is to ensure that the bid price is not too tight, thus reducing the risk of confrontational behaviours in maintaining profit margins. This was reiterated by Wolstenholme (2009).

Clearly, in consideration of the above arguments, there is no one consensus on the importance of trust, especially from the perspectives of conditionality and dependences. Furthermore, many of the participants, interviewed to provide research for this book, did not agree with academic views on the reliance, influence, and importance of trust for construction partnering. A summary of the observations and inconsistencies, in this regard, and the possible consequences and effects for projects is contained in Table 8.2.

8.5 The Interdependency between Trust, Collaborative Working and Professional Ethics

One can argue that ethics and trust are mutually dependant and the two are tied together in a fundamental way. For ethics to work well it is important for individuals and organisations to adopt practices and behaviours which do not just meet regulatory compliance benchmark to avoid prosecution and fines. Moreover, the practice of ethics should revolve around exceeding compliance levels and thus according to Fewings (2009) trust is a factor which can provide the means to move away from actions simply guided by fear of penalty or punishment. Trust in this regard should be seen as the catalyst to enact guiding principles based on belief and understanding and accordingly can be interlinked with ethical behaviour. Ethical behaviours normally revolve about 'doing the right thing' and 'keeping your word' which makes trust an essential prerequisite for loyalty, reliability and honest dealings. In the construction industry there are many cases where commitments need to be made that cannot reasonably and practically always be deal with in contracts. In addition, the outcomes for projects can be sometimes uncertain and thus having a collaborative trusting relationship to deal with other parties to resolve issues becomes increasingly important. Trust remains critical in this arena of uncertainly and unpredictable environment.

There has been views and thoughts expressed over many years in terms of whether trust is generated from good collaboration as an output (dependent variable), or whether it is trust which develops collaboration as an input (independent variable). A subcontractor's perspective below would seem to support the former case that trust is a dependent variable.

Table 8.2 The importance, reliance and influence of trust in partnering strategies.

Area of Focus	Literature Sources	Observations around These Literature Sources	Data Inconsistencies with Literature	Possible Effects on Projects
Reliance on trust	Kaluarachi and Jones (2007) Walker (2009) Lann et al. (2011) Silva et al. (2012)	More trusting relationships facilitate increased collaboration, interaction and cooperation. Trust considered to represent 'bonding agent' between collaborating partners. Degree of trust shapes relationships of project teams and influences project outcomes. Degree of trust affects integration of the whole team. More trusting relationships can foster improved communication and dialogue, increase problem solving and reduce project risks.	Trust only felt desirable to project partnering whilst essential to strategic partnering. Partnering can work effectively without trust through cooperation between project teams, as long as trust is maintained at senior manager level between organisations. Trust can be substituted for increased compliance through regulated control measures with possible remedies for breaches. Deterrence of losing repeat business could be stronger influence than trust for maintaining relationships. View that it is unrealistic to offer unconditional trust to a partner where there are considerable financial considerations at stake. Important to have robust legal agreements in place working alongside trusting relationships. Trust is not required for some lower tier supply chain subcontractors. Trust is equally as important in traditional arrangements as it is for collaborative working. The degree of trust required depends on the role of individuals within project teams.	Loss of trust can result in untenable working relationships. Trust enhances collaboration and bonds teams together. Trust underpins collaborative working and partnering strategies cannot work without it. Trust is essential for collaborative arrangements to work effectively and solve problems that arise. Working relationships built up on trust become more reliable and consistent. Loss of trust could escalate problems which could cause the collaborative environment to fail and lead to a downward cycle of collaboration. Trust has to be full, unconditional and unequivocal in order to make an important contribution to collaborative working arrangements.

Perceived importance and influence of trust	Larsen (1997) RICS (2005) Colquitt et al. (2007) Silva et al. (2012) Li (2008) Pinto et al. (2009)	Increased collaboration is a crucial factor to deliver successful construction projects. Trust can safeguard against excessive contractual provisions and/or risk adverse practices. Can reduce opportunism. Levels of trust can grow if trust reciprocated between partners.	Suspicion of realisable benefits and negative past experiences. Risk of exploitation in some cases Perception of too 'cosy' business relationships. Trust can increase opportunism. Dispute resolution processes are sometimes more important than trust.	Less antagonistic and stressful environment, improved teamwork, cooperation and collaboration. Risks especially on complex projects can be identified earlier and managed more effectively. Shared ethos based on trust, equity and fairness between partners is essential.

> **Trust as an output from collaboration. Based on an interview with a specialist subcontractor**
>
> 'I see collaboration as the way that trust is generated or more precisely earned.
> This contrasts with how a lot of others regard collaboration as an output of trust. You cannot start with a relationship of trust, but one needs to play by the rules of the game, which is the contract and after, a period, you start to build up that trust in the other party, based on the inputs.'

Notwithstanding the above opinion, recent studies conclude that trust can play a central pivotal role and perform as both an independent variable and a dependent variable (Huang and Wilkinson, 2013). This is contrary to earlier studies, which found that trust has been mostly modelled as an independent variable (cause) especially where economic outcomes are prevalent during financial negotiations (Rousseau et al., 1998). Pinto et al. (2009) called for a greater insight into how trust can be considered as a dependent variable; that is what are the independent variables that influence trust? This is within an environment where there is an overwhelming lack of empirical studies around building collaborative trust in procurement strategies.

8.6 The UK Construction Industry and Recent Reforms to Encourage Trust and Collaboration

The UK construction industry has arguably been associated, over many years, with projects which have had less than successful outcomes. This has culminated from many different reasons, and many measures and recommendations have come forward to address this on-going dilemma and act as a catalyst for improvements. Some recommendations for best practice and obtaining more cost and time predictability on projects include measures to improve partnering procurement processes and therein encourage more successful team integration and collaborative ways of working. This is a deliberate attempt to improve construction practices, which have arguably not been delivering the impact, and benefits that were intended in terms of successful collaborative project outcomes. Various government reports have reinforced this dilemma over the years. These have included Latham (1994), *Constructing the Team*, Egan (1998), *Rethinking Construction*, and Egan (2002), *Accelerating Change*. Other examples include *Construction 2025, Industry Strategy: Government and Industry in Partnership* (HM Government 2013), which identifies that fractious qualities are embedded in the UK construction industry. The report stresses that collaboration and trust across the entire supply chain are crucial to deliver successful projects. More recently, this dilemma has been reiterated and become a major feature of *Modernise or Die: The Farmer Review of the UK Construction Labour Market* (Farmer 2016). Furthermore, findings from *Low Carbon Construction Final Report* (HM Government 2010) confirm the growing need for increased collaboration and integration across the industry, especially between the supply chain and clients, in order to make greater contributions to the pursuit of efficiencies.

8.7 Traditional Procurement of Construction Projects

8.7.1 Barriers and Problems for Trust with Traditional Procurement

There is an argument that mistrust has been inherent within the UK construction industry for a long time between all parties including clients and the consultants and contractors they employ. It is important to comprehend the factors which build trust between clients, consultants and contractors in this regard. In this way, the critical factors can be realised, and clients are then more able to facilitate alignment of organisational interventions to build trust (Hawke 1994). This is especially important as the development of trust has proved problematical with only limited success owing to the different contractual interests of those involved (Lu and Yan 2007). Maurer (2010) reiterated this argument but also concluded that retaining trust, in addition to building it, can be an equally challenging task. This is an area of study which has received only limited attention within the realms of construction management. According to Maurer (2010), in addressing these challenges, further research is recommended into different factors which influence the development of trust.

8.7.2 Deficiencies with Traditional Construction Procurement

In considering the problem of trust between construction clients and those construction-related organisations they employ, it is imperative to consider the traditional adversarial nature of the construction industry which has often been attributed to creating barriers for trusting relationships to grow. Most practitioners would argue that this has stemmed from the traditional forms of procuring construction work which over recent years there have been blamed for achieving low client satisfaction levels, poor cost predictability and time certainty. Such a dilemma has largely been attributable to coordination difficulties associated with separation of design and construction and the greater need for teamwork (Latham 1994, pp. 81–83; Egan 1998, p. 18–21, 2002, p. 6). Traditional or conventional procurement methods in this sense have been regarded as the standard practice in the construction sector for many years (JCT, 2014) and normally rely on completion of design and full documentation before tender. It has been predominantly geared around stages of construction procurement which include feasibility, design, tender, construction, commissioning and handover. Traditionally contracts have in the past normally been procured with minimal if any contractor, subcontractor or supplier design input in the early stages (MacKenzie and Tuckwood 2012). This separation of design and construction is illustrated in Figure 8.1 which shows a structure chart wherein all of the design team employed by clients at all stages of the project. Furthermore Figure 8.2 illustrates under this traditional procurement route that all design stages of the RIBA Plan of Work 2013 are completed without any contractor's design or input.

Such contracts have been criticised for being awarded on lowest price tenders and having 'win' and 'lose' outcomes in the past where one party has overcome the other, normally in terms of commercial gains. This type of practice and culture that has evolved is widely considered to present an unethical and unacceptable state of affairs. Furthermore, it has also encouraged in some instances a 'blame game' when problems arise, most predominantly in the construction phase, through general lack of teamwork. This approach over many years seems

122 | *8 Ethical Considerations around Trust and Collaborative Working*

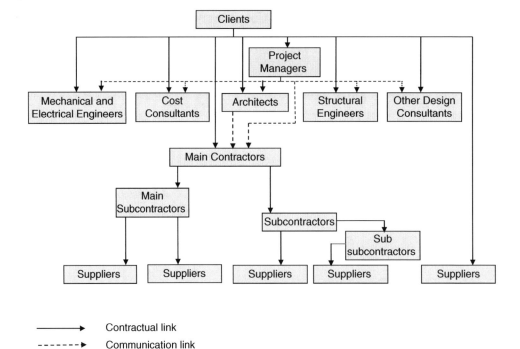

Figure 8.1 Diagram illustrating traditional procurement structure.

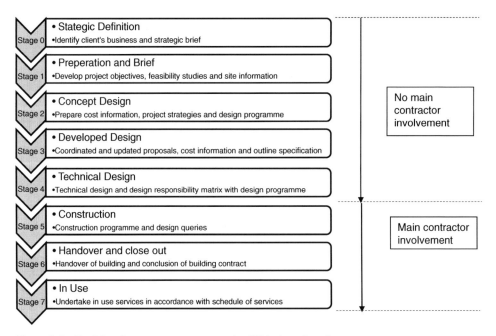

Figure 8.2 Traditional procurement route under RIBA plan of work.

rather fruitless as it has fuelled a culture and environment of mistrust between contracting parties to the extent that they are hesitant or unwilling to rely on and vest trust in the other. From the client's perspective this may manifest itself in feeling that contractors are unethical and 'out for what they can get' with maximisation of profits as the main motivator. Some clients and consultants often refer to contractors as 'claims conscious' in this regard. From the contractors' perspective, a lack of trust in traditional procurement may stem from being responsible for construction of a project which they have not had any involvement in any of the design, since they were invited to tender on full designs. As a result, they may not approve fully of some of the design elements of the project and despite feeling that they could have made improvements had they been involved earlier albeit are ultimately responsible for delivering the completed building. For these reasons, some would explain the lack of trust from the view and general perception that an embedded mentality is prevalent within the construction industry which relates to a general lack of integration of design and construction (Cartlidge 2004, p. 11). The choice of procurement strategies on projects has therefore become a long contentious issue within the construction industry to address such perspectives. Other criticisms of traditional procurement methods have focused on their adversarial nature, deficiencies of design and construction interface and the inefficiencies that these can bring for construction projects. Recommendations for changes, designed as improvement measures, emanated from such criticism and included bridging the gap between design and construction and encouraging early contractors' involvement in value management and buildability (Emmerson 1962; Banwell 1964; Latham 1994, pp. 40–51; Egan 1998, 2002). Furthermore, others have outlined that risk management under such traditional arrangements is problematic especially when complex projects create greater risks for project teams. Contracting parties in these cases may seek to preserve their own individual commercial positions, frequently causing disputes to arise. These have in some cases led to reports of programme delays, cost overruns, conflict, distrust and legal action (Chan et al. 2008). To prevent such occurrences arising there is an argument that 'collaborative working' or 'partnering' offers a more suitable context for developing cooperation and trust as less focus on price and authority (Eriksson and Lann, 2007).

8.8 Overall Context of Collaborative Working and Partnering within the Construction Industry

The aforementioned perspectives would tend to suggest that traditional procurement strategies are widely considered to be inefficient and perhaps confrontational. Collaborative working at an early stage between clients their design teams has been, post-Latham, regarded as a means to bridge the gap between design and construction to improve project outcomes. Accordingly, many have identified collaborative procurement routes through 'integrated' teams as a critical success factor on construction projects (Vaaland 2004). Such procurement routes allow construction consultant and contracting teams to interact at earlier stages in the construction cycle by allowing the overlap of design and construction. This integration of design and construction is illustrated in Figure 8.3 with reference to a design and build procurement structure wherein the contractor employs the design team directly post-novation. Furthermore Figure 8.4 illustrates under this integration of early contractors' involvement into the design and build procurement route through reference to

124 | *8 Ethical Considerations around Trust and Collaborative Working*

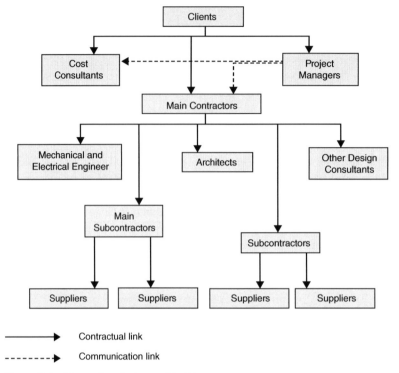

Figure 8.3 Illustrating design and build procurement structure.

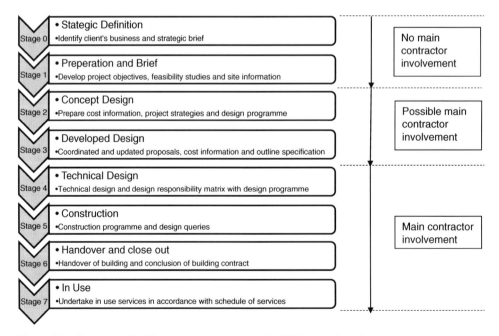

Figure 8.4 Design and build procurement route under RIBA plan of work.

the RIBA Plan of Work 2013. These have been previously heralded as the means to address calls for change and fulfil the future challenges ahead but only if parties are prepared to build trust within such relationships (Kaluarachi and Jones 2007).

8.9 Trust as a Collaborative Necessity

There is an argument that the formation stages of partnering arrangements are commonly regarded as being particularly challenging, when negotiations are progressing and interactions between the parties are complex. Notwithstanding the challenges, at this point in time, trust is considered to be a 'bonding agent' between collaborating partners and regarded as an 'essential foundation for creating relational exchange' (Langfield-Smith, 2008). Perhaps in this way it could also be regarded as a mechanism for governance of partnerships which seeks to deter cheating and opportunism especially where dependence is strong and uncertainty is great (Silva et al. 2012). Irrespective of the stage of its development, however, trust in partnering can create the willingness for further collaboration and organisational alignment, whilst reducing cost and time in managing behaviours and actions of individuals (Walker and Hampson, 2003; Rowlinson and Chueng, 2004). Accordingly, collaborative arrangements may provide the right context for trust within relationships to be embedded, and therein its ability for further development. Applying this premise in contractual terms, provisions could be incorporated into partnering agreements in terms of obligations, roles, intentions, promises and reputations (Branzei et al., 2007).

There have been concerns expressed by supply chains for negotiations in case projects do not come to fruition. To address this, an additional clause in NEC3, was introduced in January 2016. It expressly refers to contractor costs for Stage 1 (pre-commencement) and Stage 2 (construction phase when projects are on site). It deals with scenarios where contractors do not receive instructions to progress to Stage 2 when they will have incurred significant costs. The clause states that the employer will 'likely need to pay the contractor for its Stage 1 costs'. This provision has the advantage of contractors deploying time and resources to Stage 1, without concerns of being unsuccessful and losing their bidding costs. It has advantages for employers as the inputs and innovative solutions generated by unsuccessful contractors can outweigh the cost, they pay them for such valuable contributions. Within this mutually advantageous arrangement, trust can be generated between the parties without the fear from contractors of being financially disadvantaged.

Egan (2002, pp. 13–20) highlighted the need to change adversarial practices and behaviours and recommended the need for significant improvements in communications, attitudes and respect. In order to embed changes, trust amongst project teams and organisations has been identified as a factor which needs to be significantly increased (Dainty et al., 2007), especially since it is 'central to every transaction that demands contributions from the parties involved' (Cheung et al., 2011, pp. 184–196). Taking this argument, a stage further it has also been suggested that mutual trust in partnering is regarded as fundamental to relationships built-up on cooperation and collaboration (Deutsch, 1958; Deutsch, 1962; Wong et al. 2008), and based on parties not moving away from originally defined mutual objectives (Black et al., 2000). Others would advocate that it can also assist in strengthening and improving collaborative working relationships on projects, aligning partners' interests and maximising future opportunities, whilst facilitating new knowledge acquisition and enhanced business opportunities (Maurer 2010). A similar argument is supported by Fawcett et al. (2012) who found that trust is imperative to 'collaborative innovation capacity', as without trust collaborative alliances cannot be created or maintained.

There would appear from the preceding arguments to be clear consensus of the virtues of trust in construction contracts. Despite this, what is perhaps surprising is that, according to Fawcett et al. (2012), trust is the core issue as to why partnering has failed to mirror collaborative business models from successful organisations outside the construction sector. They opined that few companies have actually managed to generate the mutual trust in partnering, since they have not fully understood its true nature and how it can be best applied in developing and enhancing capacity and opportunities. Clearly this demonstrates that developing trust in construction contracting is not without its challenges.

Wong and Cheung (2004), for instance advocated the notion that developing trust is the most critical factor in influencing the success of partnering projects. Certainly Brookes et al. (2006) would appear to agree with this perspective and stressed its importance in the context of successful construction project management. Despite this, the prevalence of trusting relationships is the exception in most cases rather than the norm in construction contracting (Cheung et al., 2011, pp. 184–196), which tends to suggest that more trust building initiatives are required. For these reasons, campaigns have called for the knowledge of trust to be formally included in Construction Management curriculum programmes at Universities and Colleges (Atkinson et al., 2006; Ahrens, 2004). More recent findings from the National Construction Contracts and Law Survey supported this campaign and outlined that

> *All design and construction professionals continue to be trained in silo environments with, at best, occasional contact with the other trainee professionals. There is an inbred distrust of one another and a need to demonstrate better leadership to drive collaboration into the heart of the curriculum of the training institutions.* (emphasis added)

When considering the influence of trust in partnering strategies, one practitioner interviewed, described that his own consultancy had only managed to survive the period of austerity, 2008–2012, in the construction industry, through trust and collaboration in the widest possible sense. Examples of such collaborative practices were quoted as undertaking work without payment, sharing expertise, providing information, introducing contacts and working at risk without formal appointments being in place. Conversely another participant dismissed the notion that trust was in fact a prerequisite for partnering agreements. He explained that partnering contracts such as NEC3 have legal frameworks where roles, responsibilities, duties, management processes and procedure are clearly set down with legal redress in the event of breaches. This raises the issue of trust and whether partnering arrangements need one or both to be successfully implemented. It was accepted, however, in this context that time can be saved in not having to formally document each and every form of communication if individuals can trust and thereby rely more on each other. It is also generally acknowledged that it also creates a less antagonistic and stressful working environment which can support individuals in performing better (Wood, 2005).

Trust can provide the right environment for an 'upward cycle of trust' and encourage further collaboration to develop. For this reason, having a trust regime allows projects to move forward more effectively. This creates the conditions where problems can be solved more easily and where barriers are broken down, which is broadly consistent with the literature.

Trust is regarded by some as an output as well as an input; something that must be earned and certainly not something that should be taken for granted. Others would advocate that if trust is going to make an important contribution and be truly effective in partnering strategies it has to be 'full' and unequivocal. These would broadly underpin theories of trust from perspectives of 'deep interdependence' (Shepperd and Sherman, 1998) and 'unconditional trust' (Jones and George, 1998). Conversely in situations where this is not the case, and only partial or 'conditional' trust exists, then such benefits will not be realised. In extreme cases, where trust is lost altogether, findings indicate that it can make working relationships untenable. This can lead to a 'downward cycle of trust' and supports the theory of Thurairajah et al. (2006, pp. 1–8).

It was revealed earlier in the chapter that construction professionals have frequently revealed that trust and professional ethics are influenced by various positions and roles that exist within project teams. Accordingly, there are differences of opinion on the importance of trust between clients, design consultants, contractors and subcontractors. This may represent a constraining factor and provides further justification for the focus on trust as an integral part of construction procurement strategies. There have been reported cases of abuse which have emerged where contractors and subcontractors have been disadvantaged by more powerful clients as 'paymasters'. Such examples further reinforce the importance, for having those trust-building mechanisms in place which promote openness, fairness and equity in partnering strategies. It could also support the facilitation of workshops and introduction of 'issue' rather than 'dispute' resolution processes for addressing such concerns. The importance of trust could also be dependent on the seniority of individuals within project teams and partnering organisations. Atkin and Borgbrant (2007) found that trust at senior management level has the most influence on partnering strategies, especially if it can be transgressed downwards throughout organisations. The aforementioned examples and opinions appear to represent an unequivocal perspective that reforms and initiatives are required to approach and respond to an inherent and potentially damaging issue.

8.10 Different Context for Clients to Consider in Construction Procurement

8.10.1 Government Context: Reports, Codes of Practice and Recommendations for Change

Throughout the decades there has been much criticism directed at the UK construction industry and many government reports have been commissioned over the years. Early reports in this respect included the Simon Report (1944), Emmerson Report (1962), Banwell Report (1964) and Potts Report (1967). These reports highlighted the deficiencies with construction-related business approaches in terms of low performance, failure to meet client satisfaction levels and need for efficiency measures (Simon 1944; Emmerson 1962; Banwell 1964; Potts 1967). Perhaps, though, the most prominent report which has made recommendations on procurement in recent years is 'Constructing the Team' by Sir Michael Latham (1994). In this report Latham was critical of traditional procurement and contractual routes owing largely to a lack of coordination and integration between construction and design. He explained that in normal circumstances the main objectives of clients would include good value for money, attractive appearance, no defects, completion on time, fit for purpose, issue of guarantees, and low maintenance costs. Furthermore, he described, owing to fundamental problems of conflict between employers and contractors, why these objectives have not been realised to their full extent. Accordingly, he suggested that a change of behaviour towards partnering was required to increase fairness, encourage teamwork and enhance performance through clients and consultants engaging more with contractors and their supply chain (Latham 1994, pp. 50–57). The Egan Reports, *Rethinking Construction* and *Accelerating Change* (Egan 1998, 2002) followed on from this, calling for integration of the design and construction phases with project teams working more collaboratively. These reports referred to construction projects in the USA where modularisation, standardisation and supplier/subcontractor design innovation had facilitated greater efficiencies than those procured in the UK (MacKenzie and Tuckwood 2012). Figure 8.5 illustrates some of the perceived benefits, especially on complex projects that could emanate from the collaboration and early involvement of contractors, from findings of Latham and Egan.

To reinforce Latham's recommendations, the Construction Industry Council Strategic Forum for Construction (CIC 2002, p. 15) reported that projects which had applied principles of both Latham and Egan in the use of collaborative procurement methods, have led

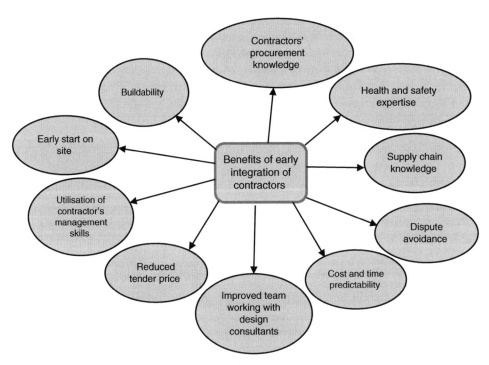

Figure 8.5 Benefits for clients in early integration of contractors.

to significant improvements in client satisfaction, cost predictability, safety and time predictability. Despite this, however, Constructing Excellence reported that there has been 'patchy' take up of recommendations for adoption of collaborative working practices, mostly attributable to strict and inflexible UK public sector procurement rules.

The Government study, *Construction 2025: Industry Strategy: Government and Industry in Partnership* (HM Government 2013), has identified low levels of innovation and investment in research and development coupled with uncertain demand, which it claims is hindering collaborative working and partnering in the UK construction industry. It has also highlighted high construction costs as a major weakness in UK construction, mostly attributable to it 'being driven by inefficient procurement and processes'. For this reason, it states Government's intention to lead the way in strategic procurement and sets down targets through both the *Infrastructure Cost Review* (HM Treasury 2010b) and the *National Infrastructure Plan 2013* (HM Treasury 2013). Such targets reflect a reduction of between 15% and 20% on construction costs over a five-year period. A further Government report: *Low Carbon Construction Final Report* (HM Government 2010), reinforced that partnering with extensive integration and collaboration within project teams remains a crucial vehicle to deliver successful construction projects. Key performance indicators linked to success, in this report, included achieving lower carbon emissions and improving predictability in terms of time and cost of projects through collaboration. The report also claimed that a downward trend in partnering has led to limited knowledge sharing, familiarisation and efficiencies being realised. In order to counteract this problem, it recommended greater collaborative working between different organisations including clients and their supply chain partners as the catalyst for reversing the trend. Further support came from The *Infrastructure Cost Review* (HM Treasury 2010b) which reported that procurement strategies particularly in the public sector have not always been efficient. It opined that 'public sector clients are more risk averse to the cost and time implications of potential challenges, and processes are overly complex and too much of a box ticking exercise'. This once again reinforces the need to reappraise procurement strategies and justifies the focus for strategies linked to increasing collaboration and trust.

The other major initiative that Government has been spearheading recently is the wider use of Building Information Modelling (BIM) and its intention to require that 3D BIM is mandatory on all public sector projects since 2016. This is part of the UK Government's four-year programme to modernise the industry with key objectives to reduce capital costs and lower carbon emissions from the construction and operation of new buildings by 20% (BIM Task Group 2013). There are already, however, concerns being raised about collaboration and integration of the whole supply chain which is arguably one of the most important prerequisites to making BIM work effectively. These relate to the notion that main contractors, subcontractors and suppliers may not have an appetite for investing up front resources in BIM, spending additional time on tenders and sharing their best ideas, if they perceive they only stand a one in six chance of winning contracts. This reflects tendering in the UK construction industry, traditionally geared around competitive bids from a number of potential contenders. However, it is unlikely that a BIM tendering process based on negotiation or fewer tenders would satisfy the market in terms of demonstrating competition. This is especially the case for public sector procurement with strict regulation and governance around such processes. Perhaps therefore a cultural change, as advocated by

Construction 2025 (HM Government 2013), following on from Latham and Egan, might be the only means to reduce the UK's reliance on the ethos of competition.

In the pursuit of the Government's drive to modernise the UK construction industry it has outlined that central to aforementioned challenges of reducing build and operating costs, the sector will be the need to adopt information technologies and reform collaborative behaviours and processes to realise more efficient ways of working. To assist in these reforms, the British Standards Institute has published a new standard *PAS 1192–2: Information Management for the Capital Delivery Phase of Construction Projects* with the aim of adopting a best practice and consistent approach around such initiatives (BSI 2014a).

8.10.2 Industry and Professional Context

In considering the problems around collaboration and trust, it is necessary for construction professionals to understand the context within which the UK construction industry operates. On reflection, the environment is unique and arguably more complex than most other sectors which include manufacturing, as projects tend to be bespoke and require creative thinking and innovation at most times (Walker 2009).

The other contributing factor which makes the construction industry unique is the fragmented nature of both the professional disciples and workforce within it. It is claimed that 192,000 independent small contractors operated in the UK in 2007, of which over 93% had fewer than 14 employees. More recently official figures published in the Business Population Estimates 2014 (BIS 2014b) reported that there were 950,000 SMEs operating in the UK construction sector at the start of 2014. Fragmentation will emanate from such a large number of relatively disparate and mostly small specialist contractors and suppliers having 'arms-length' relationships on building projects (Briscoe et al., 2001). This could hinder innovation and continual improvement from lessons learnt and experience with partnering being heralded as the means to address such a deficiency (Ikechukwu and Kapogiannis, 2014).

In consideration of the aforementioned difficulties there have been many reports of good practice and benefits arising from partnering (Rahman and Kumaraswamy 2004; Walker 2009). Unfortunately, not all organisations agree with such arguments and partnering has attracted its critics in some instances. The RICS (2005, p. 2), for example, explained that recently reported successful experiences in collaborative procurement 'are largely anecdotal and attributable to the experiences of exemplar organisations'. They argued that the focus on success rather than failure has created an unbalanced view and presented a false impression in terms of the contributions made by partnering. Accordingly, they questioned the reliability and validity of some information sources. There is also an argument that partnering is losing popularity in a climate where some construction professionals still view competitive tendering as the best means to secure best value (RICS 2012). Furthermore, Morgan (2009, p. 9) concluded that with major capital projects, procurement routes that promote alliances and partnerships are not always appropriate and open to abuse owing to the scale of the commercial interests involved. This reinforces the previous arguments that partnering could be losing its popularity.

Notwithstanding these opposing views to partnering approaches, studies conducted by the Construction Products Association (HM Government 2010) concluded that greater integration with supply chain partners could make a significant contribution to the low carbon agenda and reduce construction costs. Crespin-Mazet and Portier (2010) supported this and explained that the construction industry could benefit from a 'joined-up approach' as already experienced in other sectors. This has served to 'front-load' problem solving and lessen expensive design modifications later in the process. Furthermore, the British Standards Institution has recently developed with industry and introduced BS11000 which provides a framework for building and maintaining collaborative relationships between organisations. It is a model based on eight phases and designed to 'enable organisations to focus their efforts from concept to disengagement' (BIS 2014b). Initiatives such as these could be specifically designed to 'turn the tide' in favour of more participation in partnering practices. Perhaps these could make an impact and encourage more collaborative working in the construction industry.

8.11 Summary

Trust and professional ethics should be regarded as a collaborative necessity which construction professionals should be encouraging and closely managing on their projects. It is crucial to consider the importance of professional ethics for building trust, adopting collaborative policies and developing good working relationships, as there is a reciprocal correlation between all these factors. This is especially the case in construction management where the industry and all those professionals that are employed within it heavily rely on interaction between each other. The construction industry is composed of many different groups of individuals and disciplines; there are many interdependencies between these groups where the behaviour, successes or failures of one party could affect another in a positive or negative way. It is for this reason that professional ethics have a strong relationship and bearing on trust especially in the context of partnerships and collaboration working relationships. In this context it is therefore not surprising, owing to the fragility of trust theories, that actions of unethical behaviour or practice can have a hugely negative effect on such relationships. This can be very damaging for construction projects with parties becoming reluctant to collaborate fully, share information and trust their fellow professionals. Accordingly, this chapter has focused on the bespoke nature of the construction industry and the need for cultural and behavioural reforms to improve relationship between construction professionals. This is aimed at improving performance for the industry, maintaining professional ethics and obtaining more successful outcomes for construction projects.

The UK construction industry has arguably been associated, over many years, with projects which have had less than successful outcomes. There are many different reasons that have emerged as to why this might be the case. Some relate to the short-lived duration of construction projects whilst others to lack of continuity and fragmentation across supply chains. Various government reports have reinforced this dilemma over the years and identified that unethical behaviours and fractious qualities are embedded in the UK construction industry. Arguments have emerged that mistrust has been inherent within the UK construction industry for a long time between all parties including clients and contractors they employ. This may have stemmed from the traditional forms of procuring construction work which over recent years has been blamed for achieving low client satisfaction levels, poor cost predictability and time certainty.

To prevent such occurrences arising, there is an argument that 'collaborative working' or 'partnering' offers a more suitable alternative in creating working relationships built on ethical values and trust. The use of collaborative procurement methods and improvements to relationships that they bring about, applying the principles of both Latham and Egan, can procure more successful project outcomes. Success in this way can be measured through significant improvements in client satisfaction, cost predictability, safety and time predictability. In this context, such partnering-based approaches have been presented in this chapter as a pathway to greater cooperation and trust.

References

Ahrens, C. (2004). Ethics in the built environment: a challenge for European universities. The Socrates project. In: *Proceedings of the 2004 American Society for Engineering Education Annual Conference and Exposition*, 5281–5289.

Atkin, B. and Borgbrant, J. (2007). Development processes in construction management. In: *Proceedings of 4th Nordic Conference on Construction Economics and Organisation*. Sweden: Luleå University of Technology.

Atkinson, R., Crawford, L., and Ward, S. (2006). Fundamental uncertainties in projects and the scope of project management. *International Journal of Project Management* 24 (8): 687–698.

Banwell, H. (1964). *The Placing and Management of Contracts for Building and Civil Engineering Work*. London: HMSO.

BIM Task Group (2013). *Welcome to the BIM task group website*. Available at http://www.bimtaskgroup.org [18th March 2021].

BIS (2014b). *Department of business, innovation and skills: business population estimates for the UK and regions 2014*. Available at https://www.gov.uk/government/uploads/system/uploads/attachment_data/file/377934/bpe_2014_statistical_release.pdf [21st August 2021].

Black, C., Akintoye, A., and Fitzgerald, E. (2000). An analysis of success factors and benefits of partnering in construction. *International Journal of Project Management* 18 (2000): 423–434.

Branzei, O., Vertinsky, I., and Camp, R.D. (2007). Culture-contingent signs of trust in emergent relationships. *Organisational Behaviour and Human Decisions Processes* 104 (2007): 61–82.

Bresnen, M. and Marshall, N. (2000). Partnering in construction: a critical review of issues, problems and dilemmas. *Construction Management and Economics* 18: 229–237.

Briscoe, G., Dainty, A.R.G., and Millet, S. (2001). Construction supply chain partnerships: skills, knowledge and attitudinal requirements. *European Journal of Purchasing and Supply Management* 7 (2): 243–255.

Brookes, N.J., Morton, S.C., Dainty, A.R.J., and Burns, N.D. (2006). Social processes, patterns and practices and project knowledge management: a theoretical framework and an empirical investigation. *International Journal of Project Management* 24 (6): 474–482.

BSI (2014a). The british standards institution. *new standard for BIM to help meet 2016 government saving target*. Available at http://www.bsigroup.com/en-GB/about-bsi/media-centre/press-releases/2013/3/new-standard-for-bim-to-help-meet-2016-government-savings-target [23rd March 2021].

Cartlidge, D. (2004). *Procurement of Built Assets*. London: Buttersworth-Heinmann.

Challender, J., Farrell, P., and McDermott, P. (2019). *Building Collaborative Trust in Construction Procurement Strategies*. Oxon: Wiley.

Chan, A.P.C., Chan, D.W.M., Chiang, Y. et al. (2004). Exploring critical success factors for partnering in construction projects. *Journal of Construction Engineering and Management* 130 (2): 188–189.

Chan, A.P.C., Chan, D.W.M., Fan, L.C.N. et al. (2008). Achieving partnering success through an incentive agreement: lessons learned from an underground railway extension project in Hong Kong. *Journal of Management in Engineering* 24 (3): 128–137.

Chen, X.P., Eberly, M.B., Chiang, T.J., Farh, J.L and Cheng, B.S. (2012). Affective trust in Chinese leaders: linking paternalistic leadership to employee performance. *Journal of Management* 26: 234–243.

Cheung, S.O., Wong, W.K., Yiu, T.W., and Pang, H.Y. (2011). Developing a trust inventory for construction contracting. *International Journal of Project Management* 29: 184–196.

CIC (2002). *Construction Industry Council Strategic Forum for Construction: Accelerating Change*. London: Construction Industry Council.

Cicmil, S. and Marshall, D. (2005). Insights into collaboration at the project level: complexity, social interaction and procurement mechanisms. *Building Research and Information* 33 (6): 523–535.

Colquitt, J.A., Scott, B.A., and LePine, J.A. (2007). Trust, trustworthiness, and trust propensity: a meta-analytic test of their unique relationships with risk taking and job performance. *Journal of Applied Psychology* 92 (4): 909–927.

Crespin-Mazet, F. and Portier, P. (2010). The reluctance of construction purchasers towards project partnering. *Journal of Purchasing and Supply Management* 16 (2010): 230–238.

Das, T.K. and Teng, B.S. (1998). Between trust and control: developing confidence in partner cooperation in alliances. *The Academy of Management Review* 23 (3): 491–512.

Dainty, A., Green, S., and Bagihole, B. (2007). *People and Culture in Construction*. New York: Taylor and Francis.

Deutsch, M. (1958). Trust and suspicion. *Conflict Resolution* 2: 265–279.

Deutsch, M. (1962). Cooperation and trust: some theoretical notes. In: *Nebraska Symposium on Motivation*, 275–320. Lincoln: Nebraska University Press.

Diallo, A. and Thuillier, D. (2005). The success of international development projects, trust and communication: an African perspective. *International Journal of Project Management* 23: 237–252.

Egan, J. (1998). *Rethinking construction. The Report of the Construction Task Force*. London: DETR.TSO. 18–20.

Egan, J. (2002). *Accelerating change. rethinking construction*. Strategic Forum for Construction. London.

Elangovan, A.R. and Shapiro, D.L. (1998). Betrayal of trust in organisations. *The Academy of Management Review* 23 (3): 547–566.

Emmerson, H.C. (1962). *Emmerson report; survey of problems before the construction industries*. Ministry of Public Building and Works. London. HMSO.

Eriksson, P.E. and Lann, A. (2007). Procurement effects on trust and control in client-contractor relationships. *Engineering, Construction and Architectural Management* 14 (4): 387–399.

Farmer, M. (2016). *Modernise or Die: The Farmer Review of the UK Construction Labour Market*. London: Construction Leadership Council.

Fewings, P. (2009). *Ethics for the Built Environment*. London: Routledge.

Gadde, L.E. and Dubois, A. (2010). Partnering in the construction industry; problems and opportunities. *Journal of Purchasing and Supply Management* 16 (4): 254–263.

Hawke, M. (1994). Mythology and reality: the perpetuation of mistrust in the building industry. *Construction Papers of the Chartered Institute of Building* 41 (1994): 3–6.

HM Government (2010). *Low Carbon Construction Final Report (November 2010)*. London: HM Government. 52–62, 196–199.

HM Government (2013). *Construction 2025. Industry Strategy: Government and Industry in Partnership*. London: HM Government. 23–25, 61–71.

HM Treasury (2010b). *Infrastructure cost review: main report*. Infrastructure UK (2013). Available at https://www.gov.uk/government/uploads/system/uploads/attachment_data/file/192588/cost_review_main211210.pdf [10th July 2021].

HM Treasury (2013). *National Infrastructure Plan*. Available at https://www.gov.uk/government/uploads/system/uploads/attachment_data/file/263159/national_infrastructure_plan_2013.pdf [1st August 2021].

Huang, Y. and Wilkinson, I.F. (2013). The dynamics and evolution of trust in business relationships. *Industrial Marketing Management* 42 (2013): 455–465.

Huxham, C. and Vangen, S. (2000). Leadership in the shaping and implementation of collaboration agendas: how things happen in a (not quite) joined-up world. *The Academy of Management Journal* 43 (6): 1159–1175.

Ikechukwu, U. and Kapogiannis, G. (2014). A conceptual model for improving construction supply chain performance *In*: *Procs 30th Annual ARCOM Conference* (ed. A.B. Raiden and E. Aboagye-Nimo), 1-3 September 2014, Portsmouth, UK: Association of Researchers in Construction Management, 1029–1038.

Jones, G.R. and George, J.M. (1998). The experience and evaluation of trust: implications for cooperation and teamwork. *The Academy of Management Review* 23 (3): 531–546.

Kaluarachi, D.Y.A. and Jones, K. (2007). Monitoring of a strategic partnering process: the amphion experience. *Construction Management and Economics* 25 (10): 1053–1061.

Langfield-Smith, K. (2008). The relations between transactional characteristics, trust and risk in the start-up phase of a collaborative alliance. *Management Accounting Research* 19 (2008): 344–464.

Lann, A., Voordijk, J., and Dewulf, G. (2011). Reducing opportunistic behaviour through a project alliance. *International Journal of Managing Projects in Business* 8 (4): 660–679.

Larson, E. (1997). Partnering on construction projects: a study of the relationship between partnering activities and project success. *IEEE Transactions on Engineering Management* 44 (2): 188–195.

Latham, M. (1994). *Constructing the Team*. London: The Stationery Office.

Li, P.P. (2008). Toward a geocentric framework of trust: an application to organisational trust. *Management and Organisation Review* 4 (3): 413–439.

Lu, S. and Yan, H. (2007). A model for evaluating and applicability of partnering in construction. *International Journal of Project Management* 25: 164–170.

MacKenzie, N. and Tuckwood, B. (2012). A model to manage the water industry supply chain effectively. Management, Procurement and Law. *Proceedings of the Institution of Civil Engineers* Institution of Civil Engineers Publishing 165 (3): 181–192.

Maurer, I. (2010). How to build trust in inter-organisational projects: the impact of project staffing and project rewards on the formation of trust, knowledge acquisition and product innovation. *International Journal of Project Management* 28 (2010): 629–637.

McDermott, P., Khalfan, M.M.A., and Swan, W. (2005). Trust in construction projects. *Journal of Financial Management of Property and.*

McKnight, D.M., Cummings, L.L., and Chervany, N.L. (1998). Initial trust formation in new organisational relationships. *The Academy of Management Review* 23 (3): 473–490.

Morgan, S. (2009). The right kind of bribe. *Building Magazine*, 9th October. 8–9.

Parayitam, S. and Dooley, R.S. (2009). The interplay between cognitive and affective conflict and cognition and affect based trust in influencing decision outcomes. *Journal of Business Research* 62 (2009): 789–796.

Pinto, J.K., Slevin, D.P., and English, B. (2009). Trust in projects: an empirical assessment of owner/contractor relationships. *International Journal of Project Management* 27 (6): 638–648.

Potts (1967). *Potts report; action on the banwell report: a survey of the implementation of the recommendations of the committee under the chairmanship of sir harold banwell on the placing and management of contracts.* Economic Development Committee for Building of the National Economic Development Office. Available at http://www.designingbuildings.co.uk/wiki/Construction_industry_reports [1st August 2021].

Rahman, M.M. and Kumaraswamy, M.M. (2004). Contracting relationship trends and transitions. *Journal of Management in Engineering* 20 (4): 147–161.

RICS (2005). An exploration of partnering practice in the relationships between clients and main contractors. In: *Findings in Built and Rural Environments*, 2–3. London, UK: RICS Research.

RICS (2012). *Contracts in use. A survey of building contracts in use during 2010*. London: Royal Institution of Chartered Surveyors Publications.

Rousseau, D., Sitkin, S., Burt, R., and Camerer, C. (1998). Not so different after all: across-discipline view of trust. *Academy of Management Review* 23 (3): 393–404.

Rowlinson, S. and Cheung, F.Y.K. (2004). A review of the concepts and definitions of the various forms of relational contracting. In: *Proceedings of the International Symposium of CIB W92 on Procurement Systems* (ed. S.N. Kalindindi and K. Varghese), 227–236. January 7th-12th, Chennai, India.

Shepperd, B.H. and Sherman, D.M. (1998). The grammars of trust: a model of general implications. *The Academy of Management Review* 23 (3): 422–437.

Silva, S.C., Bradley, F., and Sousa, C.M.P. (2012). Empirical test of the trust performance link in an international alliances context. *International Business Review* 21 (2012): 293–306.

Simon, E. (1944). *Simon report; the placing and management of building contracts.* Available at http://www.designingbuildings.co.uk/wiki/Construction_industry_reports [1st August 2021].

Strahorn, S., Gajendran, T., and Brewer, G. (2014). Experiences of trust in construction project management: the influence of procurement mechanisms. In: *Procs 30th Annual ARCOM Conference* (ed. A.B. Raiden and E. Aboagye-Nimo), 463–472. 1–3 September 2014, Portsmouth, UK: Association of Researchers in Construction Management.

Thurairajah, N., Haigh, R., and Amaratunga, R.D.G. (2006). Cultural transformation in construction partnering projects. In: *Proceedings of the Annual Research Conference of the Royal Institution of Chartered Surveyors COBRA*, 7-8 September. 5–8. University College, London: RICS.

Vaaland, T.I. (2004). Improving project collaboration: start with the conflicts. *International Journal of Project Management* 22 (2004): 447–454.

Walker, A. (2009). *Project Management in Construction*, 5e. 150–158. Oxford: Blackwell Publishing Ltd.

Walker, D. and Hampson, K. (2003). Enterprise networks, partnering and alliancing. In: *Procurement Strategies: A Relationship-based Approach*. UK: Blackwell Science Ltd.

Wolstenholme, A. (2009). Never waste a good crisis; a review of progress since rethinking construction and thought for our future. London: Constructing Excellence in the Built Environment. Available at http://dspace.lboro.ac.uk/dsspacejspui/bitstream/2134/6040/1/Wolstenholme%20Report%20Oct%20Oct%202010.pdf [13th January 2014].

Wong, S.P. and Cheung, S.O. (2004). Trust in construction partnering: views from parties of the partnering dance. *International Journal of Project Management* 22: 437–446.

Wong, W.K., Cheung, S.O., Yiu, T.W., and Pang, H.Y. (2008). A framework for trust in construction contracting. *International Journal of Project Management* 26: 821–829.

Wood, G. (2005). Partnering practice in the relationship between clients and main contractors. *RICS Research Paper Series*, 1–42.

9

The Importance of Ethical Leadership in Construction and Engineering Management

To be an ethical leader is indeed to be different. This kind of leader acknowledges the complexity of running a responsible business, yet tries to do it anyway.

Andrew Leigh

9.1 Introduction to Chapter

This chapter of the book essentially links ethical leadership theory within professional practice. Construction professionals, in this regard, are seen as leaders on projects they procure. Accordingly, the purpose of forging this academic/professional practice link is intended to be a vehicle for improving leadership skills of construction professionals and especially those related to ethical practices and behaviours. With this purpose in mind, the chapter will focus on construction leadership identity and will highlight some aspects of leadership styles, broader practices and impact on individuals and organisations involved in construction projects. The review is primarily based on the leadership identity development (LID) model to provide a critical review of leadership identity knowledge and understanding. Ethics will be considered within the overall context of the LID framework and is intended to improve outcomes for construction professionals in leading and managing project teams. Furthermore, leadership definition, its transformative dynamics, the approaches to its applications, the challenges that come with it and its implications will be critically reviewed using the LID model.

The chapter will examine how to think and behave as a responsible leader and based on study by the Imperial College Business School (ICBS) in 2020. This will consider the three main characteristics and behaviours that the ICBS have developed in categorising individuals as responsible leaders. Practical examples will be given to assist readers in contextualising how such characteristics might affect decision-making for certain individuals.

Leadership challenges and the dilemmas that could present themselves for construction professionals around professional ethics will be identified and discussed. In addition, ethical leadership and the benefits that this provides for both organisations and employees from a staff development and training and Best Companies perspective will be explained and analysed. Finally, the relationship between poor leadership and project failure will be scrutinised in a concerted effort to encourage best practice from lessons learnt in the past and thus avoid repeated failures in the future.

9.2 Key Components of Ethical Leadership for Construction Professionals

The Chartered Management Institute (CMI) defined leadership as 'the capacity to establish direction, influence and align others towards a common aim, motivate and commit others to action, and encourage them to feel responsible for their performance' (CMI 2008). It is also defined by Lester (2007) as 'the ability to inspire, persuade or influence others to follow a course of actions or behaviour towards a defined goal'. Francke (2012) referred to the Centre for Creative Leadership's definition in that 'leadership is about creating the conditions for others to succeed'. This argument is also advocated by the Association for Project Management Body of Knowledge (APMBOK) in its quest for leadership to build the supportive environment for teamwork (APM 2012). Lewis (2007) explained that there are hard and soft skills (human and technical elements), which need to be integrated with leadership and management skills, for achieving successful outcomes.

Successful leaders need to be good motivators and 'bring out the best in people' (Wong 2007), and the importance of vision for leaders is paramount. Lewis (2007) believed that, as the majority of the managers' jobs often revolve around dealing with people, it is essential that people practice leadership and especially around adherence to professional ethics alongside management. Furthermore, leaders hold power and authority over team members and knowing how to apply these is a major factor in maintaining morale (Ritz 1994). Some would argue that the art of leadership is to make people to want to do things rather to be forced to do them. This could be specifically relevant for adherence to ethical codes of conduct for compliance with professional values and principles. Milosevic et al. and Strohmeier (cited in Clarke 2010) argued, that the role of leaders should be focused on trying to get the best out of a spectrum of technical experts and professionals, which signifies leadership and effective management of team relationships and emotions. Opposing arguments have stated that exercising quality and competent leadership is the difference between leaders and dictators (Lewis 2007). Along these lines, it might explain why leadership is a combination of personal characteristics and competency areas (Geoghegan and Dulewics 2008). Furthermore, Abraham (2002) linked leadership to managing change within organisations and described it as a constant process to succeed, as leaders have to keep evaluating change and make improvements for future changes.

Arguably the role of construction professionals could, in the procurement of successful project outcomes, can sometimes be too 'task focused' rather than 'people focused' which is a common misnomer when considering the aforementioned arguments around leadership. Most academic qualifications linked to construction management focus on the task nature of managing projects which are commonly referred to as 'hard skills'. Arguably, what construction professionals mostly require, moreover, is people management skills often referred to by most commentators as 'soft skills'. Walker (2007) suggested that management is more of a mechanical process and using techniques, whereas leadership is perceived to be forward thinking, charismatic and inspirational.

Buchanan and Huczynski (2017) articulated that there is a transitional period in leadership development identity. This process takes in to account that people go through a stage where there is mutual agreement of 'stepping up to the plate' and thus proving suitability to become a leader. Equally, this can be seen as an acceptance of the new leader by the group of people who are to be managed or the followers to be. Different perspectives exist, in relation to the distinction between management and leadership in this regard. Buchanan and Huczynski (2017) argued that 'managers do things right' and 'leaders do the right things'; hence, leaders are seen to have a vision. This notion is particularly relevant for professional ethics in ensuring that leaders constantly review their ethical policies and procedures and make any changes for improvements wherever possible. In this way construction professionals can 'lead by example' to gain confidence from other members of the project team that ethical actions, behaviours and decision-making are being practiced.

9.3 Ethical Leadership and Best Companies

It is essential for aspiring leaders to understand what makes a good leader, in order to perform their roles and motivate others for achieving successful outcomes. Accordingly, in those positions of leadership, construction professionals require courage, conviction and assertiveness when faced with difficult choices and decisions. Their decision-making is subject to ethics, fairness and principles which require to be considered and reflected in the decision-making process. The actions of construction professionals in this regard, can have consequences for the vested interests of their employees, communities and the wider society and decisions made can have far-reaching consequences and devastating effects. Leadership is about how to 'know'; first of all, to know themselves, their competitors, friends, enemies, available resources and their own capabilities. Once they know the way, they can lead on the construction process to benefit the build outcomes, the design team and the project environment. Construction professionals attribute this to the realisation that being a leader is far more challenging than originally envisaged.

Most responsible employers seek to comply with employment and workplace regulations set down by government and acts of parliament. Such regulations relate to such issues as non-discriminatory treatment of staff, working hours, sick pay, employment contracts, dismissal, pay and particularly minimum wage. Other regulations relate to minimum standards in the workplace around including health and safety, the disciplinary discrimination act (DDA) and making reasonable adjustments for those staff with specific needs around disabilities. There are however varying degrees of compliance in this regard with some companies choosing to comply with the minimum standards of the regulations rather than upholding the underlying spirit of the regulations. According to Fewings (2009) this could be described as a protectionist perspective, where companies are complying to avoid regulatory breaches and therein not put themselves or their organisation in a vulnerable legal position.

Ethical leadership goes beyond the notion of only satisfying minimum compliance standards. It relates to measures 'above and beyond' legal requirements and involves developing the workforce and improving the service to their clients and supply chains to make them feel valued, listened to and engaged in their business. Government acts of parliament, regulations and employment law are less proficient in dealing with ethical leadership which revolves around staff development, motivational measures, informal communications and engagement, collaboration, creation of a team spirit, empathy, integrity, organisational culture and values. These measures are arguably as important as those others covered by legal regulatory means but not always integrated and incorporated in what could be considered as the normal course of business for some organisations. Some more responsible employers have sought to initiate human relations projects tasked with engaging and developing staff more in the business and these have proved very successful in the past in improving staff morale, motivation and performance. In addition, there are many accreditation and metrics whereby companies can be scored or ranked on their efforts and outputs in this regard. In the UK, the *Sunday Times* 'Best Companies to Work For' accreditation scheme is regarded as one of the more prominent schemes. An 'A1 Star' accreditation and award from Best Companies is a significant achievement that shows an organisation is taking workplace engagement seriously. The award is the only guide to the UK's best companies to work for and is based on what the employees really think. Best Companies has developed a rigorous way to assess what it is like to work for a company, with a survey provided to each employee to determine what matters to them in the form of tally confidential feedback.

Based on the answers of employees to the survey, each organisation is given a Best Companies Index (BCI) score on a scale of 1 to 1000. Depending on an organisation's BCI, it may achieve a one, two or three-star status, which is the unique three-star accreditation system used by Best Companies Ltd. The purpose of the accreditation system is to produce a model based on employee engagement that has been developed over the last six years by Best Companies Ltd. In 2019 there were some excellent results with more than 1000 organisations applying to achieve 'star status', which represented a significant increase on applications in the previous years.

Year on year statistical analysis acquired through the employee surveys has shown that the eight main factors that are fundamental in driving employee satisfaction and engagement with employers are leadership, honesty, personal growth, my company, my team, my manager, well-being and fair deal. Being a best company is about ethical leadership and excelling in every area throughout the workplace and an organisations commitment to the workforce as it is the most important asset. Focusing on employees brings real benefits which include better staff retention, improved workforce engagement, reduced recruitment costs and greater financial performance. This is particularly important for those companies in the construction industry where competition between rival companies remains high.

Ethical leadership has traditionally not been practiced widely in developing countries which in some circumstances has led to employees' rights being compromised. This has taken the form of restrictions on workers freedom of speech, poor health and safety measures in the working environment and employers doing the bare minimum to avoid any action against them. This is particularly the case for the construction industry in those emerging economies which are more labour intensive when compared with more developed countries such as the UK. Given the labour-intensive nature of the construction industry in this context, it requires the capabilities, skills and experience of supervisors and managers to control and coordinate a large workforce of construction employees. There is less emphasis on technological and innovative enhancements to procuring construction projects such as off-site prefabrication, modularisation and use of digital processes and equipment.

One of the most important factors for improving ethical leadership revolves around staff development and training. Training and development are regarded as being the pillars which underpin a successful and motivated workforce and excel a company to achieve successful outcomes and increase growth. In this context, Fewings (2009) described development as training and preparation to deal with changes in role or responsibilities within the organisation, which distinguishes it from the definition of training which is related to imparting information to benefit the current job. There are a number of benefits for individuals and organisations from training and these are shown in Figure 9.1.

The benefits in Figure 9.1 revolve around the organisation but there are clear benefits for the employees. Training allows them to expand their knowledge base and work capabilities and this allowing them to pursue career opportunities and promotions. It also makes them more marketable, contented and comfortable in their roles and motivated to

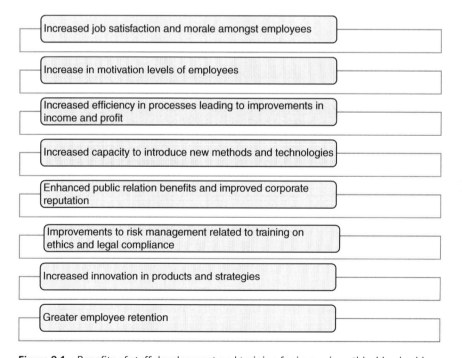

Figure 9.1 Benefits of staff development and training for improving ethical leadership.

perform, which are key ethical rationale for training. Training also has the advantage of ensuring that staff are competent to carry out the jobs assigned to them, and therein avoid any repercussions and consequences for others in not fulfilling their work in a properly manner. An example of this within a construction context could relate to a crane operator. If that individual does not have the training and right qualifications for operating the type of crane they are working with, then this could result in a major accident on site and safely has to therefore be maintained as one of the highest considerations. This underlines the importance of training for health and safety, updates and new responsibilities. Ethically there are responsibilities for companies to protect the welfare of their workforces and this includes staff development and giving them empowerment to be more productive and effective.

9.4 Leadership Identity and Relevance for Construction Professionals

> A leader is one who knows the way, goes the way, and shows the way.
> Maxwell (2006, as cited in Goodreads 2018)

Arguably, social interactions of those individuals making up construction project teams require management, regulation and leadership to enable their creation, success and survival. Whilst project managers can provide this important intervention in the construction industry, the role of construction clients and project managers, as the creators and ultimately the owners or 'project sponsors', should not be underestimated. Temporary organisations, associated with development phases of projects, may be formed for the life of projects; others may already exist and passed on to future leaders to continue their development.

Leadership for construction professionals should be focused on creating the right ethical conditions and environment for people to succeed (Francke 2012; APM 2017). According to the Leadership Identity Model (LID model), leadership grows and matures through stages. Nurturing and supporting followers are an important and a distinctive feature of successful leaders. Lopes (2016) supported this argument and articulated that there are different ways to lead and motivate people, mostly by creating the right working environment and conditions. This has particular relevance for construction teams in selecting, motivating and developing people and providing the environment for them to procure successful construction outcomes.

There is wide acceptance that most construction professionals, who perhaps have little or no previous construction management experience, knowledge or qualifications, learn from other professionals within project teams. From an ethical leadership perspective, one would hope that those professionals that are mentors for other less experienced practitioners, follow codes of conduct set down by their professional body and adhere to governance regulations of their employers. In this way therein they can then present themselves as good examples for others to follow. However, it is not unusual for those relatively inexperienced construction professionals to sometimes feel bewildered with the prospect of leading the way on projects and managing project teams. In such cases, they rely heavily on their construction consultants, especially at the early stages of projects. This may stem from the notion that less experienced individuals follow more knowledgeable people, who may be in positions of authority, power and legitimacy. In this regard, researchers such as Lord and Hall (2005) suggested that leadership in performance terms, through learning from others, progresses in stages from novice to intermediate and then to expert levels. They also claim that leadership skills are developed from micro levels through to problem and challenge-based experience, and eventually to progressively a high-level system influencing knowledge, attitude, behaviour and social perception. This 'learning pathway', influenced by lessons learnt from gaining experience, is sometimes referred to as an experiential model, gaining experience by doing using psychomotor domains, and is particularly relevant to how construction professionals learn on projects.

It is an important perspective to analyse some of the leadership skills that construction professionals need to develop in construction management. Leadership on projects can be regarded as a process whereby the interaction and activities of organised groups is planned and managed to achieve common goals (Buchanan and Huczynski 2017). In addition, it could be affiliated with Northouse (2015), who articulated that leadership is a complex process with multiple dimensions. Opinions on leadership have emerged from different disciplines, with many varying approaches to nurture the skills that are deemed necessary. Northouse also raised and challenged the subject of trait perspective from the viewpoint of 'born to be a leader' and argued that some people are born with characteristics and qualities to make them leaders. Leadership for construction professionals could also relate to having the ability to influence groups of people to achieve prescribed goals (Newell and Grashina 2004). This view was supported by Stogdill (1950, as cited in Buchanan and Huczynski 2017) who defined leadership as a process to influence achieving common goals. In order to view this aspect from psychological angle for construction professionals, Knippenberg et al. (2005) emphasised that one of the principles of leadership is 'having influence'. Therefore, it is the impact and degree to which clients influence their construction teams, that leadership can be monitored and examined from a psychological perspective.

9.5 How to Think and Behave as a Responsible Leader

According to the Imperial College Business School (ICBS 2020) there is a dilemma for business leaders when they are put under pressure and faced with making critical decisions especially when there is limited information. Ethical leadership should dictate that, irrespective of the pressures involved, leaders should not stray from ethical principles and values and make decisions which are both appropriate and moral. They explained that this sometimes can mean putting their clients or employees ahead of their own personal interests in some cases. One example of this could in the context of the coronavirus (COVID-19) pandemic where leaders chose to temporarily close their businesses rather than risk the health, safety and welfare of their staff. It is likely that decisions such as this would have been taken with the full knowledge of the harm this would cause to them and their livelihood. Another example during the pandemic could be where responsible employers allowed staff to work from home and therein trusting them to conduct their work without the level of normal supervision that they would receive if they were in an office environment. Clearly by the nature of the work in the construction industry, it does not always allow this level of flexibility and adaptability. However, for those site employees that cannot work remotely, responsible contractors have instigated COVID-19 safe and robust working procedures on construction sites for their protection. Safeguards and arrangements that construction leaders have made in this regard has included enhanced personal protective equipment (PPE), policies and arrangements for maintaining social distance and making decisions that put the health and livelihoods of their employees ahead of their own personal interests. This has added extra costs for contractors to instigate and maintain such precautionary measures that they more than likely will not be able to recoup from their clients. However, such ethical policies and undertakings have proved particularly beneficial in keeping building sites operable whilst at the same time reducing the risks of the infection spreading.

The COVID-19 crisis has proved to be a historic moment which has made responsible leaders easy to spot in orchestrating these special measures at considerable cost to them. However, outside this crisis, the qualities that underpin responsible leaders can be harder to spot. According to ICBS (2020) when considering what makes a responsible leader looks at a prerequisite requirement of those who see their responsibility as encompassing a much wider set of agents, activities and outcomes, rather than simply managing their company.

ICBS (2020) have advocated that they can categorise responsible and ethical leadership in three main ways. Firstly, responsible leaders should focus on others rather than solely themselves. They should think about how the decisions they make might affect a wide range of stakeholders, including those that are affected directly by the actions of their respective company. Furthermore, they should carefully consider how their organisation affects the communities they work in and the general public at large. With a reduced focus on themselves and an acute understanding and consideration for others, this propels such responsible leaders to make decisions that support the greater good. The environment, political systems and the economy account for a few of the factors that ethical leadership should take into account in decision-making processes. In this way a concerted focus on others coupled with lesser focus on themselves could be considered to be a moral and ethical hallmark of humility and thus a characteristic that responsible leaders share.

Secondly, responsible leadership should consider the longer-term ramifications and implications in their decision-making and thought processes. In the past there has been much criticism of less responsible leaders only focusing on short-term effects of their decision-making possibly linked to financial gain at the expense of other longer-term implications. ICBS (2020) have articulated that there is a general cognitive bias towards favouring the shorter term over the longer term. This is understandable in the context of the construction industry where random problems emerge and require urgent address to avoid costly delays. This is exacerbated even further in times of crisis wherever increasing threats are prevalent and accordingly events such as the COVID-19 pandemic should in theory result in a decrease in responsible and ethical leadership. Notwithstanding this propensity for construction leaders to have to make swift decision-making against this backdrop, due consideration should be made for the longer-term implications and outcomes of these. By way of a hypothetical example to articulate how such a dilemma of this nature might play out in practice please consider the following case.

Hypothetical example of a decision that was taken where longer-term implications were not fully considered

A construction project manager, who is overseeing a large new build for their clients, is made aware at an early stage in the construction process that there are adverse ground conditions on the site, requiring additional remedial works to address major underground contamination. The project is already on a tight budget and programme and there are no contingencies to deal with abnormal costs of this nature. The project manager has two choices, the first being to report back to their clients and request an extension to the budget where additional funds are sought and the second being to recoup the time and cost from other elements of the works. They consider that the short-term implications of the first option may escalate into them being called into account for this problem arising. For instance, they may be asked to explain why more robust surveys and due diligence of existing ground conditions were not conducted at the design stages. This would then have made them aware of the problem much earlier and potentially prior the budget being agreed. The project manager does not regard this approach as particularly attractive for them on a personal basis as they may be blamed for the delays and additional costs. Furthermore, they may consider that their clients might prefer to not use them for other contracts that will be going out to tender in the near future given the problems that have emerged.

The second option to recoup the potential delays and additional costs from other aspects of the contract through 'value engineering', is regarded by the project manager as being the most appropriate and preferable solution to overcome this problem. The building had been designed to meet a high standard of environment and sustainable construction and achieve an Excellent BREEAM accreditation rating. This has included specification of materials linked to reducing embedded carbon together with an extensive range of renewable energy measures. In the latter case this included photovoltaic energy, underground heat pumps, rainwater harvesting and wind turbines. Notwithstanding the benefits of the sustainable measures, the project manager presents a strong case to their clients to remove these 'desirable but not essential' aspects of the works which will allow them to complete the project on time and to budget. The client agrees with their recommendations and these elements of the project are simply taken out of the scope of the works. The project manager believes this has now satisfactorily resolved the issue at hand as the short-term problems of potential budget and delays have been overcome.

Author's Verdict

This is a typical scenario of individuals making spontaneous and rather hasty decisions to resolve short-term problems but failing to consider longer-term implications. The decision to remove all the energy efficient renewal elements of the project will have significant implications for the building over its life cycle. The operation costs (commonly referred to as opex costs), taking into account that the renewable energy components of the build have been removed, will increase significantly. Furthermore, these additional operational costs could account for much more than the capital cost (capex cost) savings achieved. In addition, the effects on carbon generation have been increased as the building is no longer energy efficient.

(Continued)

> **(Continued)**
>
> Taking account of the above the project manager has behaved irresponsibly and should have considered and explained the implications of the value engineering exercise to their clients for an informed decision to be made. It typifies that what can seem like a satisfactory short-term outcome can have hugely damaging consequences in the longer term. Furthermore, the project manager has put their own interests ahead of those of their client who will have to suffer the longer-term consequences. In addition, the failure to uphold environmental considerations is another issue where the project manager's ethical credentials have been brought into question.

The third factor for categorising ethical leadership is related to responsible leaders closely following a 'moral compass' in their management approaches and paying close consideration to their actions and decision-making in this pursuit. ICBS have pointed out that the average person makes endless decisions in one day and whilst some do not have a great degree of moral content the vast majority do have some a degree of social impact. An example could be that ethical leaders are more vigilant and expend more of their finite resources on sourcing sustainable products or undertaking more due diligence on the ethical stature of their supply chains. In the latter case, they may be increasingly conscious of Modern-Day Slavery when sourcing and selecting construction products or materials from overseas; taking additional precautions to ensure that their appointed suppliers meet ethical employment policies.

It is important when considering the traits of ethical leadership for people, especially in authority, to lead by example. In this way the best leaders continue to make decisions that make those around them feel safe, valued and listened to. In a construction context if professional demand certain behaviours and values from their staff that they are failing to uphold themselves, then there will be less likelihood that their demands will be met. This is a particularly critical aspect for professional ethics as any breaches or compromises by managers on what could be deemed as ethical practices could be seen as 'par of the course' by their staff. This could set a bad example and precedent and could escalate into a vicious circle of repeated and escalating unethical practices and behaviours.

When one considers ethical and responsible leadership there is a tendency to look at this from compliance and altruistic perspectives. However, in the latter case, maintaining an altruistic focus is easier said than done. Several frameworks for human motivation describe self-focused needs as more fundamental than externally focused ones. To contextualise this premise, Abraham Maslow, who created the five-level hierarchy of needs that put self-actualisation at the top, proposed a sixth level of motivation shortly before he died. This ultimate level transcends the self, and involves being motivated by truth, fairness and justice. But striving for these transcendent values often waits until more self-focused and fundamental base needs are already met. Unfortunately, according to ICBS (2020) individuals tend to focus on their base needs when facing a threat to their sense of security, self-esteem or confidence and some possess a tendency to become aggressive and adversarial when they feel their authority or competence is being compromised. Ethical leadership theory recognises this dilemma, and the most moral leaders will continue their decision-making around making others around them feel seen, valued and safe. With increasing competitiveness especially in the construction industry, bringing potentially more threats than opportunities, ethical leadership will become more important in the future. In this context there will be an ever-increasing demand for ethical leadership practices and for nurturing future leaders who take decisive action while focusing on others, the long-term, and higher order values.

9.6 Leadership Challenges for Construction Professionals around Professional Ethics

Project management within the construction industry requires the need to effectively manage people, organisations, processes, systems and finance alongside the ever-challenging goal of achieving success. In the leadership stakes, construction professionals are often required to create the vision for a particular project, and then plan and execute it. In order to realise the dream and implement the vision, it can take many years through the planning stages to get all the stakeholders on board and get their buy in. This can sometime present a daunting task for construction professionals, especially those who have little or no prior experience of project procurement processes. During this important planning stage, construction professionals frequently have to lobby politicians and government ministers to attract funding and get support and approvals through their respective local authorities. This is particularly the case for delivery of large-scale master plans, possibly linked to major regeneration, where funding is key to financial viability and affordability, and governance is paramount in ensuring best value for money is achieved. Programmes faced by construction professionals can be challenging in this regard, against an environment of political uncertainly. Normally such large-scale development involves a degree of cultural change within organisations, which presents another challenge for construction professionals in gaining commitment and 'buy in' from staff within their organisations. Accordingly, change management processes are critical to the well-being of organisations and successful outcomes for projects. This notion is reinforced by Lopes (2016) who articulated that the key to successful leadership emanates from dealing with challenging situations, possibly related to staff resisting change, and gathering support for new initiatives. Construction professionals therefore need to focus on any aspects related to change management. From an ethical standpoint one could surmise that there is a moral duty to consult and discuss change proposals with key staff and therein dispel concerns and anxieties at early stages.

9.7 Leadership Styles and Professional Ethics as Applied to Construction Professionals

In terms of leadership style, some would argue that construction professionals need to be assertive, persuasive and articulate, but at the same time reasonable. Some have argued that UK construction professionals in the past been solely focused on obtaining successful outcomes for their employers or clients. This has sometimes led to less than fair unethical practices, predicated around win-lose scenarios. Fortunately, such behaviours have reportedly improved over the years and there is largely now a more collaborative approach to working with construction partners. Accordingly, it is now generally accepted that for projects to succeed they should always be focused on 'win-win' outcomes for their respective client organisations and their contracting construction partners. Collaborative and partnering strategies, devised by construction professionals, which create the right environment for ethical behaviours amongst project teams, provide them with adequate and reasonable resources and get the stakeholders on board tend to achieve more successful outcomes (Challender et al. 2019). This argument seems to be in line with Fisher et al. (2017), which described such approaches to motivating others as 'identity based motivation theory'.

Walker (2007) argued that construction professionals need to conduct themselves in an ethical and professional manner to enable them to get the best out of people who they manage, which is where leadership skills are crucial. This view was also endorsed by Northouse (2015), who suggested that leadership is similar to management due to commonalities between the two facets. He described a skills approach to leadership which shifts individuals' characteristics to skills learning and knowledge gain. This viewpoint enforces the perception that leaders are not born, but leadership can be taught to create leaders. Nonetheless, the debate centres on the question as to whether leaders who learn the skills and the 'tricks of the trade', but are without certain traits, can make good and exceptional leaders. Since research has informed that leadership and management are similar, questions are asked whether a good manager is good enough to be a leader? Chan and Drasgow (2001) claimed that non-cognitive credentials such as personality and values are based on individual differences that are linked to leadership performance indirectly. However, Miscenko et al. (2017) argued that only skills-based approach to leadership cannot cover all aspects of leader development including adherence to professional ethics, and this is particularly the case for construction professionals. This is justified on the complex nature of leadership from a skills and identity basis, as articulated by Lord and Hall (2005).

9.8 The Importance of Communication for Construction Professionals in Ethical Leadership

Academics have long argued that to become an effective leader, it requires not only to be a good communicator but also an outstanding one. Communications, in this regard, are defined as the art of transmission of ideas from an individual to another or a group of individuals with understanding (Newell and Grashina 2004). Communication is key to understanding organisational behaviour (Buchanan and Huczynski 2017), hence strong leaders need to be able to effectively convey their messages. This is particularly applicable to construction professionals as leaders, in clearly articulating their organisational visions, as applied to project deliverables. They should also be able to persuade and influence their project teams to achieve common goals and purpose, whilst adhering to ethical standards.

Creative leadership and the act of driving change in cultures, attitudes and behaviours in all aspects of business and/or organisations can be very challenging and frustrating for construction professionals. However, if it is done well it can be very rewarding in driving transformation improvements. Knippenberg et al. (2005) debated that leadership is more effective when it is engendered in the leader's identity as well as a group identity as a collective character for the organisation. They suggested that consistent evidence exists to support that leadership is more effective when followers recognise collective identity and behaviour, and furthermore it suggests that followers as groups are more likely to endorse leaders who are group oriented.

Situational leaders can sometimes exist and emerge by a matter of choice, for example in the case of elected leaders in democratic societies. Construction professionals who 'rise through the ranks' in their respective organisations could be an example of a similar path to leadership. Alternatively, at the other extreme, leaders can emerge by default or historical traditions such as monarchy systems or other means such as coercion, which unfortunately will lead to create autocratic leaders and dictators who are still sadly classed as leaders. In a similar scenario, construction professionals can sometimes be those individuals who have inherited these roles as owners of the business where projects are being procured. It appears that the difference between different leaders may be the aspect of the use of power. Tourish (2013) presented a negative portrait of leadership as he argued that often leaders are too often regarded as heroes, charismatics and visionaries and are held responsible for organisational success or failure. Whilst it can be argued that holding leaders accountable for success and/or the failure of any organisation is logical he articulated that leaders hold and use huge powers, which may not always be used fairly and wisely. There seems to be a fine line between getting the balance right to have a constructive directive leadership style, or an authoritative and autocratic style of leadership. Other balancing acts that one has to be mindful of are how much autonomy and participation is given to have a democratic or a laissez-faire style, and how this affects motivation.

It can be argued leadership amongst construction professionals can address a wide range of management issues, and one of them is the process of change within their own organisations. The primary task for construction professionals, as leaders, is managing and implementing change within their companies and deal with the challenge of anticipating employees' reactions (Buchanan and Huczynski 2017). Leadership for clients in the construction industry in its different forms and styles is about managing, directing, influencing and achieving common objectives and goals on projects. Leadership in this sense is a concept to lead and manage organisations to achieve change. Businesses and organisations need to keep changing, adapting and fulfilling corporate social responsibilities and ethical values at the wake of globalisation and the ever-increasing pressure on resources and commodities.

9.9 Adaptability of Leadership Styles for Construction Professionals

Leadership skills are about motivating others and getting the best out of people (Wong 2007). This notion is supported by Newell and Grashina (2004, p. 128), who concluded that leadership is a managerial interrelationship between leaders and followers and defined as the 'ability to influence groups of people in order to make them work and achieve prescribed goals'. Guillen et al. (2015) articulated that leadership drives organisation success and emphasised the importance of motivational measures and leading by example.

Construction professionals are mostly senior individuals within organisations and normally accountable to executive for the projects they become involved with, taking on additional responsibilities and getting involved with a wider range of people to manage directly or indirectly. One of the challenges this position brings is dealing with adversarial people from time to time and having difficult conversations deal with them, whilst managing workplace politics and organisational governance. In this way unethical or inconsiderate practices and behaviours can be addressed before they escalate within a given organisation. This sometimes calls for construction professionals to be courageous, diverse, dynamic and adaptive to different styles of leadership.

Applications of leadership styles for construction professionals can be complex and depends on the context of the situation they are under and the type of individuals they are dealing with. For this reason, there is no one size that fits all as far as approached to the question of how to lead others. This view is attributed to the vast range of facets and complexity of leadership in leading and managing different type, structure, size, location and culture of different organisations. The findings of a study suggested that leader development processes are complex and evolve over time. Chan and Drasgow (2001) reinforced this view in presenting a theoretical framework around individual differences and leadership. Their argument introduced a theoretical model around the integration of leadership performance processes and development. They assumed when leaders bring quality and personal characteristics to certain situations, this includes knowledge-based attitudes and behaviours, learned and acquired alongside cognitive abilities and personality.

9.10 The Relationship between Poor Leadership and Project Failure

Lack of leadership displayed by construction professionals can lead to many 'pitfalls' on projects. These can ultimately lead to project failures which can result in costly and disruptive implications for organisations. The most common causes of project failures, from lack of leadership, are as follows:

- Project initiation and planning
  - Failures to plan effectively 'if you fail to plan you plan to fail'
  - Lack of clear project management and methodology
  - Failures to understand project complexity and the effect on the probability of failure or success
  - Lack of clear business objectives
  - Poor briefing or changes to the brief
  - Lack of understanding of project scope
  - Lack of whole project life perspective
- Governance
  - Lack of ownership
  - Lack of leadership
  - Procedural issues
  - Avoidance

- People
  - Behavioural issues and attitudes
  - Breakdown of client and team relationships
  - Poor communications
  - Personnel changes and a lack of continuity throughout project
  - Lack of collaboration
  - Poor engagement with stakeholders
  - Bureaucracy
  - Political aspects and barriers
  - Inexperienced project sponsor/construction client
- Technical
  - Lack of resources
  - Poor brief
  - Failure to understand risks
  - Procurement and finance difficulties

There should be a concerted attempt to support construction professionals and avoid failure of projects wherever possible. Competent leaders can ensure that measures such as construction professionals adopting the right project controls, governance, communication strategies, ethical codes of conduct and collaboration approaches are all in place.

9.11 Summary

This chapter has referenced current academic literature on theories relating to leadership and especially around professional ethics and applied these to the roles of construction professionals. There are many different definitions of leadership and what leadership actually means in practice. Leadership can be defined as the capacity to establish direction, influence and align others towards a common aim, motivate and commit others to action, and encourage them to feel responsible for their performance (CMI 2008). Accordingly, successful leaders need to be good motivators and have the ability to bring out the best in people. They should lead by example to gain the confidence from other members of the project team and give them the assurance that ethical actions, behaviours and decision-making are being practiced. Furthermore, it is essential for aspiring leaders to have courage, conviction and assertiveness when faced with difficult choices and decisions and their behaviours and actions should always be subject to fairness, integrity and a sense of responsibility. It is essential for aspiring leaders to understand what makes a good leader, in order to perform their roles and motivate others for achieving successful outcomes.

The leadership identity development (LID) and progression process as applied to construction professionals, is dependant and influenced by individual differences, cognitive capacity, personality and temperament, personal identity, personal values and emotional intelligence, driven by cultural context and personal experience. According to the LID model developing leadership identity develops in three stages. The first stage highlights the awareness the concept of leadership and the fact that some people lead, and others follow. The second stage relates to the period that individuals gain experience and develop unique leadership styles first-hand. The third stage relates to people learning the difference between management and leadership whilst at the same time experiencing leadership in practice.

The chapter has also focused on ethical leadership which goes beyond the notion of only satisfying minimum compliance standards. The benefits from measures that organisations can implement to go 'above and beyond' legal requirements have been articulated and discussed. Such benefits include developing the workforce and improving the service to their clients and supply chains to make them feel valued, listened to and engaged in their business.

In a UK organisational context, the Best Companies benchmark and accreditation scheme represents a significant achievement and shows that employers are taking workplace engagement seriously. It is a model based on ethical leadership predicated on employee interaction and an unconditional and unequivocal commitment to the workforce. Notwithstanding this premise, ethical leadership has traditionally not been practiced widely in developing countries which in some circumstances has led to employees' rights being compromised. There are many examples where restrictions on workers' freedom of speech alongside poor health and compromises of safety measures have occurred. One of the most important factors for improving ethical leadership revolves around staff development and training. In this regard training and development are regarded as being the pillars which underpin a successful and motivated workforce and excel a company to achieve successful outcomes and increased growth.

Nurturing and supporting followers are an important and a distinctive feature of successful leadership. There is a wide acceptance that most construction professionals, who perhaps have little or no previous construction management experience, knowledge or qualifications, learn from other professionals within project teams. Opinions on leadership have emerged from different disciplines, with many varying approaches to nurture the skills that are deemed necessary. ICBS (2020) advocated those responsible leaders should focus on others rather than solely themselves, consider the longer-term ramifications and implications in their decision-making and thought processes and follow a moral compass in their management approaches. Accordingly, it is important when considering the traits of ethical leadership for people in authority to lead by example and make decisions that make those around them feel safe, valued and listened to.

It is generally accepted that for projects to succeed they should always be focused on 'win-win' outcomes for their respective client organisations and their contracting construction partners. For this reason, collaborative and partnering strategies, devised by construction professionals will create the right environment for ethical behaviours amongst project teams. Notwithstanding this premise, applications of leadership styles for construction professionals can be complex and will depend on the context of the situation they are under and the type of individuals they are dealing with. Leaders can bring quality and personal characteristics to certain situations, and this includes knowledge-based attitudes and behaviours, learned and acquired alongside cognitive abilities and personality.

Finally, the relationship between poor leadership and project failure has been examined in this chapter from the main areas of project initiation and planning, governance, people and technical competent. Accordingly, leaders should ensure that measures are all in place which include adopting the right project controls, communication strategies, ethical codes of conduct and collaboration approaches.

References

Abraham, G.L. (2002). *Identification of Critical Success Factors for Construction Organization in the Architectural/Engineering/Construction (AEC) Industry*. Georgia: ProQuest.

APM 2012. *Project Management*. [Online] Available at www.apm.org.uk (accessed 19th January 2017).

Buchanan, D.A. and Huczynski, A.A. (2017). *Organisational Behaviour*, 9e. London: Pearson.

Challender, J., Farrell, P., and McDermott, P. (2019). *'How to Improve Construction Procurement through Collaborative Trust Strategies'. A New Approach and Practical Guide*. Chichester: Blackwell-Wiley.

Chan, K.-Y. and Drasgow, F. (2001). Toward a theory of individual differences and leadership: understanding the motivation to lead. *Journa of Applied Psychology* 86: 481–498.

Clarke, N. (2010). The impact of a training programme designed to target the emotional intelligence abilities of project managers. *Internation Journal of Project Management* 28: 461–468.

CMI (2008). *Managing Projects*. Corby: Chartered Management Institute. Checklist 035.

Fewings, P. (2009). *Ethics for the Built Environment*. London: Routledge.

Fisher, O., O'Donnell, S.C., and Oyserman, D. (2017). Social class and identity-based motivation. *Current Opinion in Psychology* 18: 61–66.

Francke, A., 2012. Briefing. *Professional Manager*. 05.

Geoghegan, L. and Dulewics, V. (2008). Do project managers' leadership competencies contribute to project success? *Project Management Journal* 39 (4): 58–67.

Goodreads (2018). *Goodreads*. [Online] Available at https://www.goodreads.com/quotes (Accessed 2018).

Guillen, L., Mayo, M., and Korotov, K. (2015). Is leadership a part of me? A leader identity approach to understanding the motivation to lead. *The Leadership Quarterly* 26: 802–820.

Imperial College Business School (ICBS) (2020). *How to think like a responsible leader*. Available at https://www.imperial.ac.uk/business-school/ib-knowledge/strategy-leadership/how-think-responsible-leader (accessed 2nd December 2020).

Knippenberg, B.v., Knippernberg, D.v., Cremer, D.D., and Hogg, M.A. (2005). Research in leadership, self, and identity: a sample of the present and a glimpse of the future. *The Leadership Quarterly* 16: 495–499.

Lester, A. (2007). *Project Management Planning and Control*, 5e. Oxford: Butterworth-Heinemann.

Lewis, J.P. (2007). *Fundamentals of Project Management*, 3e. New York: AMACOM, American Management Association.

Lopes, P.N. (2016). Emotional intelligence in organisations: bridging research and practice. *Emotion Review* 8: 1–6.

Lord, R.G. and Hall, R.J. (2005). Identity, deep structure and the development of leadership skill. *The Leadership Quarterly* 16: 591–615.

Miscenko, D., Guenter, H., and Day, D.V. (2017). Am I a leader? Examining leader identity development over time. *The Leadership Quarterly* 28: 605–620.

Newell, M.W. and Grashina, M.N. (2004). *The Project Management Question and Answer Book*. New York: AMACOM, American Management Association.

Northouse, P.G. (2015). *Introduction to Leadership*, 3e. Los Angeles, CA: Sage.

Ritz, G.J. (1994). *Total Construction Project Management*. Boston: McGraw-Hill.

Tourish, D. (2013). *The Dark Side of Transofrmational Leadership: A Critical Perspective*. Esat Sussex: Routledge.

Walker, A. (2007). *Project Management in Construction*. Oxford: Blackwell.

Wong, Z. (2007). *Human Factors in Project Management*. San Francisco: Jossey-Bass.

10

Professional Ethics as a Means for Motivating Project Teams

> *Nothing great was ever achieved without motivation and enthusiasm.*
> Emerson, 1867

10.1 Introduction

In consideration of the above quotation, this chapter will consider the relationship and influence that motivation of employees has on compliance and adherence to professional ethics. Furthermore, it will consider the reciprocal relationship of how fostering a culture of professional ethics can lead to raising motivation levels within organisations. In striving for an optimum relationship between the two variables of ethics and motivation, the chapter will also explain how this can lead to a virtuous circle of positive and favourable outcomes.

The chapter will explain why when professional ethics are compromised, this can lead to reputational damage and degradation of trust for organisations, which can lead to heavy fines and loss of public trust. In examining the relationship between professional ethics, trust and motivation, it will provide context relating to the bespoke and unique nature of the construction industry. This will include an examination of why the construction industry, being a project-based environment, has many challenges including unique one-off products won at short notice, transient workforce moving between different work locations, male dominated and a macho culture/climate. Thereafter factors which affect motivation levels of project teams will be discussed and the reciprocal influence that these can have for professional ethics. In this regard, such ethical factors that can raise staff self-esteem, and motivation levels will be appraised. Factors in this regard will be discussed around safety programmes and supervision, training and education and adopting welcoming cultures which are regarded as being responsible and ethical measures.

Communication as a motivational factor will then be covered and how keeping staff aware of changing practices, codes of practice, standards and values can have both motivational and ethical advantages. This will then lead into training and education related to motivation and ethical compliance and the need for responsible employers to maintain reasonable levels of continued professional development in a changing environment.

Organisational Ethics in the Built Environment, First Edition. Jason Challender.
© 2023 John Wiley & Sons Ltd. Published 2023 by John Wiley & Sons Ltd.

Cultural factors affecting levels of motivation for construction related staff will be examined in addition to the use of financial incentives as a motivational management tool. Finally, job satisfiers that can have the most prolific influence on the motivation of employees will be identified and discussed.

10.2 The Relationship between Motivation, Professional Ethics and Trust in the Construction Industry?

In considering motivational factors affecting project teams it is important firstly to consider what motivation actually means. Tabassi and Bakar outlined that:

> *Motivation may be defined as the characteristic of an individual willing to expend effort towards a particular set of behaviours.*
> (Tabassi and Bakar 2009, p. 474)

When considering what we mean by behaviours in the above quotation, this could extent to attitudes and approaches around morals and upholding values, standards and institutional codes of conduct, which are clearly all related to professional ethics.

Challender et al. (2017), concluded that collaboration, trust and successful outcomes on construction projects are strongly affected by motivation levels of project teams including clients, consultants, main contractors and construction staff. Gaining trust and maintaining reputations of individuals and organisations is heavily influenced by the adherence to professional ethics and maintaining standards. When professional ethics are compromised, this can lead to reputational damage and degradation of trust for organisations. Examples earlier in the book were given where organisations have been heavily fined and lost public trust following global cases of corruption and fraud. Such examples are typified by the quotation 'Trust takes years to build, seconds to break and forever to repair' (Amy Rees Anderson). Accordingly, this should be an important area for construction personnel in their respective leadership roles to be acutely aware of and address measures to ensure motivation levels around compliance with codes of conduct remain high throughout the life of projects. Despite this, Herzberg et al. (1959, p. 67) concluded that

> Construction contractors in particular are only interested in short-term financial incentives and maintaining morale and motivational levels accordingly will only have limited influence on the success of building projects and construction productivity.

This is a longstanding, arguably misinformed, controversial and potentially unethical view held by some within the construction industry. Such an opinion could represent the root cause of those on the lower tiers of the contracting organisation sometimes being accused of 'only being in it for the money' or even worse being typecast as lazy, having no desire to 'do a job well' and uninterested in anything other than financial recompense. This being the case, could possibly explain the reason why the practice of motivational management of project teams has traditionally not been an area for widespread reflection, consideration and focus within the construction industry. Other potential reasons could be that particularly construction contractors are sometimes perceived of only being concerned with short-term economics or simply blinkered to the plights and well-being of their blue-collar employees. After all, why would construction companies become unduly concerned for motivation levels and morale of construction operatives who are at the 'bottom of the hierarchy ladder', employed mostly on temporary contracts and who will in most cases move from one job to another in short succession? Arguably this is a misunderstood area in construction management, which should be focused on by construction senior managers, in their aspirations to secure more successful project outcomes. Such negative perceptions can produce management styles geared to aggression rather than persuasion and accordingly the importance of motivation and the practice of motivational management of the workforce do not often feature very highly on their agenda. Also, the construction industry has sometimes presented a notion that motivation is not an issue for construction teams, especially those on the lower tiers. This is based on the fact that low rates of industrial action and strikes in the construction section have suggested that workers are motivated when compared to their counterparts in other industries such as manufacturing. However, Olomolaiye and Price (1989, p. 280) argued that this historically low industrial action could be the result of fragmentation of the industry and poor construction operative cohesion rather than workers being motivated.

10.3　The Bespoke and Unique Nature of the Construction Industry

In considering motivation in the context of professional ethics, we need to understand the context and environment within which building operatives operate. In this regard, the construction industry is unique and more complex than most other sectors such as manufacturing, as projects tend to be bespoke and require creative thinking and innovation at most times. Tabassi et al. supported this notion in that

> *The construction industry is considered to be one of the most dynamic and complex environments as it is a project-based industry within which individual projects are usually built to client's needs and specifications.*
>
> Tabassi et al. (2011, p. 219)

They also explain that the construction industry, being a project-based environment, has many challenges including unique one-off products won at short notice, transient workforce moving between different work locations, male dominated and a macho culture/climate. Other challenges for the sector include short-term teams forming and disbanding, changing skill and competency requirements and increased client pressures over recent years. Tabassi et al. also highlighted that the construction sector is sensitive to organisational changes possibly brought about by changing market conditions. What this symbolises is the extremely complex environment that the construction industry operates within, with ever increasing demands and expectations for successful projects in terms of cost, time and quality. In consideration of these challenges and pressures, can construction clients rationally ignore the motivational issues affecting construction workers? Despite these obvious complexities inherent within the sector, the importance of motivating the workforce has been largely undervalued in the construction industry when compared with other sectors such as commerce and manufacturing. When considering this, it is surely worth construction clients contemplating the benefits that could ensue from having a more motivated workforce involved in construction projects. Accordingly, construction clients should be able to identify and analyse the factors which affect motivation levels of construction workers and potential benefits that motivation can bring to construction management. Such awareness should not be confined to construction managers, but extend to those individuals based on site, as they by far represent the majority of individuals making up the workforce on UK-based construction projects.

10.4 Factors Which Affect Motivation Levels of Project Teams

Table 10.1 articulates research finding by Kaming et al. (1998, pp. 133–134) on four different categories of construction staff in Indonesia. It can be determined from this research that there are many high-ranking factors which are deemed to motivate construction

Table 10.1 Rank order of job satisfiers for a range of construction personnel.

Job Satisfiers	Ranking of Importance				
	Owners	Project Managers	Site Managers	Foremen	Operatives
Financially lucrative	1st	1st	–	–	–
Satisfied customer	2nd	3rd	–	–	–
Job completed on schedule	3rd	–	–	–	–
Tangible physical structure	4th	5th	–	4th	3rd
Good workmanship	5th	2nd	2nd	3rd	1st
Owner satisfied	–	3rd	–	–	–
Good working relationship	–	–	5th	5th	–
Maintain the job	–	–	1st	–	–
Meeting the challenge	–	–	2nd	–	5th
Job costs below estimate	–	–	4th	–	–
Challenge of running the work	–	–	–	1st	–
Maintain the schedule	–	–	–	2nd	–
Productive Day	–	–	–	–	2nd
Social work relation	–	–	–	–	4th

workers other than simply monetary rewards. These include good relationships, safety programmes and supervision, which clearly can be related to professional ethics. Undoubtedly, the ethical factors affecting motivation levels of all construction staff are endless and accordingly this chapter will only specifically consider, analyse and evaluate four main factors, namely communication, training, culture and financial rewards.

Construction leaders need to be aware of alienation and repetitive work of their staff and the effects on motivation levels through monotony and boredom. They could address this through job rotation and introducing variable tasks and variety of work where practicable. In addition, the structure of organisations, communication lines, behaviours and the culture that exists can present problems for the motivation levels of site operatives. There are arguments to suggest that these issues are for contractors to manage. However, low motivation levels within construction team workforces can have a detrimental and negative effect on project outcomes in terms of conflict, productivity, poor quality of construction work, low esteem and frustration of construction staff. One could argue that having such low motivation levels need to be properly addressed by employers on an ethical level to protect the well-being of their staff.

Factors that have an influence on the motivation of construction project teams include communication strategies, management styles, training, behaviours and organisational cultures. Construction clients should be aware of how these individual factors will be managed on projects and the effects that these can have on the motivational levels of construction teams.

When considering the potential benefits that improved motivational levels could reap for the industry, Tabassi and Bakar (2009, p. 475) explained that the success of construction projects largely depends upon the quality and morale of the people employed by contracting organisations. This view was supported by Steers et al. (1996, p. 67) who outlined that motivation and communication can have a direct effect on the performance of employees. Brooks (2006, p. 66) also explained that employees need to be confident and comfortable within an organisation in order to produce successful results and engage in teamwork.

10.5   Communication as a Motivational Factor

The first factor that construction clients will consider with regard to motivation of project teams is communication. Providing good communications especially from senior management to employees can be regarded as an ethical consideration. Steers et al. (1996, p. 78) outlined that communication can have a direct effect on motivation and therein the performance of construction employees. This is supported by Carnall (1999, pp. 111–134), who explained that poor communication lines and behaviours where employees feel that they are not being consulted or made aware of decisions which affect their work can severely affect motivation levels. Furthermore, he explained that the same outcome could arise where the employees feel they are not being given the sufficient level of authority to take decisions. This when applied to the construction industry could explain why some construction industry employees feel they are not being fully consulted or 'left in the dark' leading to feelings of demoralisation. Too many times managers in the construction process fail to keep others informed of such essential matters as timescales, coordination, installation dates and quality which can leave them feeling isolated, frustrated and undervalued. These negative factors can result in poor motivation levels and a downward cycle leading to further demoralisation, lower morale and poor project outcomes.

In addition, construction staff could become resentful and anxious if they do not find out information first hand and feel they are being bypassed or undervalued through not being consulted (Brooks 2006). This can exist within organisations that employ autocratic styles of management. Conversely, Brooks does, however, concede that low self-motivation levels in such cases do not necessarily lead to poor performance. He stressed that employees could still focus on and achieve success for other reasons such as fear of reprisal, intimidation or job security rather than the desire to do a good job and job satisfaction gained in the process. This could arguably be classified as survival behaviour and an approach which could be described as satisfying the first tier of Abraham Maslow's hierarchy of needs linked to safety and security. Another argument for playing down the significance of motivation could occur where construction projects are of a minor nature leading to short durations for subcontract works especially. In such cases, applying the ideas of Brooks, one could argue that motivational levels of construction staff is less important than they would ordinarily be on longer projects where relationships and interaction between individuals is paramount to success.

Strands of motivational theory relate to the premise that all staff have desires to be treated the same as colleagues within their own organisations. Where this does not happen, it could result in employees feeling less valued than their colleagues. This in turn could lead to implications of reduced effort and productivity, and in some cases, employees leaving their organisations altogether. In the context of the construction industry, construction staff are considered by some to be at the bottom end of the hierarchy structure of projects on a par with site labourers. Accordingly, applying Adams theory (as cited in Deci 1975), this could explain why construction staff feel they are treated differently to more senior colleagues, e.g. site managers and as such not worthy of inclusion in the decision-making or consultation processes. Where this occurs construction staff can feel their specialist inputs are not welcomed or important, leading to demoralisation and low motivation levels. In such cases implications can include poor productivity, quality, teamwork and cooperation from construction staff on all aspects of projects. Construction leaders should be aware of such negative aspects and manage them accordingly.

Turner et al. (2003, as cited in Dwivedula and Bredillet p. 160), identified that motivational initiatives linked to communication include giving employees good clarity, feedback on performance and a feeling of achievement and interaction. Applied to construction projects, communication could be improved in this way by including all of project teams in regular briefings and meetings to discuss and agree how their important inputs interrelate to the project as a whole. Construction clients' interventions in enabling dialogue in this way can bring a sense of 'ownership' of 'buy in' for projects and a desire for the whole project team to want them to be completed successfully. Also, the use of regular appraisals of construction staff with their line managers could enable them to know where they stand within projects and create a sense of teamwork. Giving construction employees acknowledgement and praise for good work undertaken could prove a highly important factor leading to an upward cycle of performance and motivation. However, this communication and recognition for good performance, unfortunately, is something that is seldom offered to construction staff. The possible reasons for this could emanate from construction clients expecting good results without any positive interventions or underestimating the benefits that praise and feedback can bring to the morale and motivation of the subcontract workforce.

Nesan and Holt (1999, as cited in Tabassi and Bakar 2009, p. 474), placed emphasis on recognition as an effective means for inspiring motivation and enthusiasm amongst employees. They explained that it is particularly successful when applied to teams as opposed to individuals. Considering this perspective, in the context of the construction industry, initiatives could be introduced on projects by construction clients to acknowledge and in some cases reward construction staff or subcontract teams for outstanding work undertaken. An example of this could be 'subcontractor team of the month award'.

10.6 Training and Education Related to Motivation and Ethical Compliance

Construction leaders should be aware of the advantages in keeping project teams motivated by encouraging a culture of staff development through training and education, especially around ethical issues and compliance with professional values. The potential benefits from training and education of project teams have been much debated over recent years. Alderfer's theory of motivation is based on existence, relatedness and growth (Deci 1975) and this could be applied to all construction staff. He concluded that individuals have needs to develop and where they feel that this is not being met, they regress to satisfying lower needs such as respectability or safety needs. It should therefore be not surprising that research from the Construction Industry Council found that those construction companies that invest in training and staff development initiatives for their staff have over time benefitted from increased cooperation from them and more willingness to work collaboratively with the other members of the project team. This was supported by Dwivedula and Bredillet (2010, p. 160), who explained that providing staff with training will provide them with opportunities for further learning and growth which will create an upward cycle of motivational growth linked to career progression and further learning aspirations. Furthermore, Vroom's 'Expectancy' theory could be applied to training and staff development in that this 'investment in people' could improve the belief and self-confidence of construction staff in their individual capabilities. Vroom's theory is particularly relevant as, being multidimensional, it can be related to all levels of construction staff.

Construction leaders should be aware that one of the factors adversely affecting motivation levels of construction staff is linked to alienation, where operatives are undertaking jobs which could be classified as repetitive, routine, non-challenging or non-mentally stimulating manual work. It is sometimes forgotten that construction workers, like most other individuals, benefit from being interested in their work and where they feel they are developing their skills in other areas. Accordingly, this should be addressed wherever possible by construction clients, by instructing project teams to create opportunities to vary work by giving construction staff different tasks from time to time through job rotation means. This can raise motivation levels by workers learning new skills and has proved to reduce levels of boredom and monotony and presents an entirely ethical position for employers to take. Dwivedula and Bredillet (2010, p. 163) reinforced this view and identified facets of training which have a beneficial impact on motivating the workforce that could be applied to contractors, including job advancement, variety and level of knowledge, participative decision-making, developing competencies and increased sense of achievement. One could, therefore, surmise that construction clients should encourage the whole project team to enable tasks and job descriptions for staff to be as diverse as practicably possible to instil variety and interest.

Other factors affecting motivation levels could be that education levels of construction operatives have historically tended to be relatively low and in some cases, they have not been taught aspects of working independently, which in itself can affect their ability to motivate themselves (Tabassi and Bakar 2009, p. 218).

Tabassi and Bakar (2009) also concluded that those construction companies which demonstrated a commitment to human resource development, motivated construction staff in the process, and reaped benefits associated with increased retention of staff and overall performance of employees. This is supported by Chen et al. (2003, as cited in Tabassi and Bakar 2009, p. 214) who explained that

> Having a well organised and effective human resource programme is one of the most important assets of a company, directly impacting its fruitfulness and long-term viability as a company.

Tabassi et al. (2011, p. 222) suggested that training can work in tandem with motivation and these two factors can work together to increase construction-related productivity and prepare staff for organisational change and concluded that

> Motivation can influence the willingness of an employee to follow the training programme to exert more energy towards the programme and to transfer what they have learnt onto the job.

Tabassi and Bakar (2009, p. 472) applied this view in their studies of construction companies in Iran who invested in training, and career development for construction staff. Their findings concluded that such training and development fostered an integrated learning culture within construction organisations leading to higher levels of morale and motivation. Training for skilled construction staff 'on the job' involved job rotation to give individual transferable skills to different jobs and this increase in technical skills and training 'off the job' involving classroom lectures could be effective for developing technical and problem-solving skills. Tabassi and Bakar also explained the concept of 'training motivation' whereby training can increase motivation levels, which in itself can influence the energy and willingness that individuals will have for further training and development thus constituting a positive cycle. Cheng et al. (as cited in Tabassi and Bakar 2009) advocated this approach and found 'training motivation influences trainees training performance and transfer outcomes'.

Clearly there are many positive impacts that training, staff development and education of construction staff can have on their motivation levels. Furthermore, poor training and staff development can have potentially very harmful effects on motivation. This is evidenced in Kaming et al. (1998, p. 137) who concluded that poor management systems for coordinating construction staff and poor respect of them for one another can be a highly damaging factor on motivation levels. This would suggest that more focus through training on this element to encourage cooperation and planning between construction staff could be highly beneficial. In addition, Price et al. (2004, as cited in Dwivedula and Bredillet 2010, p. 160) explained that organisational structures on construction projects are becoming increasingly flat and thus employee empowerment through training has become an important source of work motivation in recent times.

10.7 Cultural Factors Affecting Levels of Motivation for Construction-related Staff

When considering construction industry employees, the importance of the culture of the construction site or contracting organisation can be sometimes underestimated as this can have a significant impact on job satisfaction of the workforce. Culture in this sense can be related to professional ethics in providing a safe, welcoming and positive environment to support the well-being of construction staff. In addition, having a friendly and open culture that can make employees feel confident to speak out and come forward if something is untoward is also important to maintain high levels of professional ethics. The converse argument could be made where there is a negative culture and staff are hesitant, reluctant and sometimes afraid to make their managers aware of a significant issue, e.g. health and safety hazard on a construction site.

Notwithstanding the above premise, the insecure nature of subcontract work, given the one-off nature of projects can create employment insecurity for some construction individuals. In the past this has led to some construction managers regarding certain staff including labourers as not requiring any form of motivational management. In this context they may be under the misguided view that they respond best to being simply given orders in what could only be described as a dictatorial manner. This analogy could explain why autocratic management styles with regard to communications and target setting for construction staff are so prevalent within the construction industry. It is not, however, to say that this form of management is not appropriate in all cases for managing construction staff. For instance, issues such as health and safety management, reliant on strict adherence to site rules and regulations is of paramount importance and normally areas where there are clear warnings and potential disciplinary action for contravention. In such cases, site managers cannot simply rely on motivating construction staff to comply but must be adversarial and assertive to ensure regulations and the general safety of the construction workforce are maintained at all times.

What can sometimes be forgotten or underestimated is the need for construction staff to feel a sense of pride in their work and to feel a sense of teamwork in being part of a project. On these lines, McClelland (as cited in Brooks 2006, pp. 76–89) presented a theory which promoted the need for staff to have a sense of belonging, esteem and achievement. This view is also supported by Elton Mayo (as cited in Deci 1995, pp. 25–34), who explained that employees need to feel confident and comfortable within organisations in order to produce successful results and engage in teamwork and accordingly he presented a theory which related motivation to human aspects. He advocated that wherever possible staff should enjoy their work and feel a sense of teamwork and personal satisfaction for undertaking their jobs. Having employees that feel valued, confident and belonging to an organisation one could describe as a moral and ethical requirement in any context. They should also feel valued within the organisation which in theory should incentivise them to work harder. This view is not, however, shared by Taylor (as cited in Deci 1975, pp. 76–85) who conversely believes that motivation can be induced through more scientific means. In this regard, construction leaders should encourage incentive-related motivational measures. This could include giving staff adequate resources, introducing rewards for good performance and devising processes and systems for staff to feel more competent in carrying out their duties. Perhaps one argument against this approach could be that although it may give construction staff 'the tools' to do the job well it does not address the root cause of problems relating to poor motivation and morale issues. This in itself can create conflicts within construction organisations and more especially between subcontractor operatives and construction management teams. Brooks (2006, p. 45) addressed this issue and argued that conflict management policies and strategies within companies and institutions should be introduced into organisational management at all levels to avoid motivational difficulties.

Nesan and Holt (1999, pp. 122–129) explained that a powerful motivator of construction staff is the feeling of belonging to a team and this can be particularly effective where a team is given autonomy and authority to make their own decisions, raising levels of participation and communication. Tabassi and Bakar (2009, p. 474) reinforced this point and deduced that

> Many employees are motivated when they are 'empowered' and feel that their participation is important in making the company successful.

Tabassi and Bakar (2009, p. 474) also advocated that work environments should be created to maximise individuals' commitment and enthusiasm through motivational means. They explained that this can be achieved using incentives such as recognition and team belonging other than simply extrinsic motivators such as pay. Furthermore, Bellemare et al. (2010, p. 283) found that where workers experienced high levels of what they perceived to be alienation or pressure within organisational working environments, then this could have a negative effect on their productivity through lack of motivation.

Further studies on how organisational culture can affect motivation include Kaming et al. (1998, p. 134) who emphasised that 'de-motivating factors' are not necessarily the exact opposite of motivating factors and explained that most de-motivating factors originate from apparent trivial causes, yet their impact is often significant. Their study, which looked at joiner, bricklaying and steel erection construction staff found that disrespectful supervisors, lack of recognition and lack of cooperation amongst colleagues ranked as the main de-motivating factors. These in turn can lead to frustration, discontentment and thereby have a highly damaging effect on motivation levels leading to dissatisfaction and lower performance on site.

The culture of construction organisations can influence the responsibility assigned to construction staff on site. Lam and Gurland (2008, p. 1114) explained that where construction staff are trusted to do a job and given autonomy to carry it out, this leads to an increase in self-determined work motivation leading to improved job satisfaction. Conversely, they concluded the opposite effects where construction staff are given little if any autonomy or responsibility in a more controlled environment where they may be closely monitored and scrutinised. Construction clients as project sponsors should ensure that they pursue strategies designed to motivate their staff accordingly through responsibility and trust.

The culture of construction organisations and teams can also affect the degree of collaborative working between construction staff and management. Baiden et al. (2006, p. 21), however, concluded that integrated collaborative project cultures are seldom fully realised owing to barriers created by the short-term duration of the majority of building projects. They explained that a lack of team working, emanating from this could create a culture which alienates construction staff and makes them feel that they are working as individuals rather than team members. Accordingly, they recommended changes to the construction industry to encourage integration of teams working together following the principles of Latham (1994) and Egan (1998) to change cultures and improve motivation levels of construction staff. This recommendation was supported by Dwivedula and Bredillet (2010, p. 159) who advocated 'a complete cultural change within the construction industry' to give construction staff more autonomy and extensive training and thus incentivising creativity. They explained that this concept, when adopted by other industries such as manufacturing, and learning new skills through measures such as job rotation all went towards motivating the workforce. This view was further reinforced by Reichers and Schneider (1990, p. 88–98) who describe a 'shared perceptions approach' to organisational culture where a positive climate is developed when members of working groups are interacting with each other to attain common goals.

10.8 The Use of Financial Incentives as a Motivational Management Tool

The final ethical factor for consideration in terms of its effects on motivation for construction personnel is financial incentives.

Rose and Manley (2011, p. 765) attempted to explain where and why monetary rewards are used as follows:

> Financial incentives are typically used on construction projects to invigorate motivation towards above business as usual goals and provide the contractor with the opportunity for higher profit margins if exceptional performance is achieved.

Where there are little financial incentives or where pay for undertaking a particular role in the construction industry is extremely low, this could be deemed unethical in not paying the going rate and possibly illegal if below the minimum wage in the UK. This can not only compromise motivation levels on behalf of workers, leading to poor productivity but also cause reputational damage for companies and allegations of potentially exploiting staff. There have been many cases in the past where illegal immigrants undertaking construction cleaners and labouring roles have been disadvantaged in this way, which has led to criminal charges and reputational damage for their employers.

Furthermore, Rose and Manley (2011, p. 765) explained that financial incentives are common in the construction industry in terms of bonuses for trying to encourage motivation and commitment of construction staff, although they concede that there is little evidence to prove their effectiveness.

Olomolaiye and Price (as cited in Tabassi and Bakar 2009, p. 474) apportioned much emphasis on money as a '*powerful motivator of construction employees*'. MacKenzie and Harris (1984, p. 137) further supported this view and argued that money and earnings are the sole motivating factors for construction workers. This can be related to the psychological needs level of Maslow's Needs Theory. Further endorsement came from Kaming et al. (1998, p134) who found in their studies of construction workers in Indonesia that fairness of pay was the highest-ranking motivational factor, which seems to again support the importance of financial gain. They did, however, also find that a culture that promotes good working relationships with workmates was ranked as the second most important motivator. In addition, Olomolaiye and Price (1989, p. 283) conducted a survey of various construction groups involving in building projects in Nigeria. This rather contradicts that financial gain is the sole motivator for construction foremen and operatives with variables including good workmanship, job challenge and productive day gaining highest rank. Conversely owners and project managers chose financial success as their prime job satisfier at the expense of the other variables, which rather suggests that these groups are more highly motivated by money than lower ranking groups including construction staff.

Herzberg and Mausner (1959, as cited in Tabassi and Bakar 2009, p. 215) concluded that although motivation of employees was linked to the absence or existence of certain job satisfiers. Examples of this include wages, conditions of work and holidays which could remove dissatisfaction but not greatly motivate people in their jobs. Story et al. concluded that financial incentives, if managed correctly and linked in with other motivational factors, can have beneficial results. Herzberg and Mausner (1959, as cited in Story et al.2009, p. 290) supported this view and concluded that intrinsic factors such as wanting to perform tasks well can lead to employees being satisfied at work whereas extrinsic factors, such as salaries and bonuses, typically lead to potential sources of dissatisfaction. When applied to construction staff, this could possibly explain why monetary rewards such as financial rewards for early completion may not greatly incentivise and motivate staff as would be expected in most cases. However, where they are seen to be removed from construction staff remuneration agreements, this could potentially have significant detrimental effects on demoralisation and motivation of the construction workforce. Schroder and Reich (as cited in Olomolaiye and Price 1989) concluded that real construction motivation can only be derived from satisfying the 'higher needs' other than wages, working conditions and fringe benefits. In this way they argue that this can motivate them to achieve higher production which is a similar theory to McGregor's Y theory.

Tabassi and Bakar (2009, p. 474) referred to a 'participative approach' in addressing and developing good supervisor/subordinate relationships and motivation levels through the creation of cohesive work groups. They suggested examples of how this could be applied to construction staff, whereby they could be rewarded financially for suggesting alternative ways and means to improve the productivity, quality and timescales of their construction companies' building operations. Rose and Manley (2011, p. 765) concurred with this 'blended' approach and considered that financial incentives enhance motivational levels of construction staff to participate with construction-related objectives, for instance early completion. They do, however, deem that there still need to be other factors instilled within the working environment such as unity, trust and fairness to make significant improvements to motivational levels of construction staff. Story et al. suggested that a balance can exist where construction staff can have both intrinsic and extrinsic motivations on a project, for example the desire to meet deadlines and the passion for doing a good job.

10.9 Summary and Conclusion

This chapter has considered and analysed the relationship and influence that motivation of employees has on compliance and adherence to professional ethics. Furthermore, it has explored the reciprocal relationship of how fostering a culture of professional ethics can lead to raising motivation levels within organisations.

Gaining trust and maintaining reputations of individuals and organisations is heavily influenced by the adherence to professional ethics and maintaining standards and this has been covered widely in the chapter. When professional ethics are compromised, this can lead to reputational damage and degradation of trust for organisations. Accordingly, Construction leaders should be aware of the advantages in keeping project teams motivated by encouraging a culture of staff development through training and education, especially around ethical issues and compliance with professional values. There are examples that were covered earlier in the book where organisations have been heavily fined and lost public trust following global cases of unethical behaviour, corruption and fraud.

In the construction industry there have traditionally been some longstanding, arguably misinformed, controversial and potentially unethical views held that employees are only interested in short-term financial incentives. This is a potentially unfair and unjust claim and there is no evidence to suggest that this position is widely represented across the sector. Notwithstanding this premise, there is a converse argument has been presented in the chapter that does link motivation, pay and ethics. In cases where there are no financial incentives or where pay for undertaking a particular role in the construction industry is extremely low, this could be deemed unethical in not paying the going rate and possibly illegal if below the minimum wage in the UK. This can not only compromise motivation levels on behalf of workers, leading to poor productivity but also cause reputational damage for companies and allegations of potentially exploiting staff.

To summarise the findings of this chapter, there are significant benefits that can be gained from pursuing construction management initiatives on projects to motivate staff and therein lead to more successful outcomes. There has clearly in the past been an underestimation as to the benefits of motivational management and the potential implications for poor levels of construction staff motivation. The potential benefits could include:

- Improved construction staff retention
- More positive attitudes of construction staff
- Improved supervisor/subordinate relationships

- Increased productivity, and potentially higher quality standards
- Fewer conflicts and overall better communications on site

Given these clear benefits, this chapter has analysed and evaluated the factors that influence the motivational levels of construction staff. It has found that although financial incentives are undoubtedly important to construction staff, they are by no means the sole mechanism for motivating the workforce. Perhaps the common misconception that construction staff are 'only in it for the money' has been generated from the unique environment that construction staff work within. As discussed, they are positioned at the bottom of the organisational hierarchy, their contracts are generally short term and they are working in an industry traditionally very competitive, especially in the current economic climate where financial incentives are seen as the 'quick fix' to meeting contract deadlines and targets.

Hopefully, this chapter of the book has contradicted the unethical notion that construction staff as part of overall project teams are unworthy of requiring to be motivated by any other measures other than money and that attempting to incentivise and motivate them by other means will be fruitless accordingly. In this regard, factors have been identified which construction clients should consider as they could have a significant impact on motivational levels of construction staff. These include improved communication strategies, consultation, training and development, organisational culture and opportunities should be sought by construction organisations to address these aspects wherever practicable. There is also a suggestion that the construction industry could benefit from adopting certain lessons learnt in other sectors such as the manufacturing industry where more emphasis has been on management strategies geared to raising levels of workforce motivation such as job rotation and blue-collar participation in strategic decision-making.

Construction organisations have historically asked themselves questions related to whether they can afford to procure motivational management initiatives for construction staff. Furthermore, construction clients have traditionally distanced themselves from motivational initiatives under the misguided premise that it is not within their role to encourage project teams to motivate their staff. The real question, however, they should be asking themselves is can they afford not to? After all, how long can construction staff perform their duties and obtain good outcomes if not motivated and in some cases what level of performance can be expected where poor motivation levels are experienced?

This chapter has hopefully emphasised why senior management, including construction clients, should pay special attention to motivational strategies. The justification for this is that low motivation levels within the workforce can otherwise result in poor ethical compliance, low esteem, negativity, conflict, low productivity, and frustration of construction staff. Conversely, if motivation is properly managed this can have very positive results as previously outlined and is both ethical and appropriate. Clearly, motivation is not the only factor that affects productivity and other factors, such as availability of resources within a particular situation, capability and ability to undertake a particular task, could be argued as being equally important.

Motivational management of construction staff could be used as a tool to enhance organisational performance and construction outcomes but is arguably as important to improve the mental well-being of construction staff in the process. This in itself is linked to professional ethics in providing a welcoming, comfortable, and safe environment for staff to prosper. Conversely, if motivational management is not practised, leading to construction staff becoming demoralised in some cases, construction operations may continue to work well with other 'negative' factors such as job security and fear of reprisal affecting performance on site. This could affect the well-being of staff and therein provide an unethical and unsafe environment for employees. The overriding question is, however, how long can this be maintained in such scenarios? The overwhelming academic consensus would tend to suggest that a lack of motivation coupled with autocratic management styles can provide 'short term fixes'. However, it is generally recognised that poor morale and motivation of construction staff could, in the longer term, have extremely detrimental effects on the construction industry. Such implications could include poor employee retention resulting from construction staff seeking alternative employment elsewhere. This increase in mobilisation of staff could hinder internal development of construction project teams in terms of long-standing, knowledgeable staff having the right expertise and 'knowing the businesses' they work within. This in turn could increase costs for construction clients and contractors when considering such factors as disruption, loss of continuity and training of new employees to fill vacancies left by demoralised employees.

This chapter has identified the general need for alignment of individual and organisational goals and the aim for a win-win scenario. In addition, it has highlighted the need on the part of construction clients for greater awareness of the importance of motivating construction staff. Accordingly, construction clients should champion a required change in cultures within the construction industry, adopting the recommendations of Latham (1994) and Egan (1998) to procure a more integrated and motivated project team. More emphasis by construction clients should therefore be placed on intrinsic factors rather than simply relying on financial incentives to raise motivation levels of construction staff. This by no means should involve simply removing financial rewards or any other extrinsic motivational measures that construction staff have traditionally relied upon, as this could negatively affect motivation. Instead, construction clients could look wherever possible to blend intrinsic motivational measures into extrinsic measures. Examples of such initiatives could include financial rewards for good suggestions to improve the quality or buildability of a particular work element.

In the context of the complex and dynamic environment of the construction industry, and to realise benefits associated with motivational management and ethical adherence, there is clearly a change of culture and attitude required within the sector. To enable this change construction clients should consider the integration of the following recommended initiatives on their projects:

- Introduce 'bottom up' management initiatives to instil greater construction staff empowerment through responsibility and autonomy.
- Ensure that tasks for construction staff, through measures placed by construction clients on their contractors, are challenging, varied and interesting through initiatives such as job rotation. This will seek to reduce monotony, boredom, alienation and incentivise the workforce.
- Develop informal as well as formal lines of communication through regular feedback and appraisals.

- Give praise to construction staff as individuals and teams where appropriate and introduce recognition initiatives. Examples could include 'subcontractor team of the month'.
- Ensure that training and development of construction staff is undertaken and learning achievements recognised in individuals. This could be achieved through apprenticeships, other on-the-job training and through short-term training courses off-site.
- Encourage construction staff to pursue career development through qualifications and industry accreditation.
- Develop and embed greater awareness of the importance of motivation for construction staff at senior management level.
- Break down communication and cultural barriers and treat construction staff as a more integral part of the overall project team and adopt more democratic rather than autocratic styles of management.

All these measures will hopefully contribute to provide a motivating work environment for construction staff to work within and construction clients will reap great benefits not just for their respective organisations and the building projects that they procure but for the future of the construction industry at large.

References

Baiden, B.K., Price, A.D.F., and Dainty, A.R.J. (2006). The extent of team integration within construction projects. *International Journal of Project Management* 24: 13–23.

Bellemare, C., Lepage, P., and Shearer, B. (2010). Peer pressure, incentives, and gender: anexperimental analysis of motivation in the workplace. *Labour Economics* 17: 276–283.

Brooks, I. (2006). *Organisational Behaviour, Individuals, Groups and Organisations*, 3e. Essex: Pearson Education Limited.

Carnall, C.A. (1999). *Managing Change in Organisations*, 3e. Hertfordshire: Prentice Hall Europe.

Challender, J. (2017 June). Trust in collaborative procurement strategies. *Management, Procurement and Law Proceedings of the Institution of Civil Engineers* 170 (3): 115–124.

Chen, L.H., Liaw, S.Y., and Lee, T.Z. (2003). Using an HRM pattern approach to examine the productivity of manufacturing firms. *International Journal of Manpower* 24: 299–318.

Deci, E. (1975). *Intrinsic Motivation*. New York: Plenum Press.

Dwivedula, R. and Bredillet, N.B. (2010). Profiling work motivation of project workers. *International Journal of Project Management* 28: 158–165.

Egan, J. (1998). *Rethinking Construction. The Report of the Construction Task Force*. London: DETR.TSO.

Herzberg, F., Mausner B., and Synderman, B. (1959). *The Motivation to Work*. New York: Wiley.

Kaming, P.F., Olomolaiye, P.O., Holt, G.D., and Harris, F.C. (1998). What motivates construction craftsmen in developing countries? a case study of indonesia. *Building and Environment* 33: 131–141.

Lam, C.F. and Gurland, S.T. (2008). Self-determined work motivation predicts job outcomes, but what predicts self-determined work motivation? *Journal of Research in Personality* 42: 1109–1115.

Latham, M. (1994). *Constructing The Team*. London: The Stationery Office.

Mackenzie, K.I. and Harris, F.C. (1984 25-29 May). Money the only motivator. *Building Technology and Management* 23 134–145.

Nesan, L.J. and Holt, G.D. (1999). *Empowerment in Construction: Theway Forward for Performance Improvement*. England: Baldock, Hertfordshire: Research Studies Ltd.

Olomolaiye, P.O. and Price, A.D.F. (1989). A review of construction operative motivation. *Building and Environment* 24 (3): 279–287.

Reichers, A.E. and Schneider, B. (1990). *Climate and Culture: Anevolution Ofconstructs*. San Francisco: Jossey-Bass.

Rose, T. and Manley, K. (2011). Motivation toward financial incentive goals on construction projects. *Journal of Business Research* 64: 765–773.

Steers, R.M., Porter, L.W., and Bigley, G.A. (1996). *Motivation and Leadrership at Work*, 6e. Singapore: The Graw-Hill Companies Inc.

Story, P.A., Hart, J.W., Stasson, M.F., and Mahoney, J.M. (2009). Using a two-factor theory of achievement motivation to examine performance-based outcomes and self-regulatory processes. *Personality and Individual Differences* 46: 391–395.

Tabassi, A., Ramli, M., Hassan, A., and Bakar, A. (2011). Effects of training and motivation practices on teamwork improvement and improvement and task efficiency: the case of construction firms. *International Journal of Project Management* 30: 213–224.

Tabassi, A.A. and Bakar, A. (2009). Training, motivation, and performance: thecase of human resource management in construction projects in Mashhad, Iran. *International Journal of Project Management* 27: 471–480.

11

Case Study of How Ethics Can Be Influenced by an International Crisis

A shared set of ethical values is the glue that can hold us together during an intense crisis. A key lesson from the SARS outbreak is that fairness becomes more important during a time of crisis and confusion. And the time to consider these questions and processes in relation to a threatened major pandemic is now.

Peter Singer

11.1 Introduction

This chapter is intended to provide context, by way of real-life examples, to some of the behaviours, issues and dilemmas that emerged during an international crisis. In this regard it focuses on the Coronavirus (COVID-19) pandemic, which emerged in Wuhan, China, at the start of 2020 and subsequently spread across the world. It will articulate the problems that have been created for the construction and engineering sectors around attempting to maintain progress with projects and keeping sites operational whilst balancing the ethical considerations of health and safety. Furthermore, the chapter will give examples of the positive ethical behaviours and actions that have emerged during the crisis, some of which have included cases have put the interests of their employees and supply chain ahead of their own. In such instances the fairness, integrity and moral responsibilities that organisations and individuals have shown to others have been both commendable and unprecedented in modern times. Conversely the chapter will identify and discuss various examples of unethical conduct during the pandemic. It will cover cases where organisations and individuals have sought to use the crisis to bolster their own financial positions at the expense of others in business dealings. In some instances, this has included disadvantaging their supply chains which has had severe consequences for the liquidity and survival of organisations. Such shocking and immoral behaviours and actions will be examined on a case-by-case basis from interview findings, conducted for the book.

Unethical conduct will also be examined from the perspective of the UK Jobs Retention Scheme commonly known as 'Furlough'. Cases will again be analysed where there has been a flagrant abuse of the scheme on both an organisational and individual basis and where the eligibility, rules and spirit of the scheme have been severely compromised during a national crisis. Contractual implications brought about by the pandemic will be identified and analysed and how in some cases companies have disagreed on the apportionment of additional risk brought about the virus, leading to potential dilemmas and disputes.

Finally, additional measures to ensure the health and safety measures that have been put in place of construction and engineering projects on site will be articulated. This is intended to signify the ethical commitments that organisations have taken on board to ensure the health, safety and well-being of their workforce, and supress the spread of the virus, despite the considerable financial costs incurred.

11.2   Professional Ethics in the Context of the Coronavirus (COVID-19) Pandemic

At the time the author started this chapter of the book the outbreak of the COVID-19 virus had taken a stronghold over most of the world, with most countries in lockdown, following many thousands of deaths. In the UK, a large proportion of businesses had 'Furloughed' staff as part of the Government's job retention plans and the economy had suffered a major setback with an exponential increase in unemployment, and many companies planning redundancies. Unfortunately, the UK construction sector has not escaped such adversity, with many main contractors closing sites. Those that chose to remain operational, were running at reduce operational capacity owing to shortages of supply chain labour and resources and through restrictions around reduced density of the workforce linked to 'social distancing' adherence. Construction firms were allowed to continue operating after the lockdown as Government classed them as essential businesses. However, amid criticism from the public about builders flouting social distancing rules, house builders in particular began to shut down operations as they said they were unable to operate safely under the Government guidelines. This demonstrates a high degree of professional ethics on their part in doing the right thing, putting their staff and their supply chain's health above their own commercial interests. This is especially ethical commitment and is important given that shutting down construction sites is an expensive business, with costs and bills still be incurred for site preliminaries including staff costs, overheads, security and rental of plant and equipment. One of the house builders who took this decisive and moral position was UK-based house builder Taylor Wimpey, whose Chief Executive, Peter Redfern articulated that 'whilst construction could be deemed an essential industry and important to the UK economy in the long term, he did not believe it is essential in the short term'. Heads of some of the UK's house builders had urged Government to set out an exit strategy to get the construction industry and the country moving again. Another example of an ethical position taken by some of them was to assist Government in planning a roadmap, as they believed that they owed a moral degree of responsibility to assist them. The roadmap planning was to represent something that is real, and which allows other construction companies to move in a similar way. This was designed to make a meaningful contribution for the construction industry to return to work as normal. Notwithstanding this premise, it was caveated that not every construction company will follow the same approaches, but if different businesses have alternative solutions than that should be acceptable.

11.3 Positive Examples of Ethical Behaviours during the Pandemic

In the context of the Coronavirus crisis there are many examples of professional ethics and responsible behaviours that have been applied to the construction industry. Some clients have demonstrated extremely responsible responses to construction projects which were due to start around the time of the UK lockdown on 20th March 2020. One example is a UK University that has agreed a two-year build programme for a £65m building, and for which was due to commence in April 2020. As the University had gone into lockdown, with only a handful of essential staff on campus, they considered that it would be ethically and morally wrong to allow the project to commence at this time. This decision was predicated on the notion that if Government advice and recommendations via its slogan was to 'Stay at home, Protect the NHS and Save lives' than this should apply to contractors on campus as well as staff, students and visitors. Anything less than this, was considered to raise questions around whether it would be ethically acceptable to allow work to proceed and potentially put contractors lives at risk. Clearly the University felt this was not an acceptable risk to take even though it inevitably would lead to delays on delivering the building. The consequential effects of not having the facility operational in the timescale they had planned, would bring inevitable programme disruption and a need to make alternative accommodation arrangements.

One main national UK contractor reported that the crisis had resulted in their turnover in the period between March and June 2020 being down by 60% but despite this their full preliminaries still needed to be paid by them. Clearly this had had an overarching detrimental effect on this cash flow with more cash going out of the organisation in payments and costs than what was being brought in through revenue. Notwithstanding this dilemma, they had for many previous years acted responsibly and structured their business to build cash reserves for a 'rainy day'. In this way they considered that such reserves might be critical to their survival and that of their supply chain in the future. For this main contractor, being able to pay the supply chain on time, maintaining their liquidity and therein keeping their organisation operational was explained as one of their main priorities. Their Managing Director had professed that 'the most important thing for us now is to have cash in the bank to pay our subcontractors and suppliers, as any delays to settling their accounts on time them could send them down, and our major focus is to protect them from this.' This display of loyalty and commitment to the commercial interests of other businesses was considered to represent a moral and just act and demonstrated an underlying commitment to ethical principles. To enable them to pay all their staff and the supply chain on time and honour all their commitments, this involved their senior management team taking a voluntary 20% reduction in salary. The Managing Director explained that without this salary reduction measure it would not have been extremely difficult to secure the commitment to pay all their bills and possibly lead to staff redundancies and further disruption to the project. Other measures to give their supply chain comfort and reassurance included putting out regular communications to subcontractors and suppliers. These were designed to give confirmation that their existing commitments and contracts entered into before the pandemic would not be compromised, and that they would continue to be paid on time.

Other interventions which were introduced included a range of communications via posters, newsletters and signs to make staff aware of changes to working practices to keep them safe from the spread of the virus. This proved to be a very informative and proactive measure, especially in the absence of much information or advice from Government. These new practices were formulated from bespoke risk assessments and method statements that they had undertaken by the organisation which focused on each and every aspect of the business. This led to new working arrangements which included a lower density of people on site, strict social distancing rules, managing timetabling for use of welfare facilities and the issue of vital enhanced personal protective equipment to all the workforce. All these measures demonstrated an ethical approach to their safety and preservation of life, which their Managing Director articulated as 'being their prime concern at this difficult time'. Despite this, some subcontractors had openly dismissed such measures and proclaimed it was ethically irresponsible to be operating construction sites during the lockdown period. This argument was predicated on the basis that not all construction projects are classified as 'essential' and any increased risks associated in continuing construction operations during a pandemic were totally unacceptable. The main contractor did manage to curtail this criticism in explaining the safety measures put in place and the procedures to ensure that the risk was reduced to an absolute minimum. There is also the counter argument that, with Government advice that construction should continue, it would be unethical to suspend works and potentially damage the economy further, not to mention the livelihoods of all those employed on the project. Generally speaking, according to this managing director, the workforce was appreciative of the measures to keep the site open, which amongst other factors had benefits for their mental well-being. In addition to the obvious health and safety issues, other examples emerged of more collaborative working across the supply chain to overcome other challenges. One such challenge was related to sourcing a particular roofing material where the supplier had Furloughed all their staff. In this particular case the design team worked closely with the appointed roofing contractor to find an alternative that was readily available and of equivalent quality to allow the project to continue unimpeded.

Another main contractor interviewed at the time of the pandemic advised that some of their clients that they had been working with them had taken a very helpful and collaborative approach to assisting them during the crisis. This had in at least one case manifested itself in their clients agreeing not to take retention monies from their monthly valuations but instead deducting it all at a later stage when the main contractors' cash flow was less of an issue for them. Other clients, to improve the cash flow of their main contractors had agreed to settle monthly valuations in 14 days rather than the contractual 28-day period. This again, was in a concerted attempt to underpin not only the liquidity of the main contractor concerned but the whole supply chain on the project. Other examples of concessional and perhaps charitable behaviours stemmed from clients collaborating with contractors and being understanding to inevitable disruption on their construction projects. Such disruption may have stemmed from contracting staff either being Furloughed or in isolation with the virus, and therein unavailable for work. This is particularly disruptive where specialist works such as mechanical and electrical commissioning cannot be completed on time. One example was given of a fire dampening system that was delayed due to no installation staff being available. Owing to the sequential, dependency led processes in construction, this resulted in many more elements of the building project being delayed as a consequence including the ceiling and all partitioning. Other delays experienced stemmed from reduced staffing in local authority departments such as building control leading to them not being able to conduct critical inspections. The disruption to productivity for on-going projects that have been severely hampered by the COVID-19 crisis, has for many different contracting organisations had the overall effect of inevitably increased costs as well as incurring delays on their projects. Reports that clients at this difficult time have been accommodating in 'pain-sharing' with their main contractor partners for delays and cost increases were welcomed by the industry and testament to collaborative working and partnering philosophies.

One project management consultancy director had explained that their senior management had felt a moral obligation to take a significantly large pay cut globally to protect jobs and enable them to survive. This ethical position was intended to send a positive message to their workforce and their clients that they are seeking the means to maintain the business and save jobs. This was received well by most staff but interestingly some clients chose to request that their professional consultancy fees be reduced to take account of the reduction in staff costs. This represents one example of how a minority of construction clients had attempted to manipulate a national emergency to bolster their own commercial positions.

The same consultancy director explained that UK local authorities had shown an extremely ethical approach to dealings with their contractors, consultants and suppliers, in response to Government guidance to keep payments flowing and therein not affect the cash flow and liquidity of the construction industry supply chain. This was an extremely important gesture, especially for smaller organisations that arguably are more sensitive and vulnerable to missed payments and could easily be driven out of business. The position taken by local authorities in this regard has also ensured that construction projects were not adversely affected by contractors 'downing tools' as a result of non-payment and walking off construction sites. This would have only served to exacerbate the pandemic crisis and lead to further complications of projects being left in abeyance without enough resources to be completed.

11.4 Examples of Unethical Behaviours during the Pandemic

On an ethical level some of the main contractors, interviewed for this book, reported than not all aspects of the changing industry during the pandemic had been as positive as the aforementioned examples. Cases had emerged where certain individuals and organisations had not only been less willing to work collaboratively with other firms in the industry but had deliberately tried to use the crisis to further their own commercial positions. These unfortunate scenarios demonstrated that in some cases they were deliberately trying to take advantage and profiteer from their adversaries during the Coronavirus crisis. One example was given of a developer who attempted to renegotiate contract terms, that pre-pandemic had already been agreed with a national main contractor, for the construction of an office block. In this case the developer had argued that the pandemic would lower labour costs, subcontractor tenders and supplier costs with construction organisations desperate for work to survive. The developer whether right or wrong was unethically therein attempting to renege on an agreed deal on the construction sum to save costs for themselves and increase their profits for the development.

Another interesting insight into ethical behaviours, practices and positions adopted during the COVID-19 crisis came from a Regional Director of an international multi-disciplinary construction consultancy. His company had seen 45,000 staff out of a total 50,000 working remotely in the aftermath of the virus. The other 5,000 staff were based in developing countries and had no access to remote working, and therein were expected to carry out their work in their normal offices. This in itself creates an ethical dilemma in that it could be construed that those staff were to be treated differently than their contemporaries who had remote office capabilities. In addition, there had been suggestions that those staff were being disadvantaged and their safety potentially jeopardised by having to attend an office with their colleagues and in an environment where the virus could spread. The same Director articulated that he had experienced different attitudes to professional ethics and collaboration in different parts of the country. He explained that this showed a spectrum of different responses ranging from collaborative to adversarial dispute generating approaches.

Some areas of the UK had seen clients rallying to support their consultants and contractors in their work and continuing to pay them fully for work carried out. Others had been less accommodating and used the opportunity to try to defer payments on the basis of their cash flow or even tried to renegotiate the terms of the contract on the basis of the unprecedented times. This demonstrated a differentiation in approaches depending on which part of the country projects were being procured. An example came from one client who had advised that he would pay for consultancy work already carried out but on the basis of a 10% discount. The justification for the reduction was predicated on the premise that without this there would be a long delay and his company might fold under the restrictions placed on his business to the extent the consultant might never receive payment. One could argue that this was an act of unethical 'game playing' and designed to disadvantage their consultants. Cases like this could have long-lasting effects on organisational relationships, even ones built up over many years. Similar cases after the UK recession of 2008, and the austerity that followed it for many years after, led to company's refusing to work with others that had treated them badly at a time of need and had breached professional code of conduct. For this reason, it will be interesting in the following months and years after the aftermath of COVID-19 to see the importance of loyalty in how organisations actively collaborate and work together.

Other unethical behaviours during the pandemic related to the aptly named 'Pingdemic'. This related to individuals being contacted through a National Health Service mobile application informing them that were required to isolate for 10 days, as they had come into contact with someone who had recently contracted the COVID-19 infection. In most cases, anecdotal evidence suggested that individuals contacted in this way did in fact do the responsible thing and isolated themselves for the full duration, to avoid spreading the disease to others. Unfortunately, there were, however, reports of cases where some choose to ignore such notifications and carried on attending the workplaces as normal. Such actions could have been predicated on an overriding desire not to be financially disadvantaged by isolating. This is particularly relevant to the construction industry which is made of many small companies and self-employed site-based personnel where there is no option to work from home, and no sick pay or recompense for being absent from work. Notwithstanding the reasons for failing to adhere to the isolation notification, this potentially could have exacerbated the spread of the disease, leading to increased infection rates, hospitalisations, and deaths. Such actions one could describe as selfish and irresponsible with any justifications failing to measure up against their overall impact.

11.5 Ethical Considerations around the UK Government Job Retention 'Furlough' Scheme

At the beginning of the pandemic in March 2020, the UK Government introduced its Coronavirus Job Retention Scheme (CJRS) also known as the 'Furlough' scheme. It was designed to save jobs that would otherwise have been lost through redundancies where companies had been severely affected by the virus. It had the additional benefit of giving organisations protection against insolvency during the crisis by essentially underwriting the majority of their pay costs for staff who had been Furloughed under the scheme. Alongside other sectors which included the aviation industry, hospitality, catering and retail, the construction industry did not escape such turmoil of the pandemic. There were restrictions imposed around social distancing in the workplace, a sudden drop in demand for construction-related services and a push by Government for employees to 'Stay at Home, Protect the NHS and Save Lives'. The consequential effect of these measures was to force many companies employed in the construction sector to Furlough their employees. The Furlough scheme offered employees 80% of their current salaries for hours not worked up to a maximum of £2500 whether they worked full time or half time, and this included agency workers, those on zero-hours contracts and apprentices. It was welcomed by many different organisations in the construction industry including clients, consultants, main contractors, subcontractors and suppliers who may otherwise have found it difficult to survive during this challenging period. It originally was to extend up until October 2020 but was subsequently extended twice until September 2021 due to the longevity of the crisis. Under the scheme, Furloughed staff could undertake volunteer work for another employer or organisation and training to keep their skills and learning up to date. However, Furloughed staff could not do tasks or activities that generate income for their employer, or an organisation linked with their employer. Furthermore, they could not provide a service for their employer, or an organisation linked with their employer.

In the context of the Furlough scheme, there were many contrasting examples that emerged, highlighting both positive and negative responses to the scheme throughout the built environment. In the former case an example emerged where a surveying consultancy director had been forced to Furlough the vast majority of his staff which had resulted in them having to take a substantial reduction in salary. Not wishing to cause differentiation between his staff and his own position he Furloughed himself and therein placed himself in the same predicament as his staff. This had significant financial consequences as involved him having to take a 60% pay cut, as the organisation did not adopt a policy to 'top up' Government assistance to maintain normal salary levels. This represented a moral and ethical position and was designed to share the pain that his employees had to endure whilst instilling a sense of fairness, equality and continuity across the workforce. The moral compass that was followed in this case set a good example in that he was leading from the front and therein was greatly welcomed by the organisation.

Notwithstanding the above clear example of maintaining a strong sense of ethical behaviour in relation to the Furlough scheme, there have been anecdotal soundings that not every individual and organisation have been maintaining the same moral position. In this regard, there were examples of cases of flagrant abuse of the scheme which emerged during the crisis. Such cases allegedly involved a minority of smaller consultants, contractors and suppliers benefitted from the Government job creation Furlough scheme, whilst still operating their businesses. Accordingly, during the pandemic they were receiving up to 80% of their salaries paid by the UK Government whilst continuing to receive income from their businesses. This clearly represents a fraudulent act and breach of the Furlough rules and a wholly unethical act which has no doubt, once again, harmed the reputation of the construction industry. In addition, it also highlights the greed of a minority of individuals who clearly regarded the scheme to profiteer by bolstering their incomes and profit margins on the back of Government financial aid to cash strapped businesses.

11.6 Contractual Implications Brought about by the Pandemic

It was explained that certain clients had chosen to defer construction projects which were due to start around the time of the lockdown. These decisions were predicated on them not wanting their projects to be adversely disrupted, possibly leading to time and cost claims by their contractors, and thereby avoiding any potential disputes. Some clients who adopted this strategy used the downtime to work with their respective design teams and contractors to streamline the design and management of the projects, taking into account new ways of working safely, responding to bespoke COVID-19 risk assessments and adhering to social distancing rules. In this way they could be regarded as being responsible employers by planning for their projects to start only when it is appropriate to do so and with an increased degree of preparedness. In other cases, clients had chosen to proceed with their projects but with bespoke 'Covid-19 clauses' inserted into their building contracts. Such contractual provisions were generally accepting of possible claims for extensions of time but designed to absolve clients from any monetary claims associated with disruption. These were more aligned to JCT types of contract rather than more collaboratively procured contracts such as the NEC (New Engineering Form of Contract). Nevertheless, an ethical dilemma has emerged in such cases around whether it is fair and morally reasonable to expect main contractors to carry all the risks of additional costs. This is particularly pertinent in a crisis event, such as the Coronavirus pandemic, which was not of their making and completely outside their control.

11.7 Additional Measures to Ensure Health and Safety of Workforce during the Pandemic

Examples of moral and ethical practices and behaviours during the Coronavirus (COVID-19) pandemic have included instances where leaders chose to temporarily close their businesses rather than risk the health, safety and welfare of their staff. It is likely that decisions such as this would have been taken with the full knowledge of the harm this would cause to them and their livelihood. Another example during the pandemic involved responsible employers allowed staff to work from home and therein trusting them to conduct their work without the level of normal supervision that they would receive if they were in an office environment. Clearly by the nature of the work in the construction industry, it does not always allow this level of flexibility and adaptability. However, for those site employees that could not work remotely, responsible contractors instigated COVID-19 safe and robust working procedures on construction sites for their protection. Safeguards and arrangements that construction leaders have made in this regard has included enhanced personal protective equipment (PPE), policies and arrangements for maintaining social distance and making decisions that put the health and livelihoods of their employees ahead of their own personal interests. This has added extra costs for contractors to instigate and maintain such precautionary measures that they more than likely will not be able to recoup from their clients. However, such ethical policies and undertakings have proved particularly beneficial in keeping building sites operable whilst at the same time reducing the risks of the infection spreading. The COVID-19 crisis has proved to be a historic moment which has made responsible leaders easy to spot in orchestrating these special measures at considerable cost to them.

An interview for the book with a Managing Director of a national main contractor, revealed that the pandemic had necessitated special measures and interventions being required to ensure that their construction sites could remain operational and safe. This is contained below:

> **Statement from a Managing Director of a national main contractor**
>
> *Before the lockdown was introduced it was clear that controls would need to be put in place to keep our projects and offices running through the Pandemic. Accordingly, a daily conference call was arranged to discuss how the region would manage our activities to keep our projects running safely and this was also followed up with a weekly UK Safety Team conference call to ensure consistency across the UK and spread best practice.*
>
> *The Construction Managing Director also prepared emails to all employees with updates from the business and to provide links to guidance and support and had regular calls with the regional Managing Directors. From these meetings the Maintaining 100% Safe document was produced which details how the business would control things such as maintaining the supply chain, managing visitors to site, drugs and alcohol testing, welfare facilities, training, security and travel for work. To check how the projects were coping with the virus and to provide support, a weekly call was arranged with the Managing Director and all project leaders and changes to controls and documents are discussed.*
>
> *We have also had correspondence with the Health and Safety Executive and provided them with a video tour of one of our projects to show the measures we have implemented which they were very pleased with and have asked for copies of documents to enable them to share with others who may regard their sites as not be up to the required standard.*

In addition, it was explained that there was a list of documents prepared and used across the all the UK regions to manage the pandemic and to provide advice and support. These are included in Figure 11.1.

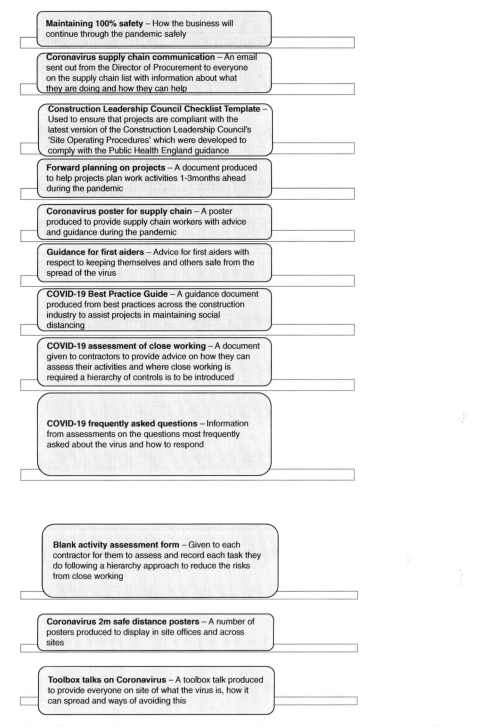

Figure 11.1 List of documents prepared and used by a national main contractor and adopted across the all-UK regions to manage the pandemic and to provide advice and support.

11.8 Ethical Leadership Throughout the Pandemic

According to the Imperial College Business School (ICBS) (2020) there is a dilemma for business leaders when they are put under pressure and faced with making critical decisions especially when there is limited information. Ethical leadership should dictate that notwithstanding the pressures involved, leaders should not stray from ethical principles and values and make decisions which are both appropriate and moral. They explained that this sometimes can mean putting their clients or employees ahead of their own personal interests in some cases. One example of this could be in the Coronavirus (COVID-19) pandemic when leaders chose to temporarily close their businesses rather than risk the health, safety and welfare of their staff. It is likely that decisions such as this would have been taken with the full knowledge of the harm this would cause to them and their livelihood. Another example during the pandemic could be to allow staff to work from home and trusting them to conduct their work without the level of normal supervision if they were in an office environment. Clearly by the nature of the work in the construction industry, does not allow this level of adaptability. However, for those site staff that cannot work remotely, responsible contractors have instigated COVID-19 safe robust working procedures for their protection. Safeguards and arrangements that construction leaders have made in this regard has included enhanced personal protective equipment (PPE), maintaining social distance and making decisions that put the health and livelihoods of their employees ahead of their own personal interests. This has proved beneficial in keeping building sites operable whilst at the same time reducing the risks of infection spreading.

The COVID-19 crisis has proved to be a historic moment which has made responsible leaders easy to spot in orchestrating these special measures at considerable cost to them. However, outside this crisis, the qualities that underpin responsible leaders can be harder to spot. According to ICBS (2020) when considering what makes a responsible leader raises the question of them seeing their responsibility as encompassing a much wider set of agents, activities and outcomes, rather than simply managing their company.

ICBS (2020) has advocated that they can categorise responsible and ethical leadership in three main ways. Firstly, responsible leaders focus on others rather than themselves. They think about how the decisions they make might affect a wide range of stakeholders, including those that are affected directly by the actions of their respective company, as well as how their organisation affects their communities and the general public at large. With a reduced focus on themselves and an acute understanding and consideration for others, this propels such leaders to make decisions that support the greater good. The environment, political systems and the economy account for a few of the factors that responsible leaders will take into account in their decision-making. In this way a concerted focus on others coupled with lesser focus on themselves could be considered to be a moral and ethical hallmark of humility and thus a characteristic that responsible leaders share.

Secondly, responsible leadership should consider the longer-term ramifications and implications in their decision-making and thought processes. In the past there has been much criticism of less responsible leaders only focusing on short-term effects of their decision-making possibly linked to financial gain at the expense of other longer-term implications. ICBS (2020) has articulated that there is a general cognitive bias towards https://science.sciencemag.org/content/306/5695/503/tab-pdf favouring the shorter term over the longer term. This is exacerbated even further in times of crisis wherever increasing threats are prevalent and accordingly events such as the COVID-19 pandemic should in theory result in a decrease in responsible and ethical leadership.

The third factor for categorising ethical leadership is related to responsible leaders closely following a 'moral compass' in their management approaches and paying close consideration to their actions and decision-making in this pursuit. ICBS has pointed out that the average person makes endless decisions in one day and whilst some do not have a great degree of moral content the vast majority do have some a degree of social impact. An example could be that ethical leaders are more vigilant and expend more of their finite resources on sourcing sustainable products or undertaking more due diligence on the ethical stature of their supply chains. In the latter case, they may be increasingly conscious of Modern-Day Slavery when sources construction products or materials from overseas and taking additional precautions to ensure that their appointed suppliers practice ethical employment policies.

It is important when considering the traits of ethical leadership for people, especially in authority to lead by example, and this has never been as important throughout the pandemic. In this way the best leaders continue to make decisions that make those around them feel safe, valued and listened to. In a construction context if projects managers demand certain behaviours and values from their staff that they are failing to uphold themselves, then there will be less likelihood that their demands will be met. This is a particularly critical aspect for professional ethics as any breaches or compromises by managers on what is deemed ethical practices could be seen as 'par of the course' by their staff. This could set a bad precedent and one which could escalate into a vicious circle of repeated and escalating unethical practices and behaviours.

When one considers ethical and responsible leadership there is a tendency to look at this from compliance and altruistic perspectives. In the latter case maintaining an altruistic focus is easier said than done. Several frameworks for human motivation describe self-focused needs as more fundamental than externally focused ones. To contextualise this premise Abraham Maslow who created the five-level hierarchy of needs that put self-actualisation at the top, proposed a sixth level of motivation shortly before he died. This ultimate level transcends the self, and involves being motivated by truth, fairness and justice. But striving for these transcendent values often waits until baser and more self-focused needs are already met. Unfortunately, according to ICBS (2020) individuals tend to focus on their base needs when facing a threat to their sense of security, self-esteem, or confidence and some possess a tendency to become aggressive and adversarial when they feel their authority or competence is being compromised. Ethical leadership theory recognises this dilemma, and the most moral leaders will continue their decision-making around making others around them feel seen, valued and safe. With increasing competitiveness especially in the construction industry during the pandemic, this brings potentially more threats than opportunities, and ethical leadership will become more important. In this context there will be more demand for ethical leadership practices and nurturing future leaders who take decisive action while focusing on others, the long term, and higher order values.

11.9 Reflections for the Future

At the time of writing this chapter the medium to long-term implications that the Coronavirus would bring to the global construction industry could only be speculated upon. It will be telling in the months and years to come whether the future of the construction industry will change in the aftermath of the pandemic. Such changes may involve potentially more staff working remotely and whether a new normal for site working encompassing social distancing will make construction costs increase significantly. This could lead to a sharp increase in the tender price index at a point in time where austerity may return to global economies and therein threaten to undermine the industry for the foreseeable future.

11.10 Summary

This chapter was written during the COVID-19 pandemic and has been aimed at portraying good and bad behaviours in the construction industry that have emerged during the crisis.

At the start the first wave of the pandemic in March 2020 the UK construction industry suffered a major setback with an exponential increase in unemployment, and many companies planning redundancies. Most of the major main contractors chose to work through the crisis, with the construction sector classified as essential and therein exempt from the lockdown. However, a large proportion of construction related-businesses, including many subcontractors and suppliers 'Furloughed' staff as part of the Government's job retention scheme, which enabled them to claim 80% of salaries for their employees. Some main contractors, however, did opt to shut down their sites, with problems experienced in sourcing labour and materials. Those that chose to remain operational, were running at reduce operational capacity owing to shortages of supply chain labour and resources and through restrictions around reduced density of the workforce linked to 'social distancing' adherence. However, amid criticism from the public about builders flouting social distancing rules, house builders began to shut down operations as they said they were unable to operate safely under the Government guidelines. In some cases, businesses made decisions to allow staff to work from home and this included many design team consultants. They saw this as doing the right thing and maintaining professional ethics on their part, putting their staff and their supply chain's health above their own commercial interests despite the inevitable disruption and financial costs. In addition, there are many other examples of professional ethics and responsible behaviours that have been applied to the construction industry during the pandemic. These included construction organisations providing many acts of loyalty and commitment to the commercial interests of other businesses. This demonstrating a collaborative teamwork approach to working through the pandemic with their contemporary organisations. These were considered to represent a moral position and one that demonstrated an underlying commitment to ethical principles. Other examples have emerged where construction business have protected and assisted their supply chain partners in navigating their way through the crisis. Such assistance included paying their supply companies on time and in some cases ahead of time to assist with cash flow at a difficult time when their revenue may have been compromised. To enable this, in some cases, this involved their senior management team taking a voluntary reduction in salary to afford to pay their outgoings.

Other ethical measures which have been reported throughout the crisis included main contractors giving their supply chain comfort and reassurance included putting out regular communications to subcontractors and suppliers. These were designed to give confirmation that their existing commitments and contracts entered into before the pandemic would not be compromised, and that they would continue to be paid on time. In addition to the obvious health and safety issues, other examples emerged of more collaborative working across the supply chain to overcome other challenges. These included clients agreeing not to take retention monies from their monthly valuations but instead deducting retention at a later stage when the main contractors' cash flow was less of an issue for them. One project management consultancy director had explained that their senior management had felt a moral obligation to take a significantly large pay cut globally to protect jobs and enable them to survive. This chapter has articulated examples where clients, to improve the cash flow of their main contractors, had agreed to settle monthly valuations in 14 days rather than the contractual 28-day period. Furthermore, some clients agreed to 'pain-sharing' with their main contractor partners for delays and cost increases brought about by the pandemic. Such moral acts of kindness and support were welcomed by the industry and testament to collaborative working and partnering philosophies throughout the construction industry.

Other examples of moral and ethical practices and behaviours during the Coronavirus (COVID-19) pandemic have included instances where leaders chose to temporarily close their businesses rather than risk the health, safety, and welfare of their staff. It is likely that decisions such as this would have been taken with the full knowledge of the harm this would cause to them and their livelihood. However, for those site employees that could not work remotely, responsible contractors instigated COVID-19 safe and robust working procedures on construction sites for their protection. Such measures for safe working included providing enhanced personal protective equipment (PPE), and orchestrating policies and arrangements for maintaining social distance and making decisions that put the health and livelihoods of their employees ahead of their own personal interests.

Notwithstanding the positive examples of ethical and responsible behaviours, cases have been articulated in this chapter where certain individuals and organisations had not only been less willing to work collaboratively with other firms in the industry but had deliberately tried to use the crisis to further their own commercial positions. These unfortunate scenarios demonstrated that in some cases these organisations were deliberately trying to take advantage and profiteer from their adversaries during the Coronavirus crisis. One example which was described involved a developer who attempted to renegotiate contract terms with a contractor to give themselves a commercial advantage. Another case included a business which sought to utilise an opportunity to try to defer payments and seek discounted rates for work already completed. One could argue that these examples constitute acts of unethical 'game playing' and ones which are designed to disadvantage the businesses that are reliant on them for settling their accounts in full. Such cases also reinforce that not all businesses take a completely moral position, especially where there is the temptation to increase profit margins.

At the beginning of the pandemic in March 2020, the UK Government introduced its Coronavirus Job Retention Scheme (CJRS) also known as the 'Furlough' scheme. It was designed to save jobs that would otherwise have been lost through redundancies where companies had been severely affected by the virus. In the context of the Furlough scheme, there were many contrasting examples that emerged, highlighting both positive and negative responses to the scheme throughout the built environment. This chapter has provided anecdotal reports of flagrant abuse of the scheme which emerged during the crisis. Such cases allegedly involved a minority of smaller consultants, contractors and suppliers benefitted from the Government job creation Furlough scheme, whilst still operating their businesses. This clearly represents a fraudulent breach of the Furlough rules and a wholly unethical act which has no doubt, once again, harmed the reputation of the construction industry.

Reference

Imperial College Business School (2020). *How to think like a responsible leader.* Available at https://www.imperial.ac.uk/business-school/ib-knowledge/strategy-leadership/how-think-responsible-leader (accessed 2nd December 2020).

12

Independent Research Study

Exploring Ethics in the Construction Industry and Understanding Its Impact on the Built Environment

> *Ethics is knowing the difference between what you have a right to do and what is right to do.*
>
> Potter Stewart

12.1　Introduction

This chapter is based on a research study carried out by Simbarashe Robertson Kanotunga, at the University of Bolton, in 2017, and was conducted via a qualitative approach using interviews to collect data.

　The purpose of this research study was to explore and evaluate the phenomenon of construction ethics within the built environment and supports the other findings and discussions in previous chapters of the book. The research was designed to improve the industry's employees' perceptions on what is ethical and what is unethical. Accordingly, the main unethical elements of the construction industry were explored and analysed. Various documented ethical approaches were linked to these approaches with the desire to find out how the image of the industry can be improved. Structured interviews for the research were conducted to objectively determine if the basis of construction ethics is understood by personnel in the industry. This research will provide readers with an overview of construction ethics and what is considered as personal values that influences the way employees behave at work.

　The research study does reiterate certain aspects already covered earlier in the book but is broadly structured into five continuant parts namely:

- Introduction
- Literature review
- Research design and methodology
- Findings and discussions
- Conclusions and recommendations

12.1.1 Aims and Objectives of the Research

The research will explore construction ethics and survey, with the aim of appraising the propensity of built environment professionals and how they act in accordance with their professional bodies' codes of conduct and to establish how personal values and corporate culture promotes or douse ethical practices. To achieve these aims, the following objectives were explored:

1) Understanding the ethics of the construction industry by conducting scenario-based interviews.
2) Analysing the results and measuring the ethical industry appropriateness from the responses obtained.
3) Establish how professional governing bodies help to shape the image of the construction industry.

12.1.2 Justification for Carrying Out the Research Study

Mason (1998) acknowledges that the incalculable value of human life, demands from building (Builders) and designing professionals (Architects), nothing less than the highest moral considerations from those who might risk it. In comparison to this fact, Martin and Schinzinger (1996) indicate that engineers, architects, project managers and contractors have the fundamental responsibility of professional conscience. Fryer (1997) argues that ethics is very important in the industry as it helps establish zero conflict between morality and good management. Exploring and understanding ethics in the industries helps reduce fatalities, minimises corruption, has positive impact on work quality, minimises dishonesty and unfairness, reduces fraud, improves reliability, integrity etc., therefore this study is driven by the contemporary discussions around ethics and how it will shape the image of the construction industry. This paper is by no means intended to belittle any cultural background but on the contrary, seeks to address the hard ground which some might deem sensitive to discuss about. It is necessary for ethics to be reviewed, to enable safer working conditions and create a profit-making industry that produces quality end-products.

12.2 Review of Existing Literature and Theory

A review of the literature was carried out based on the set objectives of this research. The areas covered include providing an understanding of construction ethics, brief analysis of two other countries' ethical elements of the construction industries and to establish how professional governing bodies help to shape the image of the construction industry.

12.2.1 The Definition of Ethics

Ethics is defined as the branch of philosophy that deals with morality (Dictionary.com 2017), whereas Oxford Dictionaries (2017) defines it as moral principles that govern a person's behaviour. Ethics plays an important role within the construction industry. The industry has come a long way in the improvement of technical practices, but still needs to do a lot of work to improve the professional practice (Mirsky and Schaufelberger 2015).

This view is also supported by Buffett (2015), whose analogy states that 'there is a lot of work to do, to push the professional practice snowball back uphill since there have been a long history filled with hundreds of examples of unethical practices'. It is important to understand that the construction industry is a service industry as well as a relationship industry, connotating that good ethics should always be applied. This statement is concurred by the views of Luder (2006) who is of the view that 'when a builder takes advantage of a client, he/ she is doing more than damaging relationship, he/she is also hurting the image of the industry and every builder'.

12.2.2 Articulation of the Problems around Ethics

In the business world, according to Arewa and Farrell (2015), fraud and corruption is arguably a pervasive trait in doing business and there is growing worldwide concern over a high level of corrupt activities among corporate organisations. Arguably, the way construction organisations operate, tend to make them vulnerable to corrupt practices. The problem being articulated by the researcher is the extensiveness of corruption in the construction industry which is nevertheless highlighted by Arewa and Farrell (2015) as an aspect that exist in the context of inappropriate activities of organisations with perhaps the knowledge of many individuals in those organisations, but with no personal gains and conversely, corruption may also involve the activities of individuals who act for personal gain without the knowledge of their companies. Hence this research aims to establish how, alongside other elements, personal values and corporate culture promote or extinguish ethical practices in the United Kingdom's built environment.

Another enunciated problem revolves around the issue of few people being professionally qualified in the construction industry, therefore highlighting the lack of commitment to professional codes of conduct. BBC (2017) concurs with this theory by mentioning that the misalignment between the number of available jobs and the number of skilled and professionals is still a problem within the industry.

Codes of conduct are arguably not well policed by professional bodies who have too little resources. It is specified that one of the distinguishing prerequisites to any business, is corporate ethics (Bowen et al. 2007), hence the importance of having professional bodies that are competent to police and enforce ethical values in the industry. In the present era of growing attention on ethical malpractice in many organisations around the world, the study of ethical code implementation is imperative. In the UK, referring to an article published by CIOB (2017), unethical problems, both at the corporate and operational levels of the construction industry have become commonplace, thus the industry is accessible to issues of ethics, hence the importance of analysing how construction professional bodies in the UK help portray the image of the industry. Apart from the issues of the professional bodies' lack of ethical policing, Doran (2012) suggests that two main reasons appear to be at the centre of ethical problems:

a) Deficiencies in code embedment in the web of organisational process and routines.
b) Ineffective code implementation.

Murray and Dainty (2009) asserted that the existence of ethical codes alone is inefficient to ensure employees' ethical behaviour and therefore there is need for them to be complemented with the assignment of functional responsibility, whereas Garrett (2011) highlighted a problem articulated in this research, by stating that the process of how the codes are implemented and maintained in the workplace remain understudied in the context of the construction industry.

Whistle-blowing is a problem in the construction industry. According to a report published by CIOB (2017), the construction industry requires the coordinated effort of all professionals. It is stated that Architects have for decades been regulated by complex code of ethics. The document (CIOB 2017) further states that construction industry employees are often required to report improper, unsafe, or illegal construction activities. However, it is documented that impeaching on colleagues, contractors, clients etc. has led to dismissal of many individuals from their jobs. The problem articulated for such, is the need to have a legislation in place that protects and therefore encourages the whistle-blowing phenomenon in the construction industry as a basis of ethical practice.

This research will be focusing and integrating ethical approaches in construction as viewed by Mirsky and Schaufelberger (2015). These approaches can also be applied to some of the most known common ethical challenges in the construction industry. These approaches are:

- *The Utilitarian Approach*: which is the ethical choice which produces the greatest good for the greatest number of people. This concept is also suggested by Arditi and Chotibhongs (2005), who mention that the Utilitarian Approach is concerned with the consequences of the action or decision and the direct or indirect impact on everyone involved. In short, what is being stipulated is that when analysing this approach, it is possible that individuals may be harmed or suffer negative consequences if the majority of those affected benefit from the decision.
- *The Rights Approach* is defined by Mirsky and Schaufelberger (2015) as the ethical choice that respects the fundamental rights of others to be treated as equals capable of making their own decisions, some of the common fundamental rights highlighted by the authors include free speech, liberty, safety and security. If these rights are ignored, then an unethical act will have prevailed.
- *The Justice Approach*: Mirsky and Schaufelberger (2015) stipulate that this ethical approach treats everyone involved in the industry fairly and in accordance with what they are due. Bowden (2010) is also of the same view and highlights that it is important to treat and compensate people in the same circumstances equally.
- *The Common Good Approach* (Mirsky and Schaufelberger 2015) mentions that it puts emphasis on the ethical choice that promotes or contributes to the common good of a society or the community, a fact which is also hailed by Bowen et al. (2007) who state that 'there is a huge assumption that there are certain services and resources either natural or human made whose existence benefit our common good as a society, these include a healthy environment, good health care system, education, housing *etc*'.
- *The Virtue Approach* is defined as an ethical approach that 'arises and demonstrates our moral virtues, which include honesty, integrity, respect, trust, and fairness' (Bowden 2010).

It should be noted that the aforementioned approaches will be critically analysed in the literature review section and linkage will be provided on how they influence the built environment from start to finish of any given project.

12.2.3 Understanding Ethics

According to Delbridge (2000), ethics broadly constitute:

a) A system of moral principles by which human actions and motions may either be judged as good or bad.
b) The rules of conduct perceived in respect of a class of human actions.
c) Moral principles of an individual.

Fewings (2009) is of the belief that ethics evolved initially from religion which originated from the Judeo-Christian tradition, a view which is also shared by Cohen and Grace (1998).

The separate study of the above mentioned (Fewings 2009) (Cohen and Grace 1998), involves the activities of examining people's moral standards of society and asking how these standards apply to their lives and, Fewings (2009) further articulates whether these standards are reasonable or unreasonable. A further study, but on same concept, by Velasquez (1998) states that the outcome suggested by Fewings (2009), emphasise on the notion of rights, as doing what will promote the best and acts, that promote the general good are one of the factors that determine whether they are right. A weakness in this fact is hailed by Johnson (1991) who states that ethics cannot be arbitrarily created but is discovered through argument and persuasion, a factor which led to the suggestion by Calhoun and Wolitzer (2001) that business and ethics should not be mixed (a view which is suggested to be out of date by many scholars). According to Velasquez (1998) business ethics was a term which was not in the business vocabulary, and it is mentioned that the mere term 'business ethics' has been called as an oxymoron. However, it is now recognised that the accustomed concepts of ethics are pertinent in business (Fledderman 1999) on the basis that business exists not solely to suit certain individuals, but because it serves society and meets collective and individual needs. There is a similarity between this view of Fledderman (1999) and the ethical approach which was highlighted by Mirsky and Schaufelberger (2015) which they mentioned as the Common Good Approach, which puts emphasis on the ethical choice that promotes or contributes to the common good of a society or the community.

12.2.4 Unethical Conduct in the Construction Industry

Sutherland (1983) was of the belief that being unethical does not eventuate from a person's upbringing, but rather, it is part of the process of learning practical business or being inducted into the practice. The significance of this is argued by the new United States of America (USA) federal sentencing guidelines for not taking proactive measures to apply and emphasise ethics in all construction phases of a given project (Ferguson 1994). In the UK, according to Fewings (2009), the reality is that little training is provided to employees in the construction workplaces, a view that is supported by Weart (2001) who stated that there is a rise in dishonest and unfair practices in the industry. It is important to understand and accept that the construction industry is a service industry as well as a relationship industry, thus implying good ethics should always be applied. In a glance, the purpose of this research is to explore construction ethics with the aim of understanding its impact on the built environment hence it is important to look at aspects that are headlined as unethical within the industry although they can be applied in different industries (disciplines).

Below is a list of the most documented ethical inappropriateness practices in the construction industry:

1) *Bribery*: Almeder and Humber (1983), cited in Johnson (1991), describe bribery as 'the offering of some goods, services, or money to an appropriate person for the purpose of securing a privileged and favourable consideration (or purchase) of one's product or corporate project'. Similarly, Whitbeck (1998) simplifies the meaning of bribery by stating that bribery is the act of offering payments or inducements to someone in a position of trust to get them to do something for the bribe payer to which the bribe payer is not entitled to. However, the weakness with the two mentioned descriptions above is the failure to address the fact that most activities with an ethical content come with what some scholars can call 'grey areas'. Fewings (2009) indicated that these are between the delineation of actions that are termed 'gift giving' and what can be defined as bribery in legal terms. It is further mentioned that this situation will be compromised when meals or gifts are also no longer of low cost and the expenses of these items are not shared equally, hence the possibility of misconduct becomes large. According to Johnson (1991), the following two actions must be satisfied to transform gift giving to the illegal practice of bribery:
    a) The person receiving the gift may deliberately, or otherwise, be predisposed, anticipated to favour the interests of the gift giver
    b) The gift must be of non-token nature that is reasonable to think that it may put the interests of the giver in an honoured status even when all else is equal. Consequently, some corporations have allowed gift giving to their clients or potential clients if these two conditions do not apply.

   Fewings (2009) argues that bribery is prevalent in all stages of the supply chain that is from tendering to commissioning. The Bribery Act of 2010 seeks to penalise any individual or groups that are found guilty of an offence under Section 1, 2 or 6, who will be liable on summary conviction, to imprisonment for a term not exceeding 12 months or to a fine not exceeding the statutory maximum or both (Legislation.gov.uk. 2017).

2) *Corruption*: according to Transparency International (2015) (T I), corruption is stipulated as the abuse of entrusted power for private gain. It is also stated that it can be classified as grand, petty and political, depending on the amount of money lost and the sector where it occurs. Jain (2011) states in his research that corruption in the construction industry occurs at a cost. It is mentioned that corruption impacts society in a multitude of ways; in worst cases it costs lives. Short of this T I (Transparency International 2015) argue that it costs people their freedom, health or money. In the UK and other developed countries, the costs of corruption can be divided into four main categories namely:
    a) *Political*: According to T I (2015), on the political front, corruption is a major obstacle to democracy and rule of law. In a democratic system, offices and institutions lose their legitimacy when they are used for private advantage. It can be argued that there are similarities between bribery and corruption as they both seek to satisfy one part through unequal/or favourable means although it can be further argued that if corruption is done at a larger scale, it can lead to loss of life.
    b) *Economically*: Transparency International (2015) is of the belief that corruption depletes national wealth. Corrupt politicians and corporate businesses invest scarcely in projects that would benefit their countries or organisations and rather will invest

in projects that will enrich them along the way. According to Arewa and Farrell (2015) it is estimated that the economic costs incurred by corruption in the EU to amount to US $860 billion per year, forecasted to rise to US $1.5 trillion by 2025.

c) *Society*: Jain (2011) understands that corruption corrodes the social fabric of the society. It undermines people's trust in the political or corporate systems.

d) *Environmental*: Transparency International (2015) argues that environmental degradation is another consequence of corrupt systems. The lack of, or non-enforcement of, environmental regulations and legislation means that special precious natural resources are carelessly exploited. This can be said will affect mainly developing countries.

Within the British constraints of the construction industry, corruption is considered a problem. A report by CIOB (2017) suggests that the construction industry has always been susceptible to corruption. The CIOB's survey in 2006 found that 51% of respondents thought that corruption was either fairly or extremely common in the UK construction industry. Like bribery, corruption can occur in all phases of a construction project.

3) *Fraud*: Doran (2012) raises an important argument on the issue of deceit, calling it fraud, whereas Johnson (1991) does not consider deceit to be fraud, but considers it to be trickery. *Oxford Dictionary* (2010) defines fraud as wrongful or criminal deception intended to result in financial or personal gain. According to the Association of Certified Fraud Examiners, 3.4% of all reported fraud cases over a two-year period up to December 2011 were attributable to the construction industry. This inflicted a median average loss of US $300 000. The UK's Chartered Institute of Loss Adjusters states that fraud in construction is rife and estimates the impact at 10% of the industry's revenues. Recent estimates (Grant Thornton ltd 2013) show that the global construction industry is worth USD 8.6 trillion and is expected to rise to USD 15 trillion by 2025. This means the global cost of fraud and corruption could be USD 860 billion, rising to USD 1.5 trillion by 2015.

There are seven main types of fraud encountered in the construction industry, per Grant Thornton Ltd's (2013) perceptions. Below is the list of the seven main types.

a) *Billing fraud*: In construction, this is purposely overstating the amount of labour, materials and other equipment required to complete a project.

b) *Bid/Contract rigging*: This occurs when supplies in the market collude to fix prices or direct customers to use certain contractors. This can include bribery.

c) *Fictitious vendors*: These are created by falsifying payment applications covering up the purchase of personal items or diverting money to a phantom company. Activity is often controlled by an employee but can also be done by external entities through falsified company documentation or email addresses.

d) *Change order manipulation*: This occurs by diverting lump-sum cost to time and material cost by initially budgeting expenses as a lump-sum then billing for time and materials related to change orders.

e) *Theft or substitution of materials*: Taking material from the work site for personal use or using lower grade material than quoted which might result in subsequent reporting or replacement.

f) *False representation*: This might involve undocumented workers falsifying minority content reports test results or insurance certificates, non-compliance with environmental regulations and misrepresentation of small business status.

g) *Money laundering /tax avoidance*: This activity is to legitimise money gained illegally. In real estate, it could include making down payments on a property and selling it later to give a legitimate origin. Tax avoidance can involve commonplace activity such as cash-in-hand for labour.

4) *Workers' rights*: On an ethical perspective, Whitbeck (1998) points out that employers and employees have responsibilities to each other, thus expecting their rights to be upheld. These rights and responsibilities are considered by Harris et al. (1995) as an instrument that relate to areas such as the provisions of terms and conditions of employment, equal opportunities, and the rights to be paid minimum wage. A breach to any of the above is then considered unethical.

5) *Health and safety*: This is an important aspect within this ethics research. Fewings (2009) mentions that ethically, health and safety measures could reduce harmful accidents or dangerous incidents which will cause immediate or future health problems. However, McDowell (1991) questioned this hypothesis and is of the view that health and safety is a 'Kantian ethic', where there is a duty to protect the employee and to act by absolute standards of safe working for all parties to the best of your ability, where the means does not justify the ends and playing fast and loose within people's well-being, health and safety to gain a commercial or any other outcome will be classified as immoral. Ethically analysing responsibilities for health and safety in the construction life cycle involves:

   a) The client: Concern for the client is to be provided with information regarding the safe delivery of the building (Fledderman: 1999).
   b) The designer: According to Fledderman (1992), if a building or tunnel collapses or creates a major health problem, there is a design duty of care making it unacceptable if precautions were not taken for foreseeable risks. It is further argued that ethically, it is a betrayal of the trust that people have in the design.
   c) The health and safety co-ordinators: Johnson (1991) is of the belief that ethically it might be unrealistic to ask an inexperienced client to carry out strict competency checks without the help of a professional. Johnson's (1991) emphasis is on the moral compass of companies on the fact that they must not just have a health and safety co-ordinator, but they will be required to have a health and safety officer who is competent.
   d) Principal contractor and contractor supply chain: Johnson (1991) argues that they have a duty of care towards everyone involved on site and emphasis is given on the ethical aspect of learning from past mistakes and work towards a programme of continuous improvements.

6) *Confidentiality*: A common breach of confidentiality is whistle-blowing, described as the act of an employee informing the public or higher management of unethical or illegal behaviour by an employer or supervisor (Johnson: 1991). In the view of Harriet et al. (1995), whistle-blowing is not always appropriate and should only be attempted when four basic requirements of need, proximity, capability, and last resort option are satisfied, a fact that is neutralised by Fledderman (1999) who acknowledges that a 'grey area' exists over the length of time an employee should withhold confidential information and where discontinued employment exists. Delbridge et al. (2000) then ascertain that negligence will arise if failure to exercise the degree of care which in the circumstances is required by the law to protect those interests of other persons which maybe injuriously affected by the need of such care.

12.2.5 The Nature of Professional Ethics

Professions have always been linked with the notion of service; thus, a profession has been described as a group of people organised to serve a body of specialised knowledge in the interests of society (Applebaum and Lawton 1990). In comparison, Whitbeck (1998) substantiates that professions are 'occupations that both require advanced study and mastery of a specialised body of knowledge and aim to promote, ensure or safeguard some matters that significantly affects other's wellbeing'. Professional ethics responsibilities have been variously described as including the satisfaction of an indispensable and beneficial social need, thus per the view of Johnson (1991), and in contrast, has been described by Murdock and Hughes (1996) as a goal of service to the public. A professional operates in a world of people with whom they work, colleagues and other specialists and people whom they serve, such as their clients and the public (Pressman 1997). Johnson (1991) recommends that professionals must not be exempt from the common ethical behaviours such as responsibilities, obligations and duties that are binding on ordinary people and are usually bound by a set of principles, attitudes or types of character dispositions that control the way the profession practices. This has been termed professional ethics (McDowell 1991) and concerns potential problems encountering members of a profession or group and their impact on society. One important aspect is that of conflict of interest, defined as an interest which if pursued, could keep professionals from meeting one of their obligations. Also relevant, is the professional right termed the 'Right of Conscientious Refusal' (Martin and Schinzinger 1996), which is the right of an employee to refuse to partake in unethical conduct when forced to do so by an employer whereas Whitbeck (1998) is of the view that 'Conscientious Refusal' may be done by either simply not participating in the activity that one sees as immoral or it may be done with the hope of making a public protest that will draw attention to the situation that one believes is wrong. Different professions however have different reputations insofar as regarded by Pressman (1997), who cites a public opinion survey which rated architects as superior in ethical behaviour as compared to lawyers and doctors.

In the UK, different disciplines are governed by different bodies, for instance builders are governed by the Chartered Institute of Building, civil engineers are governed by Institute of Civil Engineers, architects are governed by Royal Institute of British Architects etc. In terms of individual professions, it is often assumed that architects are not only talented in designing the construction of buildings, but also, they are of the highest ethical calibre in the UK.

Whilst exploring ethics in construction, Pressman (1997) cites the main types of unethical behaviour in the construction industry as:

- Concealing of construction errors and stealing someone else's ideas.
- Generally telling untruths of academic achievements in resumes and applications to get more work.
- Charging clients for work not completed costs not incurred or overstated costs.
- Misleading clients in project management.
- Involvement of conflict of interest.

For Project Managers, according to Fryer (1997), one of their critical elements of professionalism is their consideration of ethics and social responsibility, a view that is also upheld by Willouby (2004) who state that there should be no conflict between morality and good management. CIOB emphasises that construction contractors are also expected to behave in an ethical manner. A survey conducted by the body on professionals indicated that significant roles of ethical conducts had an impact on construction contracting. In contrast with architects, construction contractors have a reputation for unethical behaviours. The main problem, according to a poll conducted by the Building Research and Information (2008) cited in Pilvary and Sutherland (2008), showed high level of disputes between proprietors and builders. Their generally poor behaviour has been said to have originated from the influx of new construction companies that emerged with new people who lack building ethics. However, this fact can be arguable, citing the findings of Velasquez (1998), who stated that ethics in the past were not incorporated in the business setting meaning in the past there were rules and regulations but there were not any ethical aspects that strengthened those rules and regulations.

12.2.6 Ethical Challenges: Impacts of Construction Site Activities and Sustainability

This is an important aspect that it is concerned with the interaction of the built, intentionally organised and engineered environment, with the natural spontaneously organised environment (Fewings 2009). It is further stated that in this interaction, there is an increasing ethical concern about our ability to conserve and maintain the beauty of the natural world around us because of our lack of control of the impact of the built environment on the natural world. Fewings (2009) also account that a new and more prescriptive ethical paradigm is emerging in planning the built environment which has gathered pace in the sustainable development movement. On the same aspect, Mirsky and Schaufelberger (2015) discuss the ethics of sustainability as the drive behind the implementation of a more environmentally sensitive regime that is proactive, duty-based approach to the better management of an area that has not been considered acute, a drawback which is argued by Fewings (2009) who accentuate that environmental ethics has simply been ignored on a mass scale and that more will now be needed to bring back the balance which is needed on the environment.

According to Mirsky and Schaufelberger (2015) construction site activities frequently require the removal of vegetation and topsoil and the subsequent exposure of large areas of unprotected soil. When rain from storm events falls onto the unprotected soil, some water will infiltrate into the ground, but much will run off following the path of least resistance to the nearest storm sewer or stream. It is further mentioned that (Mirsky and Schaufelberger 2015) the release of high levels of sediment into storm sewer outlets or other receiving bodies of water has a harmful effect on water quality and on wildlife. Fewings (2009) further illustrates that the requirement for preventing soil erosion and the transport of sediment in storm water runoff from a construction site is a good method of the precautionary principle in action. Many contractors, according to a document published by Sourceable (2017), still view erosion and sedimentation control as a costly imposition. However, it is important for both contractors and clients to understand that costs associated with water-quality protection must legally and ethically be considered as one of the basic costs of construction, just like materials, equipment, and labour. Fewings (2009) is of the same ideology as Sourceable

(2017) because it is mentioned that if construction projects do not take responsibilities for protecting nearby water quality from the negative impacts of site activities some other individuals or agencies or local councils will end up paying for that protection or treatments. Fewings (2009) also states that construction projects have ethical responsibilities to bear the cost of mitigation and control of other forms of pollution associated with site activities including dust and noise.

According to Sourceable (2017) the construction process can generate substantial amounts of waste: excavated soil, scrap materials, pallets, packaging, used oil, oil, miscellaneous, rags ordinary household refuse from cabins, fuels, paint, adhesives etc. are considered hazardous waste and must be disposed of, in accordance with regulatory requirements. Demolition waste is also frequently contaminated with hazardous materials including lead, asbestos or soil containing spilled fuel etc. and should ethically be disposed of. Fewings (2009) further illustrates the importance of disposing non-hazardous waste materials which can have a major impact on local communities.

12.2.7 Understanding Individual's Values and Ethics

It is important to understand values that will then lead to positive or negative ethical mannerism in the workplace. Morals on the other hand, are defined by the Oxford English Dictionary (2010) as motivation based on ideas of right and wrong. Changing Minds states that morals have a greater social element to values and tend to have broad acceptability. Futurelearn (2017) takes a different approach when ascertaining values. They are of the view that values are in essence 'what is important to you' and add to this essence by mentioning that 'values then motivate your behaviour'. Changing Minds further illustrates that values stem from a need or a void and that need, or void will motivate you to fulfil it, which will then drive your behaviours and your actions. This is sometime referred to as moral compass which is argued by Futurelearn (2017), that part of understanding what drives our moral compass will then identify who we really are. It is important to note that our values are formed through our childhood and are influenced by our upbringing and culture (Futurelearn, 2017). It is also important to note that different people value different things, and this can change as we get older; the hierarchy of that value can also change (Futurelearn, 2017). In the business world, referring to a document published by the Businessworld.com (2017), values are the embodiment of what an organisation stands for and should be the basis of the behaviour of its members. On the same publication (BBC, 2017), being ethical is consistent with what is generally considered to be right or moral. Ethical behaviour is thus said to be the bedrock of mutual trust. Andrews (2014) contends that there are three aspects to ethical behaviour in organisations:

The development of the individual as an ethical person.
The effect of the organisation as an ethical or unethical environment.
Actions or procedures developed by the organisation to encourage ethical behaviour and discourage unethical behaviour.

12.2.8 An International Aspect on Ethics

1) **Zimbabwe case study**

   A statement of ethical principles must follow honest, fairness, fair reward, reliability, integrity, objectivity and accountability. Analysing the Zimbabwean case study, it is documented by the Construction Industry Federation of Zimbabwe, that ethics within the industry must be of high regard. This is disputed by an article published in Mafaro (2017) that stated that corruption is at its highest ever since Zimbabwe got its independence. Tenders are awarded not through merit but through nepotism, corruption etc. It is also mentioned that there is no fair reward, as some of the jobs done for the government up to five years ago, have not been paid for thus forcing some small contractors into liquidation. Mafaro.com. (2017) also explains that the National Tender Board, that is supposed to be regulating tenders is the most corrupt, hence no accountability. As of CIFOZ all their publications are arguably biased because if they state as it is, they will be implicated as an organisation.

2) **United States of America case study**

   As in the United Kingdom it is documented that America also suffers from unethical elements within their construction industry. A study conducted by Doran (2012) collected the thoughts on the ethical state of the industry from 270 architects, engineers, construction managers, general contractors, and subcontractors. When asked if they had experienced, encountered, or observed construction industry-related acts on unethical transactions, the majority were in agreement and expressing 'that it was now the norm'. Bid shopping is the biggest problem, and according to Doran (2012), this practice is predicated on main contractors disclosing competing sub-contractor's prices seeking further discounts. The American study (Doran: 2010) highlights another key issue which is people's different understanding of what the rules are and what is right and wrong in any given situation. On the contrary, Goldstock (2014) suggests that people are aware of what is happening on construction sites but are too afraid to voice their concerns. According to Goldstock (2014) extortion is America's construction industry major issue. It is mentioned that only a thin line separates extortion and bribery and a lot of people are not aware of that line, (a fact that concurs with Doran 2012). Unlike the UK, Goldstock (2014) mentioned that multifarious types of extortion are commonplace in the United State of America's construction industry. It is further discussed that many, if not all, involve the use of threats to obtain services or other things of value. On the same fact, Doran (2012) adds on by mentioning that most common form of extortion involves union officials interfering with work that is being carried on and indirectly threatening contractors with labour problems if they don't receive certain payments. This indeed can be classified as extortion. Apart from extortion, Goldstock (2014) is of the view that bribery is a major problem in America's construction industry. Like in Zimbabwe, America's bribes are frequently paid by contractors to obtain contracts, favours, and services to which they are not entitled. In variation to the above-mentioned problem of bribery, a document by Doran (2012) bases that its most unethical problems are on theft and fraud. Doran's (2012) emphasis on fraud is categorised as:
   - Fraudulent billing for work not performed or materials not used
   - Defrauding union pension and welfare funds by avoiding required payments

- Defrauding the government by failing to report taxable income or falsely reporting business expenses
- Fraudulent performance bonds (bonds that guarantee that the contractor/ subcontractor will carry out their work satisfactorily and that they will pay their workers and suppliers).

12.3 Research, Design and Methodology

12.3.1 Qualitative Analysis for Data Collection

This research seeks to explore ethics in the construction industry, what can be done and understanding its impact on the built environment. This would suggest that the outcome of this enquiry would best be satisfied by analysing the responses derived from participants who took part and their potency in answering the questions designated to the problem at hand. These factors indicate that a qualitative analysis would be suited for this research. Reflecting on Farrell et al. (2016), qualitative analysis aims to gain insights and understand people's perception of the world. It is further stated that the respondents may deliberate with themselves and give careful thought answers. To fully obtain relevant information, open ended interviews will be used which would be asked to several individuals. This method was applied due to time and geographical constraints for this study. The interviewing methods allow definitive analysis of the respondent's views of the subject area. This methodology is limited insofar as it will demonstrate the respondent's espoused beliefs and values in relation to ethics, which may differ from their depicted values and values (Schein 1996). The template used to obtain the data may contain some degree of bias from the author's culture or reference framework thus according to Schein (1996), who also believes that derivatives from this are unavoidable. These limitations have been acknowledged.

12.3.2 Limitations of the Study

Due to the nature of the study, data obtained is likely to be biased, but to minimise this the researcher did not ask open-ended questions to the respondents and used scenarios that have been compiled from construction literature articles which have been published before and during the study. This study was limited by time constraints and emphasis was given on factors that impact construction projects from tendering to commissioning.

12.3.3 Interview Design, Technique, Coding, Judgement and Calculations

It is noted that no single research method or technique that is 100% efficient, therefore the writer acknowledges the strengths and limitations of methods adopted, a fact which is further emphasised by Breach (2009), who complement the strength and limitations of all types of research. Yin (2009) is of the view that the goal is to use these methods and techniques to explore different aspects of research. This, however, enabled the writer to come to some conclusions and to draw recommendations or both for future studies and or action, that can be taken by the construction industry at large.

1) **Technique**

The Researcher reviewed different sources of literature and analysed aspects that were arguably linked to unethical examples cited by different scholars. In so doing, the Researcher managed to stage scenarios likely to be encountered in the construction industry, thus leading to the creation of the interview questions. The interview questions were categorised into different sub-headings and the responses obtained per question were placed under their respective sub-heading.

Three of the interviews took place at the Respondents places of work. Two of the interviews were conducted at the participants' homesteads and five of the interviews were telephone based. Please note: the interviews were scheduled in favour of the Participants' availability, and they were given ample notice and flexibility to withdraw from the interview if they needed to. Each interview took an average of 10 to 15 minutes, and the Researcher used an application (app) on the mobile phone to record the proceedings.

2) **Coding and Judgement**

Having completed conducting and recording the interviews, the Researcher then analysed the responses and coded them on how ethically appropriate they were considered (by the Researcher) within the industry. Responses that were of high ethics or highly acceptable actions in the industry, were coded as Definitely Appropriate (D A). Responses that were of better (and not the best) ethical appropriateness, were coded as Appropriate (A). Responses which were average, were coded as Neutral (N). Responses which were below average, were coded as Inappropriate (I) and for responses which bore no moral grounds, were coded as Definitely Inappropriate (D I). Respondents' recordings were coded per voice note as 'V'.

3) **Calculations**

Please note: Frequency counts refer to the number of respondents who perceived a scenario as either; Definitely Appropriate, Appropriate, Neutral, Inappropriate or Definitely Inappropriate. For an example, on Trustworthiness (please see Table 12.1), the Researcher judged those 3 responses from the interviews were Definitely Appropriate, 1 Respondent was considered as Appropriate, 1 Respondent's answer was perceived as Inappropriate, and 5 Respondents were viewed as Definitely Inappropriate. On the column of Industry Appropriateness, the highest value to be obtained is 4 and the lowest value to be obtained is 0 (zero). The highest value (4) is at all-time linked to Definitely Appropriate (D A), value 3; is linked to Appropriateness (A), value 2; is linked to Neutral (N), value 1; is linked to Inappropriate and value 0; is linked to Definitely Inappropriate (D I). Ten (10) people were interviewed for this research document and as mentioned before, the highest value/maximum score is four (4) which is linked to Definitely Appropriate. This implies that the 10 respondents multiply (*) by the maximum score (4) will equal (=) to 40, which is the number that is going to be the base on all the calculations to get a percentage for each category. Calculation example on trustworthiness (please see Table 12.1) will be as follows:

3 people were judged to have responded Definitely Appropriate (D A) which carries a maximum score of 4, hence 3*4=12. One person was judged to have responded Appropriately (A) which carries a score of 3, hence 1*3=3. Zero persons were Neutral(N) which carries a score of 2 hence 0*2=0. One person was judged to have responded Inappropriate (I) which carries a score of 1, hence 1*1=1 and 5 people were judged to have responded Definitely Inappropriate (D I) which carries a score of 0(zero) hence 5*0=0.

Table 12.1 Question 1: Trustworthiness.

Respondent (as per voice note recorded: V)	Summarised Responses	Industry Appropriateness
V:1	Most contractors will not disclose the payment	Definitely inappropriate
V:2	Contractors will not disclose the overpayment, but that is not the right thing to do.	Inappropriate
V:3:	They will disclose because an audit by the client will surface the overpayment and that will amount to fraud on the contractor's side.	Definitely appropriate
V:4	Should return it back to the client.	Appropriate
V:5	They can keep the money unless the client asks for it.	Definitely inappropriate
V:6	Highlight and explain the mistake to stay in good books.	Definitely appropriate
V:7	Won't mention it.	Definitely inappropriate
V:8	Pocket it.	Definitely inappropriate
V:9	If it is a long-term client will tell the client, but if it is a new client will keep quiet.	Definitely inappropriate
V:10	Return the money and explain the error.	Definitely appropriate

Next step will be to sum up: 12+3+0+1+0=16.

Bearing in mind the base value is 40, this implies that 16/40*100/1=40%. So, the percentage for trustworthiness is 40%. The same formula has been applied in Table 12.2 which gives a total calculation of 611. To calculate the Mean, the total was divided by 12 (total of categories) which gave 50.92%, taken to the nearest (per 100%) is 51%. Therefore, the Mean for Table 12.2 = 51%.

It must be noted that DA=90%–100%, A=70%–89%, N=50%–69%, I=26%–49%, and DI=0%–25%.

Tables 12.3–12.12 refer to questions related to different areas.

12.3.4 The Demography of Respondents

The demographic data of the interview respondents were as follows:
  Quantity Surveyors: 3
  Civil Engineers: 1
  Project Manager: 1
  Construction Commercial Manager: 1
  Construction Project Management Students: 3
  Design Technologist:1

Table 12.2 Question 2: Leadership balance.

Respondent (as per voice note recorded: V)	Summarised Responses	Industry Appropriateness
V:1	Must take day off if problem is severe, but if not must go and work.	Appropriate
V:2	The manager must not take problems to work. In practice, managers will offload their frustration on employees.	Definitely inappropriate
V:3	Still go to work, but evaluation will be required if the issue is off serious concern. Most managers are professional enough.	Definitely appropriate
V:4	Must not let their emotions interfere with their work.	Definitely appropriate
V:5	Must be professional and able to separate their home life and work life.	Definitely appropriate
V:6	Must not take it out on someone. Must engage the day differently.	Neutral
V:7	Must go to work and be prepared to work.	Appropriate
V:8	Engage the day differently.	Neutral
V:9	Must be able to separate home life and work life.	Definitely appropriate
V:10	Separate home life and work life.	Definitely appropriate

This selection was influenced by the Respondents diversity within the supply chain and to be acknowledged is the continuous relationship the Researcher maintained with some of the respondents and relationships created with fellow students.

This provided credibility for the survey results as the respondents are knowledgeable/learning the ethical perspectives of the industry.

12.3.5 Interview Questions and Aims

Reverting to the Literature Review, the structure of the interview questions was aimed at finding people's perspectives on the main ethical dilemmas experienced in the construction industry, hence the questions were:

Q1): examining trustworthiness.
Q2): examining leadership balance.
Q3): observing accountability dilemmas in the industry.
Q4): scrutinising the issue of collusion.
Q5): viewing ethical aspects of workers right.
Q6): inspecting knowledge on conflict of interest, which also includes corruption.
Q7) and Q8): focusing on environmental issues in relation to the industry.
Q9): examining sustainability aspects.

Q10): scrutinising quality issues experienced by some construction organisations.
Q11): observing knowledge of nepotism within the industry.
Q12): examining awareness on bribery.

12.3.6 Findings and Discussions

12.3.6.1 Research Results and Findings

This section of the study appraises the findings from the data collected by conducting interviews. The findings are evaluated to examine if congruency exists between the practical knowledge of ethics and the literature documented. The overall study is analysed in terms of the findings, objectives set, and lessons learnt (what can be done?). The 1–12 incorporate the subheadings of the questions asked in relation to ethical scenarios which are likely to be experienced in the construction industry.

In addition to the previous table, Table 12.13 represents an analytical summary of each category referenced to the literature sources.

Based on the above tables Figure 12.1 represents the data from frequency counts from the different categories of Industry Appropriateness. In addition, Figure12.2 represents the percentage of Industry Appropriateness calculated from the tables and Figure 12.3 a Pie Chart illustrating the percentage of Industry Appropriateness calculated from the tables above.

Table 12.3 Question 3: Accountability/Health and safety.

Respondent (as per voice note recorded: V)	Summarised Responses	Industry Appropriateness
V:1	Duty of the employee to make sure they adhere to the health and safety policy. It is not the manager's responsibility.	Definitely inappropriate
V:2	Supervisor's responsibility to make sure proper safety equipment is worn.	Definitely appropriate
V:3	Supervisor will admit because health and safety is not a blame game and it paves way for learning lessons so that the same accident won't repeat.	Definitely appropriate
V:4	Supervisor must take responsibility.	Appropriate
V:5	They will try to cover it up.	Definitely inappropriate
V:6	Supervisor will not want to take responsibility; it is the employee's obligation.	Inappropriate
V:7	Supervisor must accept responsibility.	Appropriate
V:8	Responsibility falls on the supervisor.	Appropriate
V:9	Supervisor must take responsibility.	Appropriate
V:10	Ethically, the supervisor must accept responsibility.	Appropriate

Table 12.4 Question 4: Collusion.

Respondent (as per voice note recorded: V)	Summarised Responses	Industry Appropriateness
V:1	Wrong to discuss if it jeopardises the tender.	Appropriate
V:2	Most people will not discuss.	Appropriate
V:3	Will not disclose in detail because of declaration to be signed.	Definitely appropriate
V:4	They can discuss because competition improves quality.	Definitely inappropriate
V:5	Surely discuss.	Neutral
V:6	Most will mention and discuss. Level of detail must differ. Must consider confidentiality.	Neutral
V:7	They can discuss everything.	Definitely inappropriate
V:8	Possibly discuss and let the friend succeed on the tender.	Definitely inappropriate
V:9	(Question not understood)	N/A
V:10	Mention but do not discuss in detail for professional reasons.	Definitely appropriate

Table 12.5 Question 5: Worker's rights.

Respondent (as per voice note recorded: V)	Summarised Responses	Industry Appropriateness
V:1	Wrong to employ and abuse someone because of desperation but should employ and offer the going rate of minimum wage.	Definitely appropriate
V:2	Advertise the vacancy and abstain from using people because of desperation.	Appropriate
V:3:	Will employ on ability.	Neutral
V:4	Must not happen.	Definitely appropriate
V:5	Will consider it as a bargain and employ the person.	Definitely inappropriate
V:6	Must not consider, will put company at risk.	Appropriate
V:7	It is against the law.	Appropriate
V:8	Employ the person and save the company money.	Definitely inappropriate
V:9	Would be difficult to employ the person because they will need to be vetted.	Neutral
V:10	By law it is illegal, but take them on a minimum wage.	Appropriate

Table 12.6 Question 6: Conflict of interest.

Respondent (as per voice note recorded: V)	Summarised Responses	Industry Appropriateness
V:1	If tender is open, everyone can apply. The shareholder must follow the company's policy and procedures, but the shareholder must not apply.	Neutral
V:2	Must declare interest for evaluation purposes.	Appropriate
V:3	Tender must be awarded without the involvement of the shareholder.	Appropriate
V:4	(Question not understood by respondent)	N/A
V:5	Must be considered because he or she won't let the company down.	Definitely inappropriate
V:6	Considered to be conflict of interest.	Appropriate
V:7	Must not be considered.	Neutral
V:8	Must not be considered because of interest.	Neutral
V:9	It is okay if the shareholder is not part of the tender process.	Appropriate
V:10	There is conflict of interest and must not apply.	Neutral

Table 12.7 Question 7: Environmental issues.

Respondent (as per voice note recorded: V)	Summarised Responses	Industry Appropriateness
V:1	Protect the birds and notify the relevant authorities.	Definitely appropriate
V:2	Destroy the birds to save on time constraints.	Definitely inappropriate
V:3	Notify the relevant authorities.	Definitely appropriate
V:4	Should call the RSPCA/relevant authorities.	Definitely appropriate
V:5	Call specialist bird removers.	Neutral
V:6	Consult environmental advisors and follow the company's policy of animals on site.	Definitely appropriate
V:7	Relocate the birds.	Inappropriate
V:8	Must proceed with their plans.	Definitely inappropriate
V:9	Contact relevant authorities.	Definitely appropriate
V:10	Contact the local authority to remove the birds.	Definitely appropriate

Table 12.8 Question 8: Environmental issues/insubordination.

Respondent (as per voice note recorded: V)	Summarised Responses	Industry Appropriateness
V:1	The subordinate must follow what he is told.	Definitely inappropriate
V:2	Do as instructed.	Definitely inappropriate
V:3	(Question not understood)	N/A
V:4	Must not dispose.	Appropriate
V:5	Will comply with the order because they will feel duty bound.	Definitely inappropriate
V:6	Unfortunately follow what they are told.	Inappropriate
V:7	Must take it to qualified disposers.	Definitely appropriate
V:8	Inform the manager that it would affect the elements of the river.	Definitely appropriate
V:9	Will do what they are told.	Definitely inappropriate
V:10	Will comply with what they have been asked to do.	Definitely inappropriate

Table 12.9 Question 9: Sustainability.

Respondent (as per voice note recorded: V)	Summarised Responses	Industry Appropriateness
V:1	Site manager must accept responsibility / mismanagement.	Inappropriate
V:2	Hide the excess material so that they are not taken to task for the mistake.	Definitely inappropriate
V:3	Store the material, but if it is a tailor made, should sell it to cover the costs.	Definitely appropriate
V:4	Store in a safe place for future use.	Appropriate
V:5	Will try to sell off to some local contractors.	Neutral
V:6	Offer material to other sites, or store for future use.	Definitely appropriate
V:7	Person responsible must take responsibility and suffer the loss from their own pocket.	Definitely inappropriate
V:8	Keep the material for future use.	Appropriate
V:9	Must be stored for future use.	Appropriate
V:10	Keep for next job or re-sale.	Appropriate

Table 12.10 Question 10: Quality.

Respondent (as per voice note recorded: V)	Summarised Responses	Industry Appropriateness
V:1	Site manager must question the contractor and refuse to use the material.	Definitely appropriate
V:2	The manager will cover his/her back by raising the issue with superiors.	Appropriate
V:3	Obliged to notify the contractor about the quality of the material.	Definitely appropriate
V:4	Should be replaced with good-quality material	Definitely appropriate
V:5	Advise the Contractor first, then they will turn a blind eye and continue.	Definitely inappropriate
V:6	Stop the job and replace.	Definitely appropriate
V:7	(Question misunderstood)	N/A
V:8	Ask for better material.	Appropriate
V:9	Must raise the concerns and put the rejections across.	Definitely appropriate
V:10	Point it out and get proper material.	Definitely appropriate

Table 12.11 Question 11: Nepotism.

Respondent (as per voice note recorded: V)	Summarised Responses	Industry Appropriateness
V:1	Most managers will award favours because of the relationship.	Definitely inappropriate
V:2	The relative will be given special treatment.	Definitely inappropriate
V:3	Must be treated fairly.	Neutral
V:4	Should be treated equally and should be competent.	Definitely appropriate
V:5	Will be treated more favourably.	Definitely inappropriate
V:6	Treat the person just like any other team member.	Definitely appropriate
V:7	Treat everyone equally.	Definitely appropriate
V:8	Likely to get favours.	Definitely inappropriate
V:9	Must maintain professional boundaries.	Definitely appropriate
V:10	There should be no favouritism.	Definitely appropriate

Table 12.12 Question 12: Bribery.

Respondent (as per voice note recorded: V)	Summarised Responses	Industry Appropriateness
V:1	Junior manager will keep quiet because they will not want to lose job.	Definitely inappropriate
V:2	Will keep quiet	Definitely inappropriate
V:3	Junior manager should notify senior management.	Definitely appropriate
V:4	Report to the relevant senior authorities.	Definitely appropriate
V:5	Will keep quiet.	Definitely inappropriate
V:6	Out of fear they will not mention anything.	Definitely inappropriate
V:7	Has to act according to the company's policy and procedures on bribery.	Definitely appropriate
V:8	They will keep quiet because it might compromise their job.	Definitely inappropriate
V:9	Look the other way.	Definitely inappropriate
V:10	Will keep quiet for fear of being called a whistle-blower.	Definitely inappropriate

Table 12.13 Analytical table of findings.

Question Category	Literature Sources	Observations, Implications, or Interpretations	Frequency Counts and Codes	Percentage Calculation
Trustworthiness	Fewings (2009), Cohen and Grace (1998)	It is perceived that a lot of contractors are not trustworthy when it comes to money. Greediness can be blamed.	DA=3Nr * value 4 = 12 A=1Nr * value 3 = 3 N=0Nr * value 2 = 0 I= 1Nr * value 1 = 1 DI=5Nr * value 0 = 0	16/40 = 40%
Leadership balance	Whitbeck (1998)	Site managers are professionals, thus being able to separate personal emotions and workload.	DA=5Nr * value 4= 20 A=2Nr * value 3= 6 N=2Nr * value 2= 4 I=0Nr * value 1 = 0 DI=1Nr * value 0 = 0	30/40 = 75%
Accountability/ Health and safety	Fewings (2009), McDowell (1991)	Supervisors are perceived to be knowledgeable on aspects of accepting responsibilities on site.	DA=2Nr * value 4 = 8 A=1Nr * value 3 = 3 N=0Nr * value 2 = 0 I=1Nr * value 1 = 1 DI=5Nr * value 0 = 0	12/40 = 30%

Table 12.13 (Continued)

Question Category	Literature Sources	Observations, Implications, or Interpretations	Frequency Counts and Codes	Percentage Calculation
Collusion	CIOB (2017), Jain (2011)	It is highlighted that employees do not know the process of acquiring tenders, and understanding the ethics involved and the professionalism required.	DA=2Nr * value 4 = 8 D=2Nr * value 3 = 6 N=2Nr * value 2 = 4 I=0Nr * value 1 = 0 DI=3Nr * value 0 = 0	18/40= 45%
Workers' rights	Harris et al. (1995), Whitbeck (1998)	Knowledge is adequate on employing people legally.	DA=2Nr * value 4 = 8 A=4Nr * value 3 = 12 I=0Nr * value 2 = 0 N=2Nr * value 1 = 2 DI2Nr * value 0 = 0	22/40= 55%
Conflict of interest/ corruption	Transparency International (2015)	Complexity of interests and ethics are perceived during tendering of jobs.	DA=0Nr * value 4 = 0 A=4Nr * value 3 = 12 N=4Nr * value 2 = 8 I=0Nr * value 1 =0 DI=1Nr * value 0= 0	20/40= 50%
Environmental issues	Mirsky and Schaufelberger (2015), Fewings 2009	There is awareness on possible aspects of discovering wildlife on construction sites and natural ethics are found to prevail.	DA=6Nr * value 4 = 24 A=0Nr * value 3 = 0 N=1Nr * value 2 = 2 I=0Nr * value 1 = 0 DI=2Nr * value 0 = 0	26/40= 65%
Environmental issues/ insubordination	Jain (2011), Fewings (2009), Martin and Schinzinger (1996)	Insubordination is linked to job dismissal.	DA=2Nr * value 4 = 8 A=1Nr * value 3 = 3 N=0Nr * value 2 = 0 I=1Nr * value 1 = 1 DI=5Nr * value 0 = 0	12/40= 30%
Sustainability	Mirsky and Schaufelberger (2015), Fewings (2009)	Human instincts of preservation are linked to construction activities.	DA=2Nr * value 4 = 8 A=4Nr * value 3 = 12 N=1Nr * value 2 = 2 I=1Nr * value 1 = 1 DI=2Nr * value 0 = 0	23/40= 58%
Quality	Pressman (1997)	A recent catastrophe has enlightened how important quality materials are on construction sites.	DA=6Nr * value 4 = 24 A=2Nr * value 3 = 6 N=0Nr * value 2 = 0 I=0Nr * value 1 = 0 DI=2Nr * value 0 = 0	30/40=75%

(Continued)

Table 12.13 (Continued)

Question Category	Literature Sources	Observations, Implications, or Interpretations	Frequency Counts and Codes	Percentage Calculation
Nepotism	Harris et al. (1995)	Favouritism is discerned as an issue on sites; however, it is anticipated that managers know the moral thing to do.	DA=5Nr * value 4 = 20 A=1Nr * value 3 = 3 N=0Nr * value 2 = 0 I=0Nr * value 1 = 0 DI=4Nr * value 0 = 0	23/40= 58%
Bribery	Almeder and Humber (1983), Johnson (1991), Whitbeck (1998)	Whistle-blowing is still considered as an unthinkable act. Employees are timid to highlight unethical practices.	DA=3Nr * value 4 = 12 A=0Nr * value 3 = 0 N=0Nr * value 2 = 0 I=0Nr * value 1 = 0 DI=7Nr * value 0 = 0	12/40= 30%

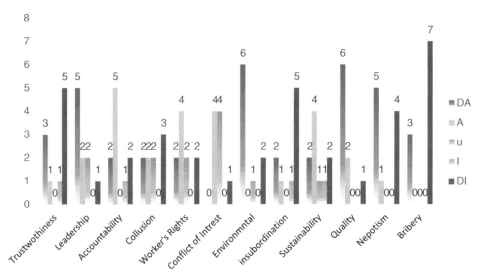

Figure 12.1 Data from frequency counts from the different categories of Industry Appropriateness.

12.3.7 Overview of the Research Findings

The collection of this qualitative data, according to Farrell et al. (2016), has attempted to give an insight into the world of others. This chapter will assimilate the ethical approaches, objectives and analysis from the derived interviews.

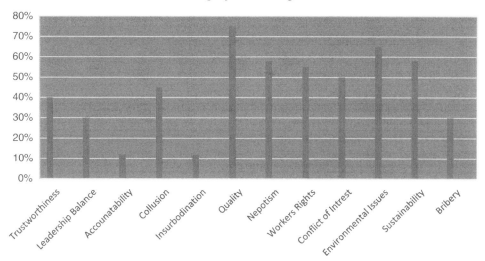

Figure 12.2 Percentage of Industry Appropriateness calculated from the tables.

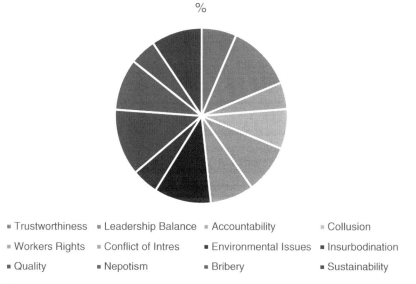

- Trustworthiness
- Leadership Balance
- Accountability
- Collusion
- Workers Rights
- Conflict of Intres
- Environmental Issues
- Insurbodination
- Quality
- Nepotism
- Bribery
- Sustainability

Figure 12.3 Pie Chart illustrating the percentage of Industry Appropriateness calculated from the tables above.

1) Analysis of construction ethics.

In the past, ethical misconduct has led to the waste of 'todays' society resources and has caused significant harm to many individuals. It is also affirmed that at the extreme end of the spectrum, in the UK, construction corporate ethical failures have led to severe injuries and death of employees and the public. This fact can be attached to the

Utilitarian Approach discussed by Mirsky and Schaufelberger (2015). This approach placed peoples' lives in danger as it is concerned with the consequences of the action or decision and the direct or indirect impact of everyone involved. In this approach, it was acceptable for a few people to suffer negative consequences, if it benefited greater number of people. Analysing this approach, it emphasises that ethical perspectives of the construction industry were then ignored because what was of paramount importance was the end-product regardless of how it was obtained.

2) How Governing Bodies help shape the image of the industry.
   a) *CIOB*: This professional body is important because it implements the codes of conduct which are then required to be used by its registered members. The CIOB institute is governed and regulated by the Royal Charter and By-laws which were approved by the Privy Council on the 23rd of February 2015. These rules are directed to institution members and the members are bound to further, to the best of their ability, the objects, interests, and influence of the institute. The rules stipulated by the CIOB include:
      i) Members to take full regard to the public interest whilst professionally undertaking their duties and responsibilities.
      ii) Members to demonstrate a level of competence and consistence with their class of membership.
      iii) Act with integrity always.
   b) *RIBA*: It is part of the construction industry structure that governs Architects. Comparable to CIOB they have three main principles namely:
      i) Integrity
      ii) Competence
      iii) Relationship

The Royal Institute of British Architects expects its members to act with impartiality, responsibly and being always truthful. It is of paramount importance that members are not improperly influenced either by their own or others self-interest. Their members should avoid conflict of interest, and if a conflict arises, they should declare it to those parties affected and either remove its cause or withdraw from that situation

   c) *Institute of Civil Engineers: ICE 2017*
      This professional body governs Civil Engineers, who are part of the construction industry. According to ICE (2017) their rules of professional conduct implemented incorporates:
      - That members must discharge their professional duties with integrity, behave with integrity and uphold the reputation and dignity of the institution and the profession of civil engineers.
      - Members to undertake work they are competent to do.
      - Members to have full regard for the public interest particularly in relation to matters of health and safety and in relation to the well-being of the future generations.
      - Members to show due regard for the environment and for the sustainable management of resources.

12.3.8 Analysis of Findings for Discussion Purposes

The research supports the view that professional bodies have an important role on the ethical structures of the construction industry and uphold the image of the industry which can eradicate the emergence of unethical practices. Nevertheless, analysing the Mean (51%) obtained after calculating the 'category percentages' from Table 15.13, it can be suggested that a lot still needs to be done as 51% is arithmetically considered as an average score.

Below are the sub-headings as per the interview questions:

i) Trustworthiness.

According to a document produced by the Construction Industry Institute (CII,2016), trust is believed to be a factor in reducing project costs, while the lack of trust in business is believed to be related to inefficiencies and increased project costs. It is further confirmed by CII (2016) that mutual trust in the construction industry is essential in prompting co-operative relationships, effective project teamwork. Referring to the interview question which was responded to for the purpose of this research, the questions asked were aimed at finding out what most contractors will do if a client overpays them by mistake. There was a mixture of responses, with some suggesting that the contractor can keep the money until discovered and some stating that it is important to disclose the mistake as failure to do so will result in bad business for that contractor. The research supports the premise that if a payment is overpaid in error, the recipient should return it back as soon as possible because ethically speaking, it is the right thing to do. Being trustworthy brings in more business for the contractor and being untrustworthy drives out business. Separate from the monetary issue, the presence or absence of trust within project teams has been highlighted in both the Latham (1994) and Egan (1998) reports as a major factor leading to the success or failure of construction projects. On the outcome of the interview question asked on trustworthiness, the percentage obtained was 40% (Inappropriate) which, arithmetically is a low score as it is under 50% thus indicating that participants judge that people in the construction industry would not act in the ethical expected manner when overpaid. The researcher analysed the responses on a large construction scale and determined that people act on things because of consequences that might arise and not on the moral principle of doing the 'right' thing. As a result of this, if a workplace does have trust issues it will have a negative impact on the quality and budget of the building projects. It is understood that people come from different backgrounds hence different values, but when it comes to trust, it must be linked to personal moral campus which must then be guided by an organisations code of conduct of which all employees should abide to. The Virtue Approach concluded by Mirsky and Schaufelberger (2015) should be implemented in all companies' ethical framework. Rowden (2010) further analyses this approach and determines that it arises and demonstrates moral virtues which include honesty, integrity, respect, trust and fairness. In contrary, review a weakness of this approach by suggesting that it is difficult to change a person's perspective, culture, belief etc. a fact which is sustained by Garret (2011), who advise that 'some cultures are taught not to trust anyone other than people whom they are close to'.

ii) Leadership balance

Leadership balance portrays key attributes of any successful business ventures. Clough and Sears (1994) argued that construction is a widely diverse industry brimming with innovation and change. Its importance and impact is substantial and critically important to the economic viability of a nation (Clough and Sears 1994). Willouby (2004) acknowledged that due to rapidly changing project load, workforce and corporate structure, the modern construction entity has experienced change over decades which included research and analysis with an attempt to integrate different technologies into construction. In the mentioned attempts, Clough and Sear (1994) argue that project management and leadership skills management have been neglected. On this note the researcher based the interview question on how a site manager must conduct a business day having had a bad morning at home. The respondents to this question had different opinions but all stimulating at the same direction. A leader must separate personal life and professional life. This suggestion concurs with Butter and Chinowsky (2006) who state that effective leaders should have a high degree of emotional intelligence. The research supports the prognosis that a leader must be able to separate personal life and professional life. However, Chan et al. (2004), emphasise that the theory of separating personal and work life does not fully incorporate the intensity of the problem the leader might be exposed to. It is further stated that some problems will require a leader to take time off and sometimes go for counselling. The score obtained under this category was 75% (Appropriate) which is a high score as it exceeds 50% approaching 100%, thus implying that the participants of this research judge that people in the construction industry are qualified enough to be able to separate their personal life and work life.

iii) Accountability/Health and safety

Accountability is arguably linked to leadership by some scholars. According to a document produced by Transparency International (2015), employees should be accountable for their decisions, and it is also added that those responsible for certain aspects during any given construction projects should also be accountable for such. The interview question asked to respondents was on, who they thought was accountable, given a scenario that an employee engaged on daily activities without the proper personal protective equipment and that employee was involved in a small accident. The site manager did not realise that the employee did not have the proper gear. Analysis of the responses from the question suggest that the supervisor must be accountable as it is his/her duty to ensure that there is no breach on the health and safety aspect on site. One respondent suggested that it is the employee's responsibility. A score of 30% (Inappropriate) was achieved implying that the participants suggest that most of the supervisors of the construction industry would not take responsibility of their subordinates' actions. Referring to a document published by Health and Safety Executive (2006) (HSE), it is proposed that everyone controlling site work has health and safety responsibilities. It is mentioned that it incorporates checking that working conditions are healthy and safe before work begins and ensuring that the proposed work is not going to put others at risk. On a revised vision of the document HSE, responsibility is placed on the employer, but emphasis is given on individuals to be aware of their responsibilities and the employer's responsibility. Analysing the two documents and the responses from the interview, the researcher propounds that although the superior is responsible, individuals also need to adhere to their company's health and safety guidelines.

iv) Collusion

This issue normally takes place during the bidding process of any construction project. According to T I (2015), collusive bidding refers to agreements by contractors to cooperate and defeat the competitive bidding process to inflate prices to artificially high levels. Some scholars are of the belief that collusion often involves corruption and bribery. A scenario given on the interview question, suggested if two friends who work for different competing companies could discuss details of a tender, they were both bidding for. Respondents again had mixed views. Some believed it was of no consequence if they discussed and some concluded that discussion was not a problem but should not discuss in detail. One respondent stipulated it was against the code of conduct of many companies as employees who work on tenders are required to sign some sort of declaration that prohibits them from disclosing any information. From this interview, the researcher believes that many construction operatives have no knowledge on how tenders are awarded, and the complete process involved.

v) Environment issues

The Common Good Approach which is explained by Mirsky and Schaufelberger (2015) as an ethical approach that promotes or contributes to the common good of a society. This explanation concurs with Bowen et al. (2007) who attest that 'there is a huge assumption that there are certain services and resources either natural or human made, whose existence benefit our common good as society.' This shows that the environment is of paramount importance within any society. According to a document published by Sourceable (2017), construction firms should design, install and maintain erosion controls to minimise the discharge of pollutants. Transparency International (2015) highlights the importance of not discharging certain components everywhere. Wastewater from washout of concrete should be managed by appropriate control. This also includes the discharge of used oils, curing components etc. Fewings (2009) authenticates that there are numerous sources of water pollution on building sites including paints, solvents, diesel, and other fossil fuels. It is further argued that even minor chemical spills leech into the ground and may enter waterways where they toxify water and harm aquatic life. For this research document two environmental scenarios were asked to respondents:

a) Scenario given was of natural habitats discovered on a new site: most respondents suggested that relevant authorities were to be called and take over the welfare of the habitats and one respondent suggested that the habitat be destroyed as it is time consuming to call authorities and wait on their action. Analysing this scenario, the researcher can ascertain that most people in the industry are aware of their company's codes of ethics when it relates to the natural habitats. It can be said that the respondent who suggested that the habitats be destroyed might either be misinformed or has seen it being done in past experiences.

b) Scenario two was in relation to dumping used oil in the stream nearby. Most respondents believe that most employees will comply with their supervisors for fear of victimisation. In the opinion of the researcher, having analysed the responses, it can be argued that most employees on construction sites cannot challenge certain aspects they are asked to do by their superiors a view which the researcher finds disturbing. On an ethical perspective, employees should be able to distinguish between what is good/legal and what is bad/illegal.

vi) Quality

According to Wong (2012), the construction industry is often criticised for its poor performance on quality, cost and safety. Clough and Sears (1994) acquiesce with Wong (2012) but further articulate that the industry has numerous parties, each of which has a role to play in ensuring the quality of the project. Poor performance of one party will negatively impact the whole project. A serious weakness of this argument, according to Fewings (2009), is that it is not due to the performance of different specialists (builders, internal walling, roofers etc.) that determines the quality of a project, but it is the balance between cost and time. It is further argued by Fewings (2009) that it is possible to have high-quality and low-cost buildings at the expense of time and conversely to have high quality and a fast project at a high cost. For this research, a scenario was given to respondents to find out if site managers would use sub-standard materials to build a project. Most of the respondents cited that most site managers will not continue but seek to address the issue. A score of 75% was achieved indicating that participants now believe employees in the construction industry will act Appropriately in challenging ethical situations of using substandard material. The researcher is of the view that due to the unfortunate incident that befell the Grenfell tower in London most people including all the respondents are now knowledgeable that construction jobs need to be undertaken using the proper material as preliminary investigations on the Grenfell Tower suggests that poor-quality cladding on the tower helped accelerate the rate of the fire spreading on the building.

vii) Nepotism

For this research document, a scenario given to respondents highlighted a supervisor in charge of a team where there was a relation to the supervisor's spouse. The respondents to this interview gave different insights. Some of the respondents suggested that there will be favouritism, and the others stipulated that the supervisor should maintain professionalism as implied by the Justice Approach. A score of 58% was obtained indicating that the respondents of the interview believe that people working in the construction industry still on average capacity in challenging ethical situations linked to nepotism.

Mirsky and Schaufelberger (2015) specified that the Justice Approach is an ethical approach that treats everyone involved in the industry fairly and in accordance with what there are due. According to a document published by Construction Week Online written by Kumar (2013) nepotism is portrayed as a controversial subject. Fewings (2009) regards the issue of nepotism to have adverse effects in organisations whereas Kumar (2013) views nepotism as natural in family or small medium businesses. The researcher believes that Kumar (2013) views will put the Justice Approach in jeopardy because the moment an individual on the same platform as others is treated unfairly or favourably then there is an act of unethical practice. It can be argued that nepotism is strongly linked to values and culture. Some cultures strongly respect their relations and do not know how to separate between relationships and professionalism.

viii) Bribery

For this thesis, the researcher gave a scenario of a junior manager witnessing a senior manager accepting a bribe.

A score of 30% (Inappropriate) was obtained by the Researcher having analysed the responses from the participants. This is a low score suggesting that participants judge that people in the construction industry would act Definitely Inappropriate when challenged with ethical situations linked to bribery.

Bribery is strongly linked to corruption, and collusion. The construction industry is often accused of fraudulent and corrupt practices. Wessing (2017), cited in Lexology. com, further states that in 2013 the Charted Institution of Building Report, which was exploring corruption in the UK construction industry, showed that 49% of respondents believed that bribery in UK was common. Fewings (2009) concurs with the report and further suggests that risks of bribery arise from the industry's tendering processes, associated licenses and permission, interaction with government officials, the complex relationships between contractors and subcontractors etc. Analysing the issue of bribery and the scenario given before of subordinates being asked to dump a drum of used oil in a stream the researcher has discovered a trend from the respondents which suggests that in the construction industry people are afraid of confronting their superiors even if they are engaging in unethical elements during the business day. It can also be said that the issue of whistle-blowing is considered to be of paramount concern, thus prompting employees to give a blind eye on unethical activities. On the UK perspective, according to a report 'Understanding the value of Professionals and Professional bribes' which surveyed employees in the construction industry, 90% acknowledged the fact that despite traditional associations of professional bodies, they have their central activity of sharing and disseminating information on how to make things better in the industry. The report finds that professional bodies in the UK offer significant value to society and their membership in areas that top current social and potential agendas. This is not the case with Zimbabwe, referring to an article published by Mafaro.com. (2017), which discussed that the professional bodies in Zimbabwe were just ceremonial bodies, they do not disseminate information that helps the industry and the bodies do not act if members of the ruling government interfere with the construction economy. As of the United States of America progress has been observed and the criminalisation of certain unethical practices in the States has realised a huge drop in the acts of unethical practices. Having mentioned that, it is important to understand how values impact people in the professional capacity. According to Fledderman (1999), values are formed through childhood and influenced by upbringing and culture. It is further mentioned that values can be positive such as fairness, honesty, creativity etc. and can also be negative such as control aggressiveness, manipulation, greed etc. Therefore, it can be formulated those parents and guardians in the UK and USA, are changing their basic perception on values that is teaching their children on what is wrong and what is right hence the development of an ethical culture in the different industries and on the contrary; in Zimbabwe, it can be analysed that it is still a huge task to accomplish.

12.4 Conclusion and Recommendations

12.4.1 Conclusions

Professional construction bodies are relevant and important in the implementation of relevant codes of conduct, but nevertheless it must be realised that codes of ethics do not just exist in absolution. A better understanding of their roles within an industry can be attained in the context of the actual realities and equivocalness confronting such an industry, as well as the people involved in the evaluation of options and decision-making (Fewings

2009). Dainty and Murray (2009) argue the great importance of business ethics in organisations owing to its potential to transform organisations' cultures, which in turn can lead to an organisational transformation towards better stakeholder management and innovation in the ways organisations do profitable business. It can be concluded that adoption of a code of ethics by an organisation is not enough for it to be ethical. Similarly, distributing the codes of ethics to all employees does not guarantee their understanding and identification with the codes. A clear difference should be established between using codes of ethics as instruments of compliance as opposed to instruments of cultivating ethical behaviour (Fewings 2009), thus, exhilarating that ethical behaviour within an organisation calls for a dire code of ethics to be engaged in organisational territory. The areas, as structured by Doran (2012), include ethical bodies, ethical tools, ethical support procedures, and internal and external ethical usage. Although Garrett (2011) discussed in detail the areas in which organisations are perceived to be engaged with ethics, the limitation is that the account provides little information about the enablers of these attributes. Personal values on the other hand are perceived to be of importance when considering aspects of moral codes. These values are enablers that help human beings make decisions about right and wrong, thus supporting the fact that issuing codes of conducts to employees is not enough to acknowledge an organisation as ethical. However, it is also documented that criminalisation of certain unethical activities (by for example, introduction of the Bribery Act of 2010) has seen a rise in development in the ethical structure of the construction industry. Current literature has highlighted the increasing progress on acceptable ethical practice and professional attitude in all forms of businesses including the construction industry.

The objectives of this research are to:

1) Understand the ethics of the construction industry by conducting scenario-based interviews.
2) Analyse the results and measure the ethical industry appropriateness from the responses obtained.
3) Establish how professional governing bodies help to shape the image of the construction industry.

This research document used unstructured interview approach resulting in the extraction of the respondents' beliefs and values on how site manager's, supervisors etc. would behave given certain work-based scenarios. The ambiguity in this type of research can be lack of cross checking to ensure that the actions of the respondents are consistent with their reporting. Further clarification of the types of unethical conduction involved is required. An example is the use of the word 'deceit' which is classed by Doran (2012) as a generalisation within itself, so another level of clarification is required to provide detailed feedback. Another example that would entail further clarification is the concept of hospitality and bribery. This research has indicated and confirmed (backed by relevant literature) the types of ethical improprieties that exists. Also confirmed as unethical, is bullying, which has frequent occurrence on construction sites (Reviewing the responses of the interviewees). However, what is new is the emergence of clients and government bodies as contenders in the ethical states. The introduction of the Bribery Act of 2010 has been hailed as a success and is a useful tool within the construction industry. The emergence of various professional government bodies has influenced the development and improved the image of the industry. History suggests that advancement of professional ethics in the construction industry is very much dependent on the implementation and policing of the ethical guidelines and policies of both professional bodies and private organisations together with the leadership of public sector procurement agencies. The response from the interviewees suggested that even though organisations have their own ethical codes of conduct, the curbing of certain unethical issues is difficult. The researcher is of the belief that all employees regardless of professional allegiance, require a familiar understanding of ethical and professional values, of which this can be achieved through continuous training (CPD). If there is lack of professionalism and lack of ethics, even the ethically good employees will have difficulty maintaining moral standards.

12.4.2 Recommendations

Ethics would improve productivity on site, lowering project cost and improve on profits. The researcher would recommend continuous ethical training and education during the construction professional's working life. It can be argued that taking personnel off their tasks to educate them about ethics can be a costly agenda, but the researcher is of the belief that there are ways of enveloping the training during the construction development period. This can be incorporated in daily morning briefs by reminding employees on what is right and what is wrong and for those seeking to be promoted they should study and be asked ethical competency-based questions for them to proceed. Posters can also be put in employee canteens and around the site. Implementing Continuing Professional Development (CPD) schemes can help drive good ethical values within a team. According to CIOB (2017), CPDs have been increasingly receiving more attention in recent decades as construction professionals are required to update themselves with new ethical developments in the expeditiously changing built environment.

Apart from training, employees should be given a sense of belongingness within any business setting. This can, however, increase productivity and help the employees behave in positive ethical manners and can be made possible by offering out shares and profit sharing. In addition, communication channels must be open for all. Some companies operate an open-door policy which enables employees to view their concerns directly with senior managers, if need be, whereas some companies use the hierarchy system of which the researcher believes to be outdated.

The researcher would recommend that the government financially help the professional bodies, as it is documented that there have too little resources.

References

Almeder, R. and Humber, J. (1983). *Professional Ethics*. Chicago: Prentice-Hall.
Andrews, K. (2014). *Ethics in Practice*. London: Longman Press.
Applebaum, D. and Lawton, S. (1990). *Ethics and Professions*. Illinois: Prentice-Hall.
Arditi, D. and Chotibhongs, R. (2005). Issues in subcontracting practice. *Journal of Construction Engineering and Management* [online] 131 (8): 875–891. Available at: doi: 10.1061/(asce)0733-9364(2005)131:8(866) (Accessed 13 Jul. 2017).

Arewa, A.O. and Farrell, P. (2015). The culture of construction organisations: the epitome of institutionalised corruption. *Construction Economics and Building* 15 (3): 59–71. doi: 10/5130/AJCEB.v.1513.4619. Accessed 2 Jully. 2017.

Bbc.com (2017). *BBC*. [online] Available at: http://www.bbc.com (Accessed 13 July. 2017).

Bowden, P. (2010). Teaching ethics to engineers. *Journal of Engineering Principles* [online] 14 (8): 124–140. Available at: http://dx.doi.org/10.1016/s0899-5362(98)00041-4 (Accessed 13 July. 2017).

Bowen, T., Hill, R., and Stott, L. (2007). *Construction Management and Economics*. Oxford: Taylor and Francis.

Breach, D. (2009). *Dissertation Writing for Engineers*. Kent: Sage.

Buffett, W. (2015). *The Snowball*. Omaha: Bantam Books.

Businessworld.com. (2017). *Business world*. [online] Available at: http://www.businessworld.com (Accessed 13 July. 2017).

Butler, C. and Chinowsky, P. (2006). *Leadership in the Construction Industry*. London: Taylor and Francis.

Calhoun, C. and Wolitzer, P. (2001). Ethics and the Cpa: building trust and value added services. *EDPACS* [online] 26 (11): 18–34. Available at: doi: 10.1201/1079/43239.26.11.19990 501/30207.4 (Accessed 13 July. 2017).

Chan, A.P.C., Chan, D.W.M., Chiang, Y. et al. (2004). Exploring critical success factors for partnering in construction projects. *Journal of Construction Engineering and Management* 130 (2): 188–189.

CIOB (2017). *CIOB*. [online] Available at: http://www.ciob.org (Accessed 13 Jul. 2017).

Clough, R. and Sears, G. (1994). *Construction Project Management*. London: John Wiley and Sons.

Cohen, S. and Grace, D. (1998). *Business Ethics: Australian Problems and Cases*. Sydney: Skirball Cultural Centre.

Construction-Institute (2016). *Construction-institute*. [online] Available at: http://www.construction-institute.org (Accessed 13 July. 2017).

Dainty, A. and Murray, M. (2009). *Corporate Social Responsibility in the Construction Industry*. London: Taylor and Francis.

Delbridge, A. (2000). *The Macquarie Dictionary*, 3e. Sydney: Macquarie Library.

Dictionary.com (2017). *Dictionary.com - the world's favourite online dictionary!* [online] Available at: http://www.dictionary.com (Accessed 13 July. 2017).

Doran, R. (2012). Fundamentals of ethics. *Business Knowledge* [online] 11 (4): 122–126. Available at: doi: 10.2307/854310 (Accessed 13 July. 2017).

Egan, J. (1998). Rethinking construction. *The Report of the Construction Task Force*. London: DETR.TSO. 18–20.

Farrell, P., Richardson, A., and Sherratt, F. (2016). *Writing Built Environment Dissertations and Projects: Practical Guidance and Answers*, 2e. Chichester: Wiley Blackwell.

Ferguson, W. (1994). Building ethical foundation in business. *The Executive Speaker* [online] 95 (65): 463–510. Available at: doi: 10.2307/949473 (Accessed 13 July. 2017).

Fewings, P. (2009). *Ethics for the Built Environment*. London [Angleterre]: Taylor & Francis.

Fledderman, C. (1999). *Engineering Ethics*. New Jersey: Prentice-Hall.

Fryer, B. (1997). *The Practice of Construction Management*. Oxford: Blackwell Science.

FutureLearn (2017). *Future learn*. [online] Available at: http://www.futurelearn.com (Accessed 13 July. 2017).

Garrett, J. (2011). The basics of ethics. *Construction Morals* [online] 18 (7): 86–92. Available at: doi: 10.1136/bmj.326.7400.1217 (Accessed 13 Jul. 2017).

Goldstock, R. (2014). *Corruption and Racketeering in the New York City Construction*. New York: Routledge.

Grant Thornton, U.K.L.L.P. (2013). *Grant thornton UK LLP*. [online] Available at: http://www.grantthornton.co.uk (Accessed 13 June. 2017).

Harriet, G., Gregory, P., and Jones, J. (1995). *Modern Business Management*. London: Blackwell publishing.

Harris, C., Pritchard, M., and Robins, M. (1995). *Engineering Ethics*. Michigan: Wadsworth Publishing Company.

Hse.gov.uk (2006). *HSE: information about health and safety at work*. [online] Available at: http://www.hse.gov.uk (Accessed 13 July. 2017).

Institution of Civil Engineers (ICE) (2017). *ICE: the home of civil engineering*. [online] Available at: http://www.ice.org.uk (Accessed 13 July. 2017).

Jain, A. (2011). *Corruption: A Review*. New Delhi: Management Pub. Co.

Johnson, D. (1991). *Ethical Facts in Morden Engineering*. Windsor, Berkshire: BDA.

Kumar, V. (2013). *Nepotism in construction* [online] construction week. Available at: http://www.constructionweek.com (Accessed 13 July. 2017).

Latham, M. (1994). *Constructing the Team*. London: The Stationery Office.

Legislation.gov.uk. (2017). *Legislation.gov.uk*. [online] Available at: http://www.legislation.gov.uk (Accessed 13 July. 2017).

Luder, O. (2006). *A Guide to Keeping Out of Trouble*, 3e. London: RIBA Publications.

Mafaro.com. (2017). *mafaro.com*. [online] Available at: http://www.mafaro.com (Accessed 10 May. 2017).

Martin, M. and Schinzinger, R. (1996). *Ethics in Engineering*, 3e. New York, NY: McGraw-Hill.

Mason, R. (1998). Ethics: a professional concern. *American Society of Civil Engineers* [online] 68 (12): 28–54. Available at: doi: 10.1002/chin.198224045 (Accessed 13 July. 2017).

McDowell, B. (1991). *Ethical Conduct and Professionalism*. New York: Quorum Books.

Mirsky, R. and Schaufelberger, J. (2015). *Professional Ethics for the Construction Industry*. Oxford: Prentice-Hall.

Murdock, J. and Hughes, W. (1996). *Construction Contracts, Law and Management*, 4e. London: Taylor and Francis.

Murray, M. and Dainty, A. (2009). *Corporate Social Responsibility in the Construction Industry*. London: Taylor & Francis.

Oxford Dictionaries (2017). *Oxford dictionaries | our story, products, technology, and news*. [online] Available at: http://www.oxforddictionaries.com (Accessed 13 July. 2017).

Oxford Languages (2010). *Oxford English Dictionary*, 3e. OUP Oxford.

Pilvang, C. and Sutherland, I. (2008). Environmental project management. *International Research* [online] 26 (2): 106–119. Available at: doi: 10.20314/als.696a9ca459 (Accessed 13 July. 2017).

Pressman, A. (1997). *A Compedium of Business and Management Strategies in Architecture*. Florida: John Wiley and Sons Publishing.

Schein, G. (1996). *Research Techniques*. Minneapolis, Minn: Augsburg Pub. House.

Sourceable (2017). *Environment and construction*. [online] Available at: http://www.sourceable.com (Accessed 13 July. 2017).

Sutherland, E. (1983). *White Collar Crime*. Binghampton: Yale University Press.

Transparency International (2015). *Transparency International*. [online] Available at: http://www.transparencyinternational.org.uk (Accessed 13 July. 2017).

Velasquez, M. (1998). *Business Ethics, Concepts and Cases*, 4e. New Jersey: Prentice-Hall.

Weait, M. (2001). Ethics at work. *Management Today* [online] 112 (13): 24–38. Available at: doi: 10.2139/ssrn.2128015 (Accessed 13 July. 2017).

Wessing, T. (2017). [online] Lexology. Available at: http://www.lexology.com (Accessed 13 July. 2017).

Whitbeck, C. (1998). *Ethics in Engineering Practice*. Washington, D.C.: Five and Ten Press.

Willouby, N. (2004). *Project Management in Construction*. Edmonton: NWN ltd.

Wong, A. (2012). Quality in construction. *International Journal of Environmental Research* [online] 13 (5),p.486. Available at: doi: 10.3390/ijerph13050486 (Accessed 13 July. 2017).

Yin, K.Y. (2009). *Case Study Research: Design and Methods*. Sage.

13

Implications in Practice for Organisational Ethics in the Built Environment

If ethics are poor at the top, that behaviour is copied down through the organisation.
                                                                    Robert Noyce

13.1 Introduction

The above quotation clearly underpins the importance of ethical standards and principles in society. This chapter is largely focused on considering the implication of the proceeding chapters of the book, underpinning the main themes and looking at potential recommendations for future practice. It will reiterate the importance of reputation and image for organisations in being regarded as 'ethical' or 'non ethical' in their business dealings, alongside the influence this can have on their long-term performance.

The relationship between ethics and quality for organisations will be discussed and the various initiatives and measures to improve organisational ethics debated and analysed. This will include the following areas:

- Raising environmental sustainability credentials for organisations
- Increasing corporate social responsibility (CSR)
- Ethical human resource (HR) considerations
- Ethics around construction health and safety
- Promoting trust and collaborative working
- Ethical leadership in construction and engineering management

Finally, education, training and continued professional development (CPD) around organisational ethics will be covered and an articulation of how construction professionals can learn to adopt ethical behaviours and act with professional integrity and a duty of care to their clients.

13.2 The Importance of Reputation and Image for Organisations in the Built Environment

When unethical practices and behaviour come to light and especially when they are exposed in the public domain, this can be extremely damaging not just for the individuals and organisation involved but for the reputation of the industry (Fewings 2009). Certainly, in the UK this has been a big issue for the built environment over many years. Previously published reports and high-profile press coverage have highlighted cartels, collusion on tenders, blacklisting of subcontractors and other illegal practices this has led to widescale public condemnation of the construction and engineering industries. Whilst these reports and negative press coverage provide good news for newspapers, they destroy the image of the built environment sector and present a very poor perception of it. Notwithstanding the fact that such practices may only represent a small minority of case, there is the perception that such behaviours are commonplace in the sector. To coin a phrase the construction industry and all the organisations employed within it are 'tarred by the same brush' by association. One could reach the conclusion that such widespread damage to the whole industry is unfair on those good organisations who are doing everything right to rebuild public confidence and act ethically. This has led to a poor image of the industry in the past and the perception that there are unregulated 'cowboy' builders operating widely and offering substandard services.

There are views that there are a few excellent contracting organisations providing outstanding services to their clients but many mediocre contractors who are still operating in a less than ethical way with poor working practices and behaviours. There have been reports that clients have been left with latent defects on work carried out, owing to poor quality of design, workmanship, and materials. These problems may cause them long-standing difficulties in building performance and increases in maintenance and running cost in some instances. In extreme cases where contractors have refused to rectify faults and defects, clients have had no other option than to engage other contractors at additional cost to remedy the problems. In such cases this has also led to legal proceedings which have attracted negative publicity and further damaged the reputation and image of the construction industry. Other problems that the sector has faced in the past emanate from organisations not employing policies linked to satisfactory human resource management. This can potentially lead to 'toxic' cultures and leave employees feel unsupported, exposed and at risk from negative forces within the organisation. Unfortunately, this can result in low staff morale, well-being and retention rates.

It is clear that unethical practices coupled with poor organisational cultures can take a negative toll on the performance and ultimately the outcome of construction and engineering projects. Negative consequences associated with poor outcomes are numerous and lead to assertions that the industry is inefficient, unsafe, wasteful, compromised on quality, and cannot deliver to time and budgetary constraints. This can severely impact on the reputation of the industry. There have been many ideas and opinions from clients, consultants and main contractors on the 'ways and means' through internal and external intervention mechanisms to curb such unethical practices. It is widely accepted that construction professional bodies, construction regulatory bodies and governments to adopt remedies, and such measures should always be embraced by construction professional in their respective roles.

Unethical conducts can, especially in developing countries, undermine public confidence which can in turn hinder economic development, investment with a consequential loss of finance and in some cases human lives (Inuwa et al. 2015). For these reasons it has become an important priority for governments to address such practices to curb practices and cultures around unethical practices. This has put more centrally governed focus on creating policies designed to implement and police ethical reforms, and this could be through public procurement agencies.

13.3 The Relationship between Ethics and Quality for Organisations

Quality and ethics are inextricably linked through a common care premise to do the right thing and therein a proven way to improve competitiveness, reduce costs and increase client satisfaction. Despite this, a study by Rahman et al. (2005) found that only a third of organisations surveyed had any ethical training programmes. Possibly this emphasises that those that did not have programmes are unaware of the importance of professional ethics for their staff. Notwithstanding this premise, three quarters of those organisations in the study by Rahman et al. (2005) agreed that they felt the construction industry was tainted by unethical behaviours. One of the main examples of unethical conduct that the study revealed was associated with failures during bidding processes. Reports of lowest price awards, at the expense of quality, were commonplace. More serious examples of bribery and corruption were also reported.

In terms of teaching and upholding ethical practices, Helgadottir (2007) concluded that there are theories to consider, and these can be both process and outcome oriented. In the former case, ethics are considered from behaviours and actions are respectful to the duties and rights of others. In the latter case, ethics are viewed from the perspective that one's actions should lead to the best outcome for as many people as possible. Accordingly, Helgadottir (2007) advocated that teaching students to analyse their own work from an ethical perspective is a valuable and effective exercise that students will find useful. Academic programmes should incorporate elements of ethical training and incorporate inventive ways and means to help students understand how to be competent in ethical issues. Organisations should look at their social responsibly in the context of managing projects.

13.4 Initiatives and Measures to Improve Organisational Ethics

The earlier chapters of the book have identified various ways that organisations can improve ethics within their respective businesses. These have been reflected upon and recommendations for the future such measures have been captured below as a model of good practice:

13.4.1 Raising Environmental Sustainability Credentials for Organisations

Organisations in the engineering and construction industries should embrace all aspects of environmental sustainability and make commitments to reduce waste, lower energy consumption, develop carbon management plan and adopt sustainable transport policies. When procuring projects, there should be a growing awareness of the importance of environmental ethics, as part of their organisations' social responsibilities. In this sense all responsible organisations should uphold policies, strategies and values linked to corporate social responsibility. This in turn will have far reaching benefits for their businesses in terms of public perception, marketing, operational efficiency, adherence to ever increasing regulations and maximising future government funding opportunities. Other initiatives may include devising and implementing credible recycling programmes and deploying state of the art technologies to reduce energy and water consumption. Use of renewable sources of heat and power are particularly appropriate especially where grant funding is available and having a minimum standard of BREEAM for all new buildings commissioned.

Reduction in single use plastic waste is particularly important in this regard. Organisations that have shown responsibility in 'doing their bit' for the sustainability and environmental agenda have benefitted from positive media coverage and gaining environmental accreditation. Such accreditation could include ISO 14001 Certification, Environmental Management Systems and ISO 50001 Certification, and Energy Management.

Organisations should also look at their supply chains and have influence on them for responsible procurement of supplies and services. In addition, adopting sustainable transport policies can have many beneficial outcomes for businesses in reducing carbon emissions and can be linked to cycling schemes, car sharing and subsidies for using public transport. All the above measures, if implemented, should benefit organisations when they are tendering for works by being able to demonstrate that they are responsible and aware of the environmental sustainability agenda. This is particularly the case in the public sector where bid evaluation criteria linked to appointments of consultants and contractors are more frequently focusing on environmental sustainability aspects.

13.4.2 Increasing Corporate Social Responsibility (CSR)

Organisations should be considering what corporate social responsibility (CSR) means in practice and how CSR philosophies can be embedded into organisational cultures, policies and procedures. In this pursuit they should focus on the legal, economic, philanthropic and volunteering perspectives as previously articulated in Chapter 4. Furthermore, they should consider implementing some of the initiatives such as joining the Considerate Constructors Scheme in helping them to improve their ethical credentials. The principles of social value and the different types of intervention that companies can make can be measured through cost-benefit analysis across projects and programmes. For such initiatives to work well, corporate responsibility should be robustly embedded into their policies, culture and management approaches. This can allow them to introduce and deliver successful outcomes, not only for their own purposes, but for communities and the public at large. Organisations in undertaking such commitments should be balanced against the reputational benefits and the positive public relations that could be generated.

13.4.3 Ethical Human Resource (HR) Considerations

Ethical human resource best practices should be deployed to enable organisations to uphold policies and strategies that cover aspects including inclusivity, equality and diversity. On reflection of how the construction and engineering industries can improve diversity and inclusion, Chapter 6 of book suggested many ways that organisations can deter breaches and bad practices. This chapter also offered solutions to how organisations can improve the culture of their workplaces through a toolkit for change. Management policies and protocols should also be in place for dealing with complaints and potential breaches and violations.

Organisations should be focused on maintaining cohesive, inclusive and safe working environments for the creation of a truly inclusive environment. Accordingly human relations management should seek to build and maintain respect and trust between disparate groups and individuals. Raising awareness of the different perspectives of people is fundamental and the building blocks for the development and maintenance of good relations. Furthermore, organisational policies and strategies should be linked to complying with the Equality Act 2010 and therein reducing discrimination on race, gender, age, disability in the workplace and avoiding the risk of any breaches. Other HR initiatives that organisation should consider include:

- Allowing freedom of speech and open dialogue
- Eliminating victimisation, harassment and bullying
- Introducing whistle-blowing processes
- Developing and maintaining a culture which makes diversity and inclusion a reality
- Deploying human relations policies for their own organisations and the companies that make up their supply chains to avoid cases where individuals will be exposed to modern slavery practices
- Improving professional ethics related to data protection

13.4.4 Ethics around Construction Health and Safety

Health and safety in the UK's construction and engineering industries have improved greatly over the last decades. Notwithstanding this premise the fatal injury rate is 1.62 per 100,000 workers which is around 4 times the all-industries rate. According to the HSE in 2020–2021 142 workers were killed on construction sites in the UK, and accordingly there is still room for improvement in health and safety management. Those ethically motivated construction organisations that are striving to reduce injuries and fatalities should consider not simply complying with health and safety legislation but exceeding regulations. In practice they can achieve this by introducing enhanced safety measures and initiatives on their construction sites. Introducing near miss reporting with essential improvement measures for each reportable incident is one example of such an initiative and has proved very effective in the past in reducing hazards on sites. Other initiatives that companies could adopt include becoming members of the Construction Skills Certification Scheme and ensuring that all construction staff are properly trained and gained accredited to the scheme.

13.4.5 Promoting Trust and Collaborative Working

Organisations could use trust and collaborative working as the way and means to encourage improved ethical practices and behaviours whilst improving business relationships. The extent and degree of trust has also been shown to underpin relationships of key stakeholders which in turn can lead to project success (Pinto et al. 2009). The emergence of these arguments appears to affirm that trust as one of the most important factor to risk minimisation on projects. The use of collaborative procurement methods and improvements to relationships that they bring about, applying the principles of both Latham (1994) and Egan (1998, 2002), can procure more successful project outcomes. Success in this way can be measured through significant improvements in client satisfaction, cost predictability, safety and time predictability. In this context, such partnering based approaches have been presented in Chapter 8 of the book as a proven pathway to greater cooperation and trust. For this reason, organisations in the construction and engineering sectors should conduct strategies and continued professional development (CPD) linked to promoting collaboration.

13.4.6 Ethical Leadership in Construction and Engineering Management

In Chapter 9, the book has referenced current academic literature on theories relating to leadership and especially around professional ethics and applied these to the roles of construction professionals. Accordingly, successful leaders need to be good motivators and can bring out the best in people. Notwithstanding this premise, individuals becoming leaders is not always something that comes naturally. For most people leadership is a reflective process, where they learn to shape their behaviours and practices based on previous experience of what was successful or unsuccessful. In this way lessons learnt from such experiences play an important role in determining such things as ethical decision-making and their individual styles of management. This should assist them in leading by example to gain the confidence from other members of the project team and give them the assurance that ethical actions, behaviours and decision-making are being practiced. Furthermore, it is essential for aspiring leaders to have courage, conviction and assertiveness when faced with difficult choices and decisions and their behaviours and actions should always be subject to fairness, integrity and a sense of responsibility. It is essential for aspiring leaders to understand what makes a good leader, in order to perform their roles and motivate others for achieving successful outcomes.

The three-stage leadership identity development (LID), discussed in Chapter 9, could be useful as a progression process applied to construction professionals to assist the creation of ethical leaders. However, this is dependant and influenced by individual differences, cognitive capacity, personality and temperament, personal identity, personal values and emotional intelligence, driven by cultural context and personal experience.

Ethical leadership which goes beyond the notion of only satisfying minimum compliance standards. The benefits from measures that organisations can implement to go 'above and beyond' legal requirements have been articulated and discussed. Such benefits include developing the workforce and improving the service to their clients and supply chains to make them feel valued, listened to and engaged in their business. For this reason, perhaps the best companies to 'nurture' ethical leaders are those where the culture of the organisation permits an environment to learn, collaborate, communicate and freely exchange different ideas with their colleagues.

The relationship between poor ethical leadership and project failure has been examined in this book from the main areas of project initiation and planning, governance, people and technical competent. Accordingly, leaders should ensure that measures are all in place which include adopting the right project controls, communication strategies, ethical codes of conduct and collaboration approaches. Organisations could pursue education and training through continued professional development to promote such ethical initiatives and this will be covered in the next section.

13.5 Education, Training and Continued Professional Development (CPD) around Organisational Ethics

In consideration of the above reflections and recommendation for improving ethics, organisations should promote continued professional development in the workplace for the teaching and learning around ethics in practice to improve quality in construction outcomes and raise the reputation of the industry. In this way, construction professionals can learn to adopt ethical behaviours and act with professional integrity and a duty of care to their clients. Notwithstanding this premise, only when the practice of ethics is commonplace and placed centre stage within organisations will professionalism be enhanced leading to improved construction outcomes. Furthermore, parties within the construction industry should be vigilant, alert and proactive to avoid unethical behaviours and report any unethical practices of others to the relevant authorities. For this reason, the main public agencies, especially in developing countries, should adopt strict rules and regulations to enforce the values of ethical standards and outlaw those deemed by be unethical.

Notwithstanding these assertions, the learning of ethics should not cease for individuals in further or higher educations but should continue and be developed during their careers and professional working lives. Such individuals will possess different values of bad and good which in theory can result in a vast array ethical or unethical practice. This can in most cases be promoted and governed by professional institutions, via Continued Professional Development (CPD) in their long-term role in ensuring professional standards of their respective institution. In this regard, such issues as maintenance and development of skills and knowledge, awareness of ethical considerations and maintaining public status all feature very highly. This is especially important within the context and environment of the built environment where things, regulations and developments are dynamic and constantly changing. The RICS approach this by imposing mandatory requirements for its members to submit written documented proof of at least 20 hours of CPD per calendar year, with penalties and sanctions applied for noncompliance. Some would argue that a 'carrot rather than the stick approach' would prove more successful and perhaps CPD participation in such areas of ethics could be better promoted through motivational measures rather than what could be regarded as old-fashioned penalties designed to punish. The implementation of CPD training may be best achieved through workshops, conferences, mentoring, lectures, webinars and by studying a new qualification. Other ways to disseminate CPD material could be through working forums, talks, discussions, and seminars designing to endorse professional ethics.

With the right degree of education, training and CPD, construction professionals can learn to adopt ethical behaviours and act with professional integrity and a duty of care to their clients. Notwithstanding this premise, only when the practice of ethics is commonplace and placed centre stage within organisations will professionalism be enhanced leading to improved construction outcomes.

13.6 Summary

There are many implications for the construction industry from unethical practices, which will include reputational damage and image considerations for organisations. Furthermore, it is clear than unethical practice can take a negative toll on the performance and ultimately the outcome of construction projects. Negative consequences associated with poor outcomes are numerous and lead to assertions that the industry is inefficient, unsafe, wasteful, compromised on quality, and cannot deliver to time and budgetary constraints.

When unethical practices and behaviours come to light and especially in the public domain, this can be extremely damaging. In this regard, some high-profile press coverage in the past have highlighted cartels, collusion on tenders, blacklisting of subcontractors and other illegal practices. This has led to widescale public condemnation and a poor image of the construction and engineering industries and a perception that they are not sufficiently regulated. Other problems that the sector has faced in the past emanate from organisations not employing policies linked to satisfactory human resource management. This can potentially lead to 'toxic' cultures within organisations and breaches of regulations and ethical standards.

Quality and ethics are inextricably linked through a common care premise to do the right thing and therein a proven way to improve competitiveness, reduce costs and increase client satisfaction. Despite this only a third of organisations surveyed had any ethical training programmes. Professional bodies and construction regulatory bodies can adopt training programmes to remedy this educational dilemma, and in this way, measures can be embraced by their members.

When procuring projects, there should be a growing awareness of the initiatives and measures to improve organisational ethics. One important area around initiatives is linked to environmental ethics, as part of their organisations' social responsibilities. In this regard they need to consider not only their own organisation but their supply chains and have influence on them for responsible procurement of supplies and services. They should also factor in sustainable transport into their policies and encourage such initiatives as cycling, car sharing and greater use of public transport to reduce carbon generation. In addition, organisations should be considering what corporate social responsibility (CSR) means in practice and how CSR philosophies can be embedded into organisational cultures, policies and procedures. Implementing measures such as joining the Considerate Constructors Scheme could help them to improve their ethical credentials in this pursuit.

Ethical human resource best practices should be deployed to enable organisations to uphold policies and strategies covers aspects including inclusivity, equality and diversity. Management policies and protocols should also be in place for dealing with complaints and potential breaches and violations. Organisational policies and strategies should be linked to complying with the Equality Act 2010 and therein reducing discrimination on race, gender, age, disability in the workplace and avoiding the risk of any breaches.

Ethically motivated construction organisations that strive to reduce injuries and fatalities should consider not simply complying with health and safety legislation but exceeding regulations wherever possible. Furthermore, companies could use trust building mechanisms and collaborative working as the way and means to encourage improved ethical practices and behaviours whilst improving business relationships. The use of collaborative procurement methods and improvements to relationships that they bring about, applying the principles of both Latham (1994) and Egan (1998, 2002), can procure more successful project outcomes.

Ethical leadership in construction and engineering management is another important factor for organisations to uphold. Accordingly, they should ensure that certain management measures are all in place which include adopting the right project controls, communication strategies, ethical codes of conduct and collaboration approaches.

It is important to promote teaching and learning around ethics in professional practice. For this reason, organisations should promote continued professional development in the workplace which could be via seminars, workshops or guest speakers. Investing in this type of training will ensure that staff are kept up to date on latest developments and changes in regulations. In addition, it has been suggested that colleges and universities should tailor more academic modules and courses to the subject of professional ethics to address this deficiency in knowledge and education. With the right degree of education, training and CPD, construction professionals can learn to adopt ethical behaviours and act with professional integrity and maintain a duty of care to their clients.

References

Abdul-Rahman, H., Rahim, M., Hamid, M., and Zakaria, N. (2005). Beyond basic: the potential roles and involvement of the QS's in public Projects-an observation. In: *QS National Convention 2005: Sustaining the Profession-Towards Diversification*, 10–11 August 2005, Kula Lumpur, Malaysia:10–18.

Egan, J. (1998). *Rethinking Construction. The Report of the Construction Task Force*. London: DETR.TSO. 18-20.

Egan, J. (2002). *Accelerating Change. Rethinking Construction. Strategic Forum for Construction*. London.

Fewings, P. (2009). *Ethics for the Built Environment*. London: Routledge.

Helgadottir, H. (2007). The ethical dimension of dimension of project management. *International Journal of Project Management* 26 (2008): 743–748.

Inuwa, I.I., Usman, N.D., and Dantong, J.S.D. (2015). The effects of unethical professional practice on construction projects performance in Nigeria. *African Journal of Applied Research (AJAR)* 1 (1): 72–88.

Latham, M. (1994). *Constructing the Team*. London: The Stationery Office.

Pinto, J.K., Slevin, D.P., and English, B. (2009). Trust in projects: an empirical assessment of owner/contractor relationships. *International Journal of Project Management* 27 (6): 638–648.

14

Summary of Key Points, Reflections, Overview and Closing Remarks

> *There is no such thing as business ethics. In an organisational context you have to adhere to the highest standards.*
>
> Marvin Bower

14.1 Introduction

This final section will summarise each chapter of the book and extrapolate the key findings and issues raised. Following on from what has been articulated and discussed, it will present some reflections and recommendations for the future of organisational ethics in the built environment, taking account of the inherent dilemmas and challenges that are faced.

14.2 Summary of the Key Issues Raised throughout the Book

14.2.1 An Introduction to Organisational Ethics

Within an organisational context it is very important to understand why ethics are relevant and important. To help us in this pursuit there is a consensus that ethical principles and practices within organisations are largely built upon behaviours, trust and values, in the work environment. Notwithstanding this premise, such qualities have been compromised in the past within the built environment. Examples of bad practices which have emerged and been widely publicised have brought with them reputational damage not just for those organisations involved but the built environment sector at large. To address this dilemma, education, personal development and teaching in the construction and engineering sectors has sought to turn the tide around addressing unethical practices and hopefully promote policies to embed a minimum standard of ethical compliance for the future.

Ethics especially when considered from an organisational perspective can create the bedrock on which all our working relationships are built. When reviewing the importance and relevance of ethics within organisations we should consider the decision-making processes and there are three basic areas of managerial decision-making. The first is around choices about the law and whether to follow, the second relates to choices about economics and social issues outside of law and the third is the priority of self over the company's interests.

In this chapter a significant relationship has been established between not upholding ethical standards, around such aspects as employment relations, and low productivity. This has underpinned why employment relations in any working organisation is important for productivity, stability, profitability and the well-being of all employees and stakeholders. Employment relations in this regard were found to be influenced by such factors as proper management of employees at work including human resources (HR) policies and procedures, adherence to ethical values, standards and principles, teamwork and collaborative working approaches. In addition, there are many other ethical behaviour factors at an organisational level which have improved construction productivity, and these include honesty, effective communication, reliability, fairness, transparency and training/education.

Organisational culture has over many years been recognised as a positive construct in organisational science and in this sense could be regarded as the way individuals behave and approach their roles and responsibilities. For this reason, it has become the important focus for the way organisations are run and perform against their aims and ambitions. Accordingly, there should be a concerted effort to encourage organisations to focus more on improvements to organisational culture as part of their human resources management strategies. There has been industry encouragement to pursue such policies and strategies and accreditation schemes have been introduced to support these initiatives. The UK's Best Companies is one example of such an initiative and is predicated on staff surveys to score their employers on how they measure up against predetermined success criteria and metrics. The importance of culture within an organisational context can possibly explain why those more responsible employers have sought to invest time and resources as part of their human relations policies.

14.2.2 Environmental Ethics

In understanding what sustainability is, there have been many different definitions over previous years. For this reason, sustainability could mean different things to many people. This considers the wide-ranging contexts and environments which construction management operates within. One more prolific definition is that environmental ethics concerns formulation and moral obligations regarding the environment.

The UK Government has committed to delivering significant reductions in greenhouse gas emissions through net zero carbon targets for the future years ahead. In doing so this will strongly influence the way buildings are designed and constructed but also used throughout their life cycle. Accordingly, if the Government pledge and commitments are to be achieved, buildings must become significantly more efficient, not just in the construction process but in terms of energy usage during their operation. For this reason, construction professionals in their respective leadership roles, when procuring projects, should be aware of the growing importance of environmental ethics, as part of their organisations' social responsibilities. Accordingly, there is a strong argument that construction professionals should be increasing political influence in this area, and therein drive the agenda forward. This could take the form of instructing their project teams to adopt environmental and sustainable design measures into their projects as one example.

The role of environmental ethics is to identify the moral obligations in terms of concerns of effects on the atmosphere, climate change, depletion of natural resources, loss of wilderness, dwindling animal and plant biodiversity. Given the increasing concern for environmental issues and the impact that our actions will have on the environments that we live in, the field of environmental ethics will become more important in the future. Political efforts will seek to limit or in some cases reverse the environmental effects of certain delirious actions that have caused harm to the environment. In this regard, it is important for all responsible organisations to uphold a sense of corporate social responsibility. This in turn will have far reaching benefits for their businesses in terms of public perception, marketing, operational efficiency, adherence to ever increasing regulations and maximising future government funding opportunities. However, the global and national agenda on sustainability has in the past had little direct influence on most organisations. Notwithstanding this premise, Government in recent years has contributed to encouraging environmental ethics into the design and construction of projects in the UK. One example of a UK Government initiative in this regard is the introduction by the Department of Trade and Industry (DTI) for Sustainable Progress Reports. This covers principles such as waste, energy efficiency, carbon generation, safety, skills, climate change and costs which have been used in the past for measuring how sustainable projects are. Furthermore, in looking at local sustainability agenda in the UK, and in particular local authority policies, there is strong emphasis on the participation of organisations to encourage reduction of waste, increased recycling, and the adoption of sustainable transport policies. In some cases, county councils have previously had to rely on enforcement of such policies through the planning system (Town and Country Planning Act 1990) for major developments. In addition, it is important and productive for organisations to work more closely with local authorities in partnership, through liaison and consultation, to promote a more robust sustainability strategy for their local area. This would improve the external relations and the public perception of businesses.

Environmental social responsibility is referred to as environmental conservation social responsibility, and curtails to the long-term problem of global warming, increased waste production and contamination on an international scale. Such has been the prominence and exposure of environmental problems, and global climate change that it has become increasing important for organisations to align themselves in concerted efforts to help the situation. The questions that have been posed for business organisations across the world are whether they have a credible recycling programme in place, and have they programmes and technologies in place for reducing energy consumption. This could take the form of creation of cleaner energy creation through renewable technology such as solar power through photovoltaic technologies and wind power through wind turbines. There are many 'greening initiatives' that companies can adopt which could make a difference and some of which carry funding opportunities. In the UK construction industry, there is much research being conducted in reducing embodied carbon as part of the construction processes. This has called for innovative solutions and challenges to more tradition construction methods and building materials such as concrete and steel which have been commonly associated in being carbon intensive in their production. Notwithstanding the global push for 'carbon friendly' buildings, the promotion and commitment to environmental and sustainability building technologies can sometimes create ethical dilemmas for designers especially. Such dilemmas have revolved around short-term client focus on reducing capital costs at the expense of longer-term outcomes. Accordingly, consultants should be articulating the implications and disadvantages of ignoring the climate emergency and how this could specifically be regarded as irresponsible, leading possibly to reputational damage for organisations. In this way clients could then be persuaded to adopt environmentally sustainable policies and not simply discount them on the grounds of affordability in the short term. This drive to increase client awareness is important as they may not be fully aware of the longer-term implications of not pursuing a policy of environmental sustainability and simply opting for solutions that generate the lowest capital costs in the construction phase.

Two case studies have been articulated and discussed in this chapter relating to environmental sustainability plans and strategies which have been adopted by both a UK college and university. Measures undertaken by these further and higher education establishments have included the introduction of effective environmental management systems, waste reduction initiatives, sustainable and ethical procurement processes, sustainable construction and engineering designs and technologies, community engagement and education around environmental issues. Successful outcomes on delivering environmental sustainability in these case studies have included reduction in energy and water consumption, the award of many environmental accreditations, improvements to travel and transport policies, reduction in waste in carbon generation and waste, decrease in the use of single use plastics, and the introduction of more robust environmentally sustainable procurement policies.

14.2.3 Corporate Social Responsibility

Corporate social responsibility (CSR) has become a highly topical area in recent years and can be defined as a corporation's obligation to its stakeholders including suppliers, employees, customers and those communities around the business. CSR focuses on optimising the positive impacts on these stakeholders whilst reducing the negative impacts. On this basis, there is a growing requirement that businesses must be responsible from a social perspective rather than being purely being focused on financial success and achieving targeted profit levels for their shareholders. Accordingly, for this reason, in today's economy it is imperative for organisations to embed corporate responsibility into their policies, culture and management approaches, and therein bringing reputational benefits from enhancement of their brands and positive public relations. Conversely, where there are breaches of corporate social responsibilities, these can have extremely negative consequences for organisations, especially those larger businesses that rely heavily on brand and reputation. There have been many controversial global cases which have contextualised this potential dilemma. Recent examples include the Deepwater Horizon oil spill in 2010 which became the worse environmental disaster in American history with the loss of 11 lives and which caused untold reputational and financial damage for British Petroleum.

Social value can be affiliated with corporate social responsibility and is a process whereby organisations meet their needs for goods, services and utilities in a way that achieves value for money on a whole life basis; therein generating benefits and positive outcomes. These benefits are regarded as contributing to increasing and enhancing social value and therein benefitting the well-being of people, communities and the general public at large. Organisations should also incorporate social value initiatives and sustainability in their management approaches which should allow them to address social, economic and environmental issues in their local community. The Social Value Portal provides a standardised approach to measurement of the benefits against costs. Notwithstanding this premise, it is important for main contractors to understand local needs before embarking on social value initiatives. Accordingly, it is important for main contractors to undertake a consultation process with the local community to ascertain needs and aspirations and therein gain their 'buy in'. Community consultees in this regard would normally include ward councillors, local business owners, elected officials, residents, colleges, universities and group leaders.

There are different types of corporate social responsibility, and these relate to legal, ethical, economic, environmental, philanthropy, volunteering and diversity and inclusivity requirements in the workplace. These can be brought together in a business ethics model which can be underpinned by transparent processes, governance and controls. Policies need to be instigated and managed by organisations to ensure that targets associated with each corporate social responsibility area are tracked and achieved through established key performance indicators and properly reported on.

In the context of the UK construction industry one example of an intervention to uphold the principles of the corporate responsibility model could include the Considerate Constructors Scheme which is a not-for-profit independent organisation. Construction sites, companies and suppliers voluntarily register with the Scheme and agree to abide by the Code of Considerate Practice, designed to encourage best practice beyond statutory requirements. It is predicated on an eight-point code of practice which includes cleanliness, good neighbourly practices, the environment, alongside being respectful, responsible, safe and accountable. It seeks to improve the image of the construction industry by striving to promote and achieve best practice. Each section of the Code contains an aspirational supporting statement and four bullet points which represent the basic expectations of registration within the Scheme and designed to encourage best practice beyond statutory requirements.

It is especially important for large organisations, to market their companies and brands on corporate codes and ethical statements as part of their corporate social responsibility and this frequently includes commitments to environmental sustainability. For many years the construction industry has been associated with environmental issues related to excessive carbon and waste generation from what some would describe as antiquated and extremely wasteful building processes. The industry has been tasked to work with research and development partners such as universities to spearhead cleaner, more energy efficient construction techniques, possibly linked to modular construction technologies to improve environmental standards as part of its corporate social responsibility.

14.2.4 Relationship between Ethics and Other Factors

This chapter has identified and discussed the relationship between ethics and several other factors including risk, quality, reputation, trust, leadership, and motivation.

When considering the relationship between ethics and risk, the acceptability of taking risk is based on likelihood of a particular risk materialising. The ethical debate in a construction context comes in the project planning stages when considering the probability of the risks emerging and their overall effects on projects. For example, if the probability of a risk materialising is low and its effects not that significant, organisations may choose to accept a particular risk, especially where mitigation measures are expensive and potentially disruptive. One could debate whether this is ethically the right thing to do when a potential problem area has been identified, especially when it could lead to devastating outcomes including loss of life. Several examples were provided in this chapter to identify catastrophes that have occurred in the past to reinforce this point. To add further complexities around risk decision-making, even those companies that have ethical strategies, processes and procedures cannot always guarantee that some of their staff will simply ignore the risks. Accordingly, companies should have both robust ethical standards and contingencies to deal with unexpected events to respond proactively to these. In this regard, most public sector organisations have disaster recovery strategies in place which covers a multitude of potentially catastrophic events.

In the past risk has sometimes been 'passed down the line' with clients passing most risks to their contracted main contractor and those main contractors passing it on to their subcontractors. Such placing of risk on those organisations who are sometimes clearly do not have the necessary resources, expertise, or experience to manage those risks could be regarded as unethical in this sense. Ideally risk workshops at early project stages should through discussion and collaboration with the wider supply chain should identify which parties are best able to manage risks identified on projects and this is considered the most ethical approach. During the construction stages of projects there may be ethical considerations associated with one party not giving another adequate warning of an avoidable issue. This raises the question of a 'duty to warn' where there might not be a contractual duty to do so, but more an ethical and moral obligation to do so. Other ethical issues could apply to those construction professionals such as architects and project managers who hold the balance of power between clients and contractors in terms of certifying various aspects of construction management. The significance of the certifying process has led to disputes and adjudication proceedings in extreme cases. Ethical issues can also emanate after construction projects have been completed and can involve risks around financial claims for events and variations which can invariably lead to disputes and litigation. An example could be where a contractor submits financial claims for a critical incident that occurred many weeks or months before, and for which their clients were unaware of.

When considering the relationship between ethics and quality, the issue of poor quality of construction projects, linked with questionable professional ethics has been widely publicised. This is especially in those developing countries around the world construction where quality is strongly influenced by ethical practice and behaviours. The correlation between ethics and quality could be possibly explained from the position that they are both built upon the promise of doing the right thing. In this sense, good ethical conduct can create the right conditions for trust to build and to optimise value for money, improve competitiveness and increase client satisfaction levels. Conversely, low ethical standards can have the reverse affect and lead to low quality outcomes. Two real-life examples to reinforce this dilemma have been given in this chapter.

Research carried out by Besterfield et al. (2003), focused on the potential impact of professional ethics on construction quality, concluded that the quality of service is dependent on ethical behaviour. Furthermore, their studies confirmed that ethics and quality have an aligned premise built up on a desire to do 'right things right'. Analysis from the study by Abdul-Rahman et al. (2010) determined that quality and unethical behaviours and practice are positively correlated. From this study most respondents (93%) agreed that acts of unethical behaviour or practices contribute to quality related problems in the construction industry. In addition, findings showed that a shift to unethical conducts of one form or another could lead to poor quality service delivery and project outcomes, alongside monetary loss for their clients. This correlation result gives strength to the prognosis that quality and ethics are not only related but can both have a strong impact on the quality of construction outputs.

The reputation of organisations and individuals can be heavily affected by how well or badly they maintain ethical standards and adhere to codes of conduct within their professional practice. Reputational damage, emanating from acts of fraud and corruption that they had been party to, can cause them negative publicity leading to loss of trust by the public. The converse situation can arise where organisations commit to ethical practices and have policies and procedures geared around upholding moral values and professional codes of conduct. In such cases organisations have gained positive publicity and enhanced their reputations. The Best Companies benchmark and accreditation scheme represents a significant achievement and shows that employers are taking workplace engagement seriously. It is a model based on ethical leadership predicated on employee interaction and an unconditional and unequivocal commitment to the workforce.

Trust and professional ethics should be regarded as a collaborative necessity which construction professionals should be encouraging and closely managing on their projects. According to Challender (2019), trust built upon partnership and collaboration working relationships has a strong reciprocal correlation with professional ethics. In addition, there have been many different academic theories that have forged various links, corelations and relationships between ethics and leadership. When leadership is particularly good, certainly within a construction context, this can gain the confidence from other members of the project team and give them the assurance that ethical actions, behaviours, and decision-making are being practiced. Leadership decisions and their behaviours and actions should always be subject to fairness, integrity, and a sense of responsibility. Benefits which can derive from good ethical leadership include developing the workforce and improving the service to their clients and supply chains whilst maintaining the reputation of their respective organisations. Conversely, the relationship between poor leadership and ethical adherence can result frequently in project failure especially in the areas of project initiation and planning, governance, people and technical competent. Furthermore, leaders should be aware of the advantages in keeping project teams motivated by encouraging a culture of staff development through training and education, especially around ethical issues, and compliance with professional values.

14.2.5 Ethical Human Resource Considerations

Inclusivity, equality and diversity in the workplace is not only beneficial but, in most countries, governed by law with severe penalties for breaches in employment law. The statutory requirements of the Equality Act 2010 play a major part in maintaining diversity and equality in the workplace. Policies, procedures and instruments should be carefully monitored by human resource departments to ensure that these are applied to all employees, both existing and new, and at every level with organisations. Management policies and protocols should also be in place for dealing with complaints and potential breaches and violations.

Equality relates to ensuring that everyone has an equal opportunity and is not discriminated against or treated any differently as a result of their background or characteristics. Diversity revolves around how we think, look and behave. Inclusion is about the creation of an environment where differences are regarded as an advantage, where differences and perspectives are shared, leading to better decision-making and working cultures. Most medium to large-scale organisations in the construction industry have many different policies, strategies and initiatives linked to embracing equality, diversity and inclusion and some of these are underpinned by training workshops and continuous professional development. Open communications routes are fundamental to making progress on diversity and inclusion communication routes and creating dialogue with people based on their narratives and experiences in terms of disability, faith and beliefs, sexual orientation or transgender identity. Construction personnel may be diverse in nature in terms of religion, nationality, gender, ethnicity, age disability, sexuality, background, education, social class and many other respects. It is important to embrace these differences and acknowledge the diversity of the workforce.

The Equality Act 2010 created nine 'protected characteristics' that cover age, disability, sex, religion and belief, race, sexual orientation, pregnancy and maternity, marriage and civil partnership and gender reassignment. The Act makes it unlawful to discriminate on any of these protected characteristics. There are many different types of discrimination and these include direct, indirect, associative and perceptive discrimination and construction organisations have an overiding responsibility to protect their employee from discrimination. However, individual employees can be held responsible for acts of unlawful discimination provided their organisations can prove that they have provided their employees with a resonable level of ability to carry out their duties and an awareness of their ethical and legal responsibilities.

Organisations have an ethical, moral and legal responsibility to address equality issues in the workplace with policies, procedures and processes in place to promote best practice. They should minimise or remove disadvantages suffered by individuals who share a relevant protected characteristic and devise and implement processes and procedures that are consistent, accessible, transparent and fair. The most ethical of companies have robust policies, procedures and management approaches linked to reducing and eliminating victimisation, harassment and bullying in the workplace. Such measures should seek to ensure that it is safe and easy for people to raise sensitive matters and concerns without reprisals from other people within the organisation and sometimes whistle-blowing processes assist in this regard. In addition, managers need to be also mindful that when recruiting staff, processes allow their respective organisations to employ the best person for the job, reap the benefits of a diverse workforce, and avoid allegations of unlawful discrimination.

Ethical supply chain management is becoming an ever-increasing area for construction organisations especially those where international suppliers are adopting principles linked to ethical procurement of labour. Notwithstanding this premise, modern slavery is an increasing problem across the world with an estimated 13,000 slaves working in the UK, many of these working in the construction industry. Under the Modern Slavery Act 2015, all businesses with over £36 m turnover are required to publish an annual statement demonstrating that they have taken adequate steps to combat modern slavery. In addition, the Act, lays down requirements to identify whether modern slavery is present within an organisation, and this entails a risk assessment of current suppliers. If companies identify high or medium risk suppliers within their supply chain, they should complete an on-site audit to uncover potential abuses. Having completed investigations, the audit report should be shared with their suppliers and used as the basis for a 'Correct Action Plan' (CAP).

Professionals who work in the construction industry, as in other sectors, need to be aware of ethical and regulatory requirements related to the General Data Protection Regulations (GDPR). The GDPR is the European Union (EU) regulations covering the personal data of individuals and are designed to protect data in an era of mass digital use. The GDPR applies to organisations anywhere in the world that use or stores the personal data of anyone in an EU country, when offering goods or services or monitoring their behaviour. There are several points of compliance as part of the GDPR which allows organisations to collect and process personal data. Organisations must document and evidence how they are protecting data and using it fairly, ethically, lawfully and transparently. Under the GDPR there are strict penalties for breaches to the rules on how data is processed and used. These normally include large fines which are normally dependant on the severity of the breaches.

14.2.6 Ethical Considerations around Trust and Collaborative Working

Trust and professional ethics should be regarded as a collaborative necessity which construction professionals should be encouraging and closely managing on their projects. It is crucial to consider the importance of professional ethics for building trust, adopting collaborative policies and developing good working relationships, as there is a reciprocal correlation between all these factors. This is especially the case in construction management where the industry and all those professionals that are employed within it heavily rely on interaction between each other. The construction industry is composed of many different groups of individuals and disciplines; there are many interdependencies between these groups where the behaviour, successes or failures of one party could affective another in a positive or negative way. It is for this reason that professional ethics have a strong relationship and bearing on trust especially in the context of partnerships and collaboration working relationships. In this context it is therefore not surprising, owing to the fragility of trust theories, that actions of unethical behaviour or practice can have a hugely negative effect on such relationships. This can be very damaging for construction projects with parties becoming reluctant to collaborate fully, share information and trust their fellow professionals. Accordingly, this chapter has focused on the bespoke nature of the construction industry and the need for cultural and behavioural reforms to improve relationship between construction professionals. This is aimed at improving performance for the industry, maintaining professional ethics and obtaining more successful outcomes for construction projects.

The UK construction industry has arguably been associated, over many years, with projects which have had less than successful outcomes. There are many different reasons that have emerged as to why this might be the case. Some relate to the short-lived duration of construction projects whilst others to lack of continuity and fragmentation across supply chains. Various government reports have reinforced this dilemma over the years and identified that unethical behaviours and fractious qualities are embedded in the UK construction industry. Arguments have emerged that mistrust has been inherent within the UK construction industry for a long time between all parties including clients and contractors they employ. This may have stemmed from the traditional forms of procuring construction work which over recent years has been blamed for achieving low client satisfaction levels, poor cost predictability and time certainty.

To prevent such occurrences arising, there is an argument that 'collaborative working' or 'partnering' offers a more suitable alternative in creating working relationships built on ethical values and trust. The use of collaborative procurement methods and improvements to relationships that they bring about, applying the principles of both Latham and Egan, can procure more successful project outcomes. Success in this way can be measured through significant improvements in client satisfaction, cost predictability, safety and time predictability. In this context, such partnering-based approaches have been presented in this chapter as a pathway to greater cooperation and trust.

14.2.7 The Importance of Ethical Leadership in Construction Management

This chapter has referenced current academic literature on theories relating to leadership and especially around professional ethics and applied these to the roles of construction professionals. There are many different definitions of leadership and what leadership actually means in practice. Leadership can be defined as the capacity to establish direction, influence and align others towards a common aim, motivate and commit others to action, and encourage them to feel responsible for their performance (CMI 2008). Accordingly, successful leaders need to be good motivators and have the ability to bring out the best in people. They should lead by example to gain the confidence from other members of the project team and give them the assurance that ethical actions, behaviours and decision-making are being practiced. Furthermore, it is essential for aspiring leaders to have courage, conviction and assertiveness when faced with difficult choices and decisions and their behaviours and actions should always be subject to fairness, integrity and a sense of responsibility. It is essential for aspiring leaders to understand what makes a good leader, in order to perform their roles and motivate others for achieving successful outcomes.

The leadership identity development (LID), and progression process as applied to construction professionals, is dependant and influenced by individual differences, cognitive capacity, personality and temperament, personal identity, personal values and emotional intelligence, driven by cultural context and personal experience. According to the LID model developing leadership identity develops in three stages. The first stage highlights the awareness the concept of leadership and the fact that some people lead, and others follow. The second stage relates to the period that individuals gain experience and develop unique leadership styles first-hand. The third stage relates to people learning the difference between management and leadership whilst at the same time experiencing leadership in practice.

The chapter has also focused on ethical leadership which goes beyond the notion of only satisfying minimum compliance standards. The benefits from measures that organisations can implement to go 'above and beyond' legal requirements have been articulated and discussed. Such benefits include developing the workforce and improving the service to their clients and supply chains to make them feel valued, listened to and engaged in their business.

In a UK organisational context, the Best Companies benchmark and accreditation scheme represents a significant achievement and shows that employers are taking workplace engagement seriously. It is a model based on ethical leadership predicated on employee interaction and an unconditional and unequivocal commitment to the workforce. Notwithstanding this premise, ethical leadership has traditionally not been practiced widely in developing countries which in some circumstances has led to employees' rights being compromised. There are many examples where restrictions on workers freedom of speech alongside poor health and compromises of safety measures have occurred. One of the most important factors for improving ethical leadership revolves around staff development and training. In this regard training and development are regarded as being the pillars which underpin a successful and motivated workforce and excel a company to achieve successful outcomes and increased growth.

Nurturing and supporting followers are an important and a distinctive feature of successful leadership. There is a wide acceptance that most construction professionals, who perhaps have little or no previous construction management experience, knowledge or qualifications, learn from other professionals within project teams. Opinions on leadership have emerged from different disciplines, with many varying approaches to nurture the skills that are deemed necessary. Some would advocate those responsible leaders should focus on others rather than solely themselves, consider the longer-term ramifications and implications in their decision-making and thought processes and follow a moral compass in their management approaches. Accordingly, it is important when considering the traits of ethical leadership for people in authority to lead by example and make decisions that make those around them feel safe, valued and listened to.

It is generally accepted that for projects to succeed they should always be focused on 'win-win' outcomes for their respective client organisations and their contracting construction partners. For this reason, collaborative and partnering strategies, devised by construction professionals, will create the right environment for ethical behaviours amongst project teams. Notwithstanding this premise, applications of leadership styles for construction professionals can be complex and will depend on the context of the situation they are under and the type of individuals they are dealing with. Leaders can bring quality and personal characteristics to certain situations, and this includes knowledge-based attitudes and behaviours, learned and acquired alongside cognitive abilities and personality.

Finally, the relationship between poor leadership and project failure has been examined in this chapter from the main areas of project initiation and planning, governance, people and technical competence. Accordingly, leaders should ensure that measures are all in place which include adopting the right project controls, communication strategies, ethical codes of conduct and collaboration approaches.

14.2.8 Professional Ethics as a Means for Motivating Project Teams

This chapter has considered and analysed the relationship and influence that motivation of employees has on compliance and adherence to professional ethics. Furthermore, it has explored the reciprocal relationship of how fostering a culture of professional ethics can lead to raising motivation levels within organisations.

Gaining trust and maintaining reputations of individuals and organisations is heavily influenced by the adherence to professional ethics and maintaining standards and this has been covered widely in the chapter. When professional ethics are compromised, this can lead to reputational damage and degradation of trust for organisations. Accordingly, Construction leaders should be aware of the advantages in keeping project teams motivated by encouraging a culture of staff development through training and education, especially around ethical issues and compliance with professional values. There are examples that were covered earlier in the book where organisations have been heavily fined and lost public trust following global cases of unethical behaviour, corruption and fraud.

In the construction industry there have traditionally been some long-standing, arguably misinformed, controversial and potentially unethical views held that employees are only interested in short-term financial incentives. This is a potentially unfair and unjust claim and there is no evidence to suggest that this position is widely represented across the sector. Notwithstanding this premise, there is a converse argument that has been presented in the chapter that does link motivation, pay and ethics. In cases where there are no financial incentives or where pay for undertaking a particular role in the construction industry is extremely low, this could be deemed unethical in not paying the going rate and possibly illegal if below the minimum wage in the UK. This can not only compromise motivation levels on behalf of workers, leading to poor productivity but also cause reputational damage for companies and allegations of potentially exploiting staff.

To summarise the findings of this chapter, there are significant benefits that can be gained from pursuing construction management initiatives on projects to motivate staff and therein lead to more successful outcomes. There has clearly in the past been an underestimation as to the benefits of motivational management and the potential implications for poor levels of construction staff motivation. The potential benefits could include:

- Improved construction staff retention
- More positive attitudes of construction staff
- Improved supervisor/subordinate relationships
- Increased productivity, and potentially higher quality standards
- Fewer conflicts and overall better communications on site

Given these clear benefits, this chapter has analysed and evaluated the factors that influence the motivational levels of construction staff. It has found that although financial incentives are undoubtedly important to construction staff, they are by no means the sole mechanism for motivating the workforce. Perhaps the common misconception that construction staff are 'only in it for the money' has been generated from the unique environment that construction staff work within. As discussed, they are positioned at the bottom of the organisational hierarchy, their contracts are generally short term and they are working in an industry traditionally very competitive, especially in the current economic climate where financial incentives are seen as the 'quick fix' to meeting contract deadlines and targets.

Hopefully, this chapter of the book has contradicted the unethical notion that construction staff as part of overall project teams are unworthy of requiring to be motivated by any other measures other than money and that attempting to incentivise and motivate them by other means will be fruitless accordingly. In this regard, factors have been identified which construction clients should consider as they could have a significant impact on motivational levels of construction staff. These include improved communication strategies, consultation, training and development, organisational culture and opportunities should be sought by construction organisations to address these aspects wherever practicable. There is also a suggestion that the construction industry could benefit from adopting certain lessons learnt in other sectors such as the manufacturing industry where more emphasis has been on management strategies geared to raising levels of workforce motivation such as job rotation and blue-collar participation in strategic decision-making.

Construction organisations have historically asked themselves questions related to whether they can afford to procure motivational management initiatives for construction staff. Furthermore, construction clients have traditionally distanced themselves from motivational initiatives under the misguided premise that it is not within their role to encourage project teams to motivate their staff. The real question, however, they should be asking themselves is can they afford not to? After all, how long can construction staff perform their duties and obtain good outcomes if not motivated and in some cases what level of performance can be expected where poor motivation levels are experienced?

This chapter has hopefully emphasised why senior management, including construction clients, should pay special attention to motivational strategies. The justification for this is that low motivation levels within the workforce can otherwise result in poor ethical compliance, low esteem, negativity, conflict, low productivity, and frustration of construction staff. Conversely, if motivation is properly managed this can have very positive results as previously outlined and is both ethical and appropriate. Clearly, motivation is not the only factor that affects productivity and other factors, such as availability of resources within a particular situation, capability and ability to undertake a particular task, could be argued as being equally important.

Motivational management of construction staff could be used as a tool to enhance organisational performance and construction outcomes but is arguably as important to improve the mental well-being of construction staff in the process. This is itself is linked to professional ethics in providing a welcoming, comfortable, and safe environment for staff to prosper. Conversely, if motivational management is not practised, leading to construction staff becoming demoralised in some cases, construction operations may continue to work well with other 'negative' factors such as job security and fear of reprisal affecting performance on site. This could affect the well-being of staff and therein provide an unethical and unsafe environment for employees. The overriding question is, however, how long can this be maintained in such scenarios? The overwhelming academic consensus would tend to suggest that a lack of motivation coupled with autocratic management styles can provide 'short term fixes'. However, it is generally recognised that poor morale and motivation of construction staff could, in the longer term, have extremely detrimental effects on the construction industry. Such implications could include poor employee retention resulting from construction staff seeking alternative employment elsewhere. This increase in mobilisation of staff could hinder internal development of construction project teams in terms of long-standing, knowledgeable staff having the right expertise and 'knowing the businesses' they work within. This in turn could increase costs for construction clients and contractors when considering such factors as disruption, loss of continuity and training of new employees to fill vacancies left by demoralised employees.

This chapter has identified the general need for alignment of individual and organisational goals and the aim for a win-win scenario. In addition, it has highlighted the need on the part of construction clients for greater awareness of the importance of motivating construction staff. Accordingly, construction clients should champion a required change in cultures within the construction industry, adopting the recommendations of Latham (1994) and Egan (1998) to procure a more integrated and motivated project team. More emphasis by construction clients should therefore be placed on intrinsic factors rather than simply relying on financial incentives to raise motivation levels of construction staff. This by no means should involve simply removing financial rewards or any other extrinsic motivational measures that construction staff have traditionally relied upon, as this could negatively affect motivation. Instead, construction clients could look wherever possible to blend intrinsic motivational measures into extrinsic measures. Examples of such initiatives could include financial rewards for good suggestions to improve the quality or buildability of a particular work element.

In the context of the complex and dynamic environment of the construction industry, and to realise benefits associated with motivational management and ethical adherence, there is clearly a change of culture and attitude required within the sector. To enable this change construction clients should consider the integration of the following recommended initiatives on their projects:

- Introduce 'bottom up' management initiatives to instil greater construction staff empowerment through responsibility and autonomy.
- Ensure that tasks for construction staff, through measures placed by construction clients on their contractors, are challenging, varied and interesting through initiatives such as job rotation. This will seek to reduce monotony, boredom, alienation and incentivise the workforce.
- Develop informal as well as formal lines of communication through regular feedback and appraisals.
- Give praise to construction staff as individuals and teams where appropriate and introduce recognition initiatives. Examples could include 'subcontractor team of the month'.
- Ensure that training and development of construction staff is undertaken and learning achievements recognised in individuals. This could be achieved through apprenticeships, other on-the-job training and through short-term training courses off-site.

- Encourage construction staff to pursue career development through qualifications and industry accreditation.
- Develop and embed greater awareness of the importance of motivation for construction staff at senior management level.
- Break down communication and cultural barriers and treat construction staff as a more integral part of the overall project team and adopt more democratic rather than autocratic styles of management.

All these measures will hopefully contribute to provide a motivating work environment for construction staff to work within and construction clients will reap great benefits not just for their respective organisations and the building projects that they procure but for the future of the construction industry at large.

14.2.9 Professional Ethics in the Context of the Coronavirus (COVID-19) Pandemic

This chapter was written during the COVID-19 pandemic and has been aimed at portraying good and bad behaviours in the construction industry that have emerged during the crisis.

At the start the first wave of the pandemic in March 2020 the UK construction industry suffered a major setback with an exponential increase in unemployment, and many companies planning redundancies. Most of the major main contractors chose to work through the crisis, with the construction sector classified as essential and therein exempt from the lockdown. However, a large proportion of construction-related businesses, including many subcontractors and suppliers 'Furloughed' staff as part of the Government's job retention scheme, which enabled them to claim 80% of salaries for their employees. Some main contractors, however, did opt to shut down their sites, with problems experienced in sourcing labour and materials. Those that chose to remain operational, were running at reduce operational capacity owing to shortages of supply chain labour and resources and through restrictions around reduced density of the workforce linked to 'social distancing' adherence. However, amid criticism from the public about builders flouting social distancing rules, house builders began to shut down operations as they said they were unable to operate safely under the Government guidelines. In some cases, businesses made decisions to allow staff to work from home and this included many design team consultants. They saw this as doing the right thing and maintaining professional ethics on their part, putting their staff and their supply chain's health above their own commercial interests despite the inevitable disruption and financial costs. In addition, there are many other examples of professional ethics and responsible behaviours that have been applied to the construction industry during the pandemic. These included construction organisations providing many acts of loyalty and commitment to the commercial interests of other businesses. This demonstrated a collaborative teamwork approach to working through the pandemic with their contemporary organisations. These were considered to represent a moral position and one that demonstrated an underlying commitment to ethical principles. Other examples have emerged where construction business has protected and assisted their supply chain partners in navigating their way through the crisis. Such assistance included paying their supply companies on time and in some cases ahead of time to assist with cash flow at a difficult time when their revenue may have been compromised. To enable this, in some cases, this involved their senior management team taking a voluntary reduction in salary to afford to pay their outgoings.

Other ethical measures which have been reported throughout the crisis included main contractors giving their supply chain comfort and reassurance included putting out regular communications to subcontractors and suppliers. These were designed to give confirmation that their existing commitments and contracts entered into before the pandemic would not be compromised, and that they would continue to be paid on time. In addition to the obvious health and safety issues, other examples emerged of more collaborative working across the supply chain to overcome other challenges. These included clients agreeing not to take retention monies from their monthly valuations but instead deducting retention at a later stage when the main contractors' cash flow was less of an issue for them. One project management consultancy director had explained that their senior management had felt a moral obligation to take a significantly large pay cut globally to protect jobs and enable them to survive. This chapter has articulated examples where clients, to improve the cash flow of their main contractors, had agreed to settle monthly valuations in 14 days rather than the contractual 28-day period. Furthermore, some clients agreed to 'pain-sharing' with their main contractor partners for delays and cost increases brought about by the pandemic. Such moral acts of kindness and support were welcomed by the industry and testament to collaborative working and partnering philosophies throughout the construction industry.

Other examples of moral and ethical practices and behaviours during the coronavirus (COVID-19) pandemic have included instances where leaders chose to temporarily close their businesses rather than risk the health, safety, and welfare of their staff. It is likely that decisions such as this would have been taken with the full knowledge of the harm this would cause to them and their livelihood. However, for those site employees that could not work remotely, responsible contractors instigated COVID-19 safe and robust working procedures on construction sites for their protection. Such measures for safe working included providing enhanced personal protective equipment (PPE), and orchestrating policies and arrangements for maintaining social distance and making decisions that put the health and livelihoods of their employees ahead of their own personal interests.

Notwithstanding the positive examples of ethical and responsible behaviours, cases have been articulated in this chapter where certain individuals and organisations had not only been less willing to work collaboratively with other firms in the industry but had deliberately tried to use the crisis to further their own commercial positions. These unfortunate scenarios demonstrated that in some cases these organisations were deliberately trying to take advantage and profiteer from their adversaries during the Coronavirus crisis. One example which was described involved a developer who attempted to renegotiate contract terms with a contractor to give themselves a commercial advantage. Another case included a business which sought to utilise an opportunity to try to defer payments and seek discounted rates for work already completed. One could argue that these examples constitute acts of unethical 'game playing' and ones which are designed to disadvantage the businesses that are reliant on them for settling their accounts in full. Such cases also reinforce that not all businesses take a completely moral position, especially where there is the temptation to increase profit margins.

At the beginning of the pandemic in March 2020, the UK Government introduced its Coronavirus Job Retention Scheme (CJRS) also known as the 'Furlough' scheme. It was designed to save jobs that would otherwise have been lost through redundancies where companies had been severely affected by the virus. In the context of the Furlough scheme, there were many contrasting examples that emerged, highlighting both positive and negative responses to the scheme throughout the built environment. This chapter has provided anecdotal reports of flagrant abuse of the scheme which emerged during the crisis. Such cases allegedly involved a minority of smaller consultants, contractors and suppliers benefitted from the Government job creation Furlough scheme, whilst still operating their businesses. This clearly represents a fraudulent breach of the Furlough rules and a wholly unethical act which has no doubt, once again, harmed the reputation of the construction industry.

14.2.10 Independent Research Study: Exploring Ethics in the Construction Industry and Understanding Its Impact on the Built Environment

Professional construction bodies are relevant and important in the implementation of relevant codes of conduct, but nevertheless it must be realised that codes of ethics do not just exist in absolution. A better understanding of their roles within an industry can be attained in the context of the actual realities and equivocalness confronting such an industry, as well as the people involved in the evaluation of options and decision-making (Fewings 2009). Dainty and Murray (2009) argued the great importance of business ethics in organisations owing to its potential to transform organisations' cultures, which in turn can lead to an organisational transformation towards better stakeholder management and innovation in the ways organisations do profitable business. It can be concluded that adoption of a code of ethics by an organisation is not enough for it to be ethical. Similarly, distributing the codes of ethics to all employees does not guarantee their understanding and identification with the codes. A clear difference should be established between using codes of ethics as instruments of compliance as opposed to instruments of cultivating ethical behaviour (Fewings 2009), thus, exhilarating that ethical behaviour within an organisation calls for a dire code of ethics to be engaged in organisational territory. The areas, as structured by Doran (2012), include ethical bodies, ethical tools, ethical support procedures, and internal and external ethical usage. Although Garrett (2011) discussed in detail the areas in which organisations are perceived to be engaged with ethics, the limitation is that the account provides little information about the enablers of these attributes. Personal values on the other hand are perceived to be of importance when considering aspects of moral codes. These values are enablers that help human beings make decisions about right and wrong, thus supporting the fact that issuing codes of conducts to employees is not enough to acknowledge an organisation as ethical. However, it is also documented that criminalisation of certain unethical activities (by, for example, introduction of the Bribery Act of 2010) has seen a rise in development in the ethical structure of the construction industry. Current literature has highlighted the increasing progress on acceptable ethical practice and professional attitude in all forms of businesses including the construction industry.

The objectives of this research were to:

1) Understand the ethics of the construction industry by conducting scenario-based interviews.
2) Analyse the results and measure the ethical industry appropriateness from the responses obtained.
3) Establish how professional governing bodies, help to shape the image of the construction industry.

This research document used unstructured interview approach resulting in the extraction of the respondents' beliefs and values on how site managers, supervisors etc. would behave given certain work-based scenarios. The ambiguity in this type of research can be lack of cross checking to ensure that the actions of the respondents are consistent with their reporting. Further clarification of the types of unethical conduction involved is required. An example is the use of the word 'deceit' which is classed by Doran (2012) as a generalisation within itself, so another level of clarification is required to provide detailed feedback. Another example that would entail further clarification is the concept of hospitality and bribery. This research has indicated and confirmed (backed by relevant literature) the types of ethical improprieties that exists. Also confirmed as unethical, is bullying, which has frequent occurrence on construction sites (Reviewing the responses of the interviewees). However, what is new is the emergence of clients and government bodies as contenders in the ethical states. The introduction of the Bribery Act of 2010 has been hailed as a success and is a useful tool within the construction industry. The emergence of various professional government bodies has influenced the development and improved the image of the industry. History suggests that advancement of professional ethics in the construction industry is very much dependent on the implementation and policing of the ethical guidelines and policies of both professional bodies and private organisations together with the leadership of public sector procurement agencies. The response from the interviewees suggested that even though organisations have their own ethical codes of conduct, the curbing of certain unethical issues is difficult. The researcher was of the belief that all employees, regardless of professional allegiance, require a familiar understanding of ethical and professional values, of which this can be achieved through continuous training (CPD). If there is lack of professionalism and lack of ethics, even the ethically good employees will have difficulty maintaining moral standards.

Ethics would improve productivity on site, lowering project cost and improve on profits. The researcher would recommend continuous ethical training and education during the construction professional's working life. It can be argued that taking personnel off their tasks to educate them about ethics can be a costly agenda, but the researcher is of the belief that there are ways of enveloping the training during the construction development period. This can be incorporated in daily morning briefs by reminding employees on what is right and what is wrong and for those seeking to be promoted they should study and be asked ethical competency-based questions for them to proceed. Posters can also be put in employee canteens and around the site. Implementing Continuing Professional Development (CPD) schemes can help drive good ethical values within a team. According to CIOB (2017), CPDs have been increasingly receiving more attention in recent decades as construction professionals are required to update themselves with new ethical developments in the expeditiously changing built environment.

Apart from training, employees should be given a sense of belongingness within any business setting. This can, however, increase productivity and help the employees behave in positive ethical manners and can be made possible by offering out shares and profit sharing. In addition, communication channels must be open for all. Some companies operate an open-door policy which enables employees to view their concerns directly with senior managers, if need be, whereas some companies use the hierarchy system of which the researcher believes to be outdated.

The researcher would recommend that the government financially help the professional bodies, as it is documented that there have too little resources.

14.2.11 Implications in Practice for Organisational Ethics in the Built Environment

There are many implications for the construction industry from unethical practices, which will include reputational damage and image considerations for organisations. Furthermore, it is clear that unethical practice can take a negative toll on the performance and ultimately the outcome of construction projects. Negative consequences associated with poor outcomes are numerous and lead to assertions that the industry is inefficient, unsafe, wasteful, compromised on quality, and cannot deliver to time and budgetary constraints.

The following areas linked to organisational ethics have been identified and analysed throughout the book as being pertinent for companies to uphold good ethical standards. These are:

- Raising environmental sustainability credentials for organisations
- Increasing corporate social responsibility (CSR)
- Ethical human resource (HR) considerations
- Ethics around construction health and safety
- Promoting trust and collaborative working
- Ethical leadership in construction and engineering management

In order to take on board these considerations and embed them into organisational practices, it is important to promote teaching and learning around ethics in the workplace. For this reason, organisations should promote continued professional development for their staff which could be via seminars, workshops or guest speakers. Investing in this type of training will ensure that staff are kept up to date on latest developments and changes in regulations. In addition, it has been suggested that colleges and universities should tailor more academic modules and courses to the subject of professional ethics to address this deficiency in knowledge and education. With the right degree of education, training and CPD, construction professionals can learn to adopt ethical behaviours and act with professional integrity and maintain a duty of care to their clients.

14.3 Final Reflections, Overview and Closing Remarks

This book has discussed many different aspects and issues that influence and affect ethics in an organisational context. Measures to improve the practice of professional ethics in the built environment, such as professional codes of conduct, have gone some way to improve the way the industry works but there are still far too many cases emerging of unethical and immoral practices that are still blighting the sector. Although arguably these practices are emerging from a small minority of the sector, they are creating a bad press for the whole industry and further measures should be instigated to address this dilemma. Traditional responses in the past, at an institutional level, have been based on governance, regulations and punishment for non-compliance and clearly these have had only limited success. Hopefully this book has highlighted that a more proactive and collaborative approach is required to address some of the challenges in promoting organisational ethics. Accordingly, leaders of organisations and professional institutions should be leading the way for a cultural change in the built environment to train, educate and motivate construction individuals and organisations in what professional ethics entail, measures to ensure compliance and the benefits that they can bring for the sector. The book has identified and recommended many different measures, in this regard, which include human resource initiatives, motivational management, leadership, and pursuing organisational goals linked to environmental sustainability, health and safety and corporate social responsibility. These could be achieved through more focus on further education and higher education course modules linked to professional ethics and continued professional development (CPD) through workshops and training events in the workplace. These measures will hopefully contribute to providing a more ethical environment for the industry to work within and reap great benefits not just for clients, all construction- and engineering-related organisations and the projects that they procure but for the future of the built environment industry at large. It is accepted, however, that to bring about these cultural changes will take resources, conviction, integrity and in some cases courage not to engage in established unethical practices. These improvements once ingrained within the built environment could then reap massive rewards in providing a safer, honest, trusting and more enjoyable working environment for all.

References

Abdul-Rahman, H., Wang, C., and Yap, X.W. (2010). How professional ethics impact construction quality: perception and evidence in a fast-developing economy. *Scientific Research and Essays* 5 (23): 3742–3749.

Besterfield, D.H., Michna, C.B., Besterfield, G.H., and Sacre, M.B. (2003). *Total Quality Management*, 3e. USA: Prentice Hall.

Challender, J. (2019). *Building Collaborative Trust in Construction Procurement Strategies*. Oxon: Wiley.

CIOB (2017). *CIOB*. [online] Available at http://www.ciob.org (accessed 13 Jul. 2017).

CMI (2008). *Managing Projects*. Corby: Chartered Management Institute. Checklist 035.

Dainty, A. and Murray, M. (2009). *Corporate Social Responsibility in the Construction Industry*. London: Taylor and Francis.

Doran, R. (2012). Fundamentals of ethics. *Business Knowledge* [online] 11 (4): 122–126. Available at doi: 10.2307/854310 (accessed 13 July 2017).

Egan, J. (1998). *Rethinking Construction. The Report of the Construction Task Force*. London: DETR.TSO.

Fewings, P. (2009). *Ethics for the Built Environment*. London [Angleterre]: Taylor & Francis.

Garrett, J. (2011). The basics of ethics. *Construction Morals* [online] 18 (7): 86–92. Available at doi: 10.1136/bmj.326.7400.1217 (accessed 13 July 2017).

Latham, M. (1994). *Constructing The Team*. London: The Stationery Office.

Appendix A

Energy, Water and Carbon Management Plan

1   Executive Summary

In 2011, we launched our first Carbon Management Plan with targets to reduce scope 1 and 2 carbon emissions by 43% by September 2020 compared to a 2005/06 baseline. As of end of academic year 2017/18 we had already reduced our scope 1 and 2 carbon emissions by 64%. This has been achieved partly through reductions in energy use on campus through investment, partly through reductions in grid electricity emissions factor, and partly through the divestment from our student accommodation in 2007/08.

Our new Energy, Water and Carbon Management Plan builds on our significant progress in carbon reduction to ensure we can achieve our objective of a scope 1 and 2 81% reduction by 2030 and work towards net zero carbon by 2038 in line with the Greater Manchester carbon reduction strategy. Environmental sustainability is a core element of our ambitious University Campus Masterplan, launched in the last year. Central to the plans is an energy strategy that takes a major step towards a zero-carbon future, while providing high-quality spaces for residential, teaching, research and commercial uses that are cost-effective to run. University of Salford buildings will undergo refurbishment to make them highly energy efficient, with maximum potential for renewable energy generation. In the meanwhile, we will carry out extensive audits on our existing buildings to identify opportunities for improvements, implement where technology, scale and finances allow and develop and submit businesses cases for funding for larger projects.

The 2005/06 baseline is to be used for absolute carbon emissions, and this has therefore been adopted under the Energy, Water and Carbon Management Plan (EWCMP). Other relevant baselines are included in the Objectives, Targets and Action Plan include the academic year 2016/17 as a baseline for energy, gas and water consumption, to be used in the University's ISO50001 Energy Management System.

Carbon emissions are broken down into three categories by the Greenhouse Gas Protocol in order to better understand the source.

Scope 1 – All Direct Emissions from the activities of an organisation or under their control. Including fuel combustion on site such as gas boilers and fleet vehicles.

Scope 2 – Indirect Emissions from electricity purchased and used by the organisation. Emissions are created during the production of the energy and eventually used by the organisation.

Scope 3 – All Other Indirect Emissions from activities of the organisation, occurring from sources that they do not own or control.

2 Background and Context

2.1 Background

Situated in Salford, to the North West of Manchester City Centre, we have three campuses with 27 principal buildings providing teaching to around 16,000 students. In the 2016/17 Academic Year (the 'audit' year selected for this Plan), we spent around £2.1 million on energy and water, of which electricity is the dominant cost at around 73%, with gas making up around 19%, and water and district heating and cooling the remainder at around 8%. Total carbon emissions from energy use for the same period were 8,901 tCO_2.

Energy costs are expected to rise significantly over the next three years (and beyond), with electricity in particular expected to rise as a result of measures announced under the Electricity Market Reform, and gas as a result of the closure of the CRC Energy Efficiency Scheme following the 2018/19 compliance year, and the recovery of the lost revenue by HM Treasury by the loading of the Climate Change Levy, particularly onto gas. We calculate that CCL changes alone will increase costs by around 11% in the next 3 years, and by at least 15% up to 2025. Upwards pressure on commodity prices because of investment decisions in new electricity generation and distribution systems and gas resource depletion and geopolitical factors make it likely that energy prices will rise still further over the period.

We launched our first Carbon Management Plan in 2011, which included a commitment to reduce Scope 1 and 2 carbon emissions by 43% by 2020 compared to a 2005/06 baseline. As of 2016/17 we had already reduced our carbon emissions by 55% over the period and hence initiated the development of a new Energy Water and Carbon Management Plan (EWCMP).

We have recently developed and published a Masterplan in conjunction with Salford City Council. Our reason for developing the new Masterplan are to improve the student experience and placemaking. Research shows that today's students are environmentally-engaged and seek learning environments which are sustainable and efficient. This has been incorporated into the Campus Masterplan, particularly through an Energy Strategy and recognition of a physically green campus. The Campus Masterplan Energy Strategy recommends an all-electric approach for the campus by 2030 in order to achieve the targeted carbon reduction emissions. This will require an increase in efficiency in new and existing buildings, use of low carbon and renewable energy technologies on campus as well as taking advantage of decarbonisation of the grid.

2.2 Methodology

In developing this Plan, we have partnered with an external consultancy. First, we held an Energy Risk and Opportunity Workshop in July 2018 together involving colleagues from across the University. We also conducted a PESTLE (Political, Economic, Social,

Technological, Legal and Environmental) analysis to determine the internal and external issues that can affect our ability to achieve the intended outcomes of our Energy Management System, which are:

- To enhance the energy performance of all activities of the University of Salford Estates and Facilities Division.
- To fulfil compliance obligations.
- To comply with the policy and achieve environmental objectives.

These activities have helped us develop a list of energy-related issues and associated risks and opportunities, together with actions to address them. In total over 60 issues were identified, of these 20 were prioritised as potentially having a significant impact (see Table below).

Internal Issues	External Issues
Systems and Procedures: Budget/funding process for energy saving initiatives Setting of Utilities Budget Procurement systems and frameworks Consultation on technological issues	Compliance Issues: Minimum Energy Efficiency Standards (MEES) Requirement for EPCs/DECs Changing legislation and awareness of legislation (existing, new and amended)
Development Activities and New Projects: Implementation of Masterplan Communication and visibility of projects Consultation process during a building design Ability of new buildings to achieve expected energy performance	
Buildings and their Use: Managing building user's expectations Building manager's engagement in energy performance Old systems within buildings, e.g. heating, lighting, IT. Heating and Cooling Expectations/Policy Old inefficient 1960s buildings	
Building Management System Issues: Access to BMS for relevant parties BMS Performance BMS sensors in wrong place, e.g. on West- or East-facing walls or near heat sources Training and competency of staff on installed building systems, e.g. BMS	

We also developed an Energy and Water Review tool to review all of our current and historic energy use, costs and associated carbon emissions.

The purpose of the energy review was:

- To establish an appropriate baseline for measurement of energy use and carbon emissions.
- To derive suitable energy, water and carbon benchmarks and key performance indicators (KPIs) which can be used to set targets and measure performance going forward.

- To accurately quantify energy use, water use and carbon emissions for the University's main campuses, buildings and end users, and to use this as the basis of derivation of areas of significant energy use and opportunities for energy performance improvement in line with the requirements of ISO 50001:2018.
- To establish relationships between energy and water use and key driving variables, in order to develop a forecasting methodology. This in turn will be used to calculate future energy and water use and carbon emissions, and then used to establish what level of reduction is required to meet any future targets.

3 Energy and Water Review

3.1 Baseline and Key Targets

Through the Higher Education Funding Council for England (HEFCE) carbon management programme we initially adopted the sector target published in January 2010. HEFCE advocated targets for the higher education sector for reduction in Scope 1 and 2 carbon emissions of 43% by 2020 when compared to a 2005 baseline.

We adopted the 2005/06 academic year as our baseline for carbon emissions accordingly. To the 2018/19 Academic Year, we had achieved an actual reduction of around 64%. This has been achieved partly through reductions in energy use on campus through investment, partly through reductions in grid electricity emissions factor, and partly through our divestment of our student accommodation in 2007/08.

The 2005/06 baseline is to be used for absolute carbon emissions, and this has therefore been adopted under the Energy, Water and Carbon Management Plan. Other relevant baselines are included in the Objectives, Targets and Action Plan (see Section 4) and include the academic year 2016/17 as a baseline for energy, gas and water consumption, to be used in our ISO50001 Energy Management System.

Our Campus Masterplan Energy Strategy suggested a target to reduce carbon emissions by 81% by 2030, through adoption of an energy strategy which would see the Campus move to an all-electric model, with heat pumps to provide heating and cooling, and with grid electricity supplemented by the limited implementation of building-mounted solar PV. This target and an aspiration to move towards net zero carbon by 2038 was adopted and published via the Mayor of Manchester's Green Summit in March 2018.

Our Carbon Reduction Objective:

To reduce scope 1 and 2 carbon emissions from 2005/6 baseline by 81% by 2030 and work towards net zero carbon by 2038.

3.2 Benchmarks and Key Performance Indicators

Potential benchmarks and key performance indicators (KPIs) have been explored. Obvious KPIs are those that relate energy and water use and carbon emissions to a driving variable, such as floor area or student numbers. This subject is more complex than perceived as there is more than one variable affecting energy and water use. This is explored further in Section 3.5.

Data has been obtained from the Higher Education Statistics Authority (HESA) and used to compare our 2016/17 performance against that of our peers using suitable normalised KPIs, i.e.:

- Energy use/floor area/year (kWh/m^2/year)
- Water use/full-time equivalent (staff and students)/year (m^3/FTE/year)
- Carbon emissions/floor area/year (kgCO$_2$e/m^2/year).

As these are the accepted KPIs used within the sector, they have been adopted as the overarching KPIs within our Environmental and Energy Management System.

For energy use, on a total kWh/m^2 (gross internal floor area) basis the University of Salford performs at around 193 kWh/m^2/year (119th out of 161) as shown in Figure 1.

For water, the University's performance is 5.7 m^3/FTE/year (123rd out of 161) as shown in Figure 2.

Document Ref.: Author: Version No. and Date:
EWCMP Bec Bennett/Nia Prys-Williams V4.0 03/03/2020 $_8$

For Scope 1 and 2 carbon emissions, the University's performance is 54.8 kgCO$_2$e/m^2/year (97th out of 161) as shown in Figure 3.

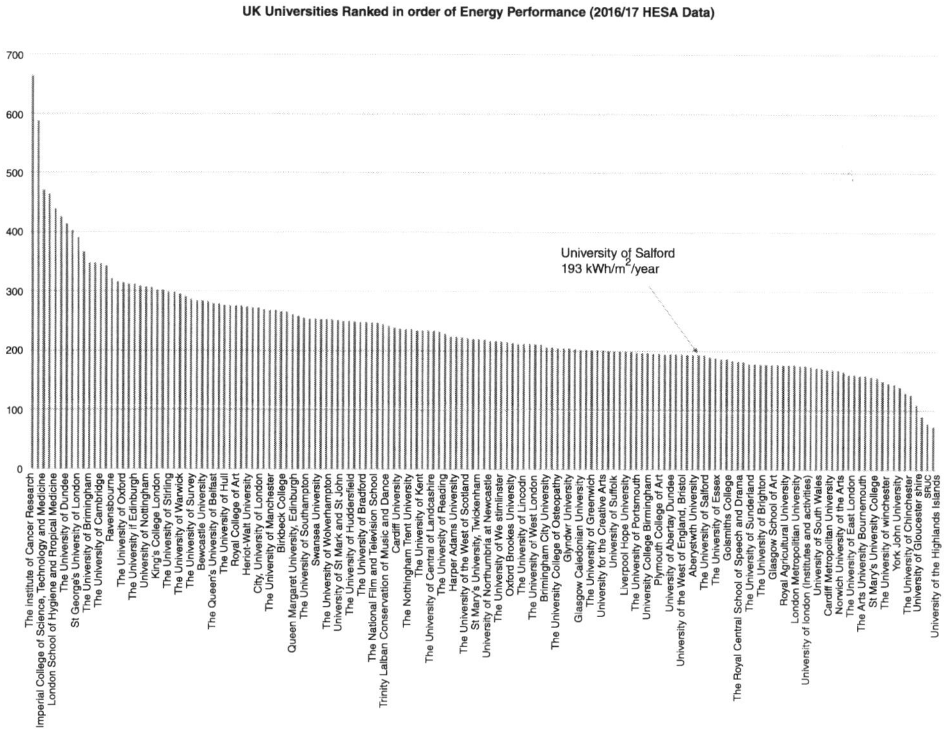

Figure 1 Comparative energy performance of UK Universities (HESA Data (www.hesa.ac.uk) / CC BY 4.0 / HESA).

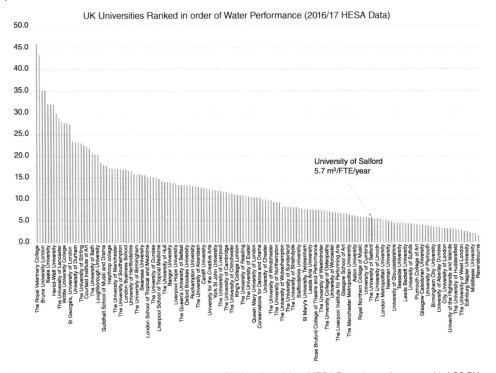

Figure 2 Comparative water performance of UK Universities (HESA Data (www.hesa.ac.uk) / CC BY 4.0 / HESA).

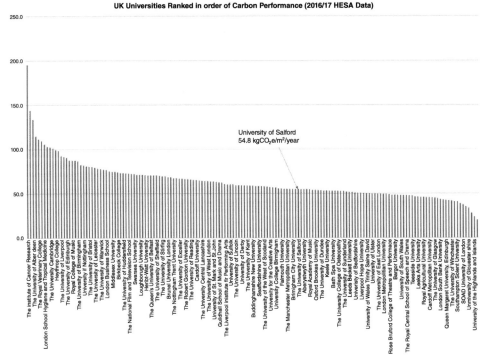

Figure 3 Comparative carbon performance of UK Universities (HESA Data (www.hesa.ac.uk) / CC BY 4.0 / HESA).

3.3 Energy and Water Review Results

In line with the requirements of ISO 50001:2018, an Energy and Water Review has been carried out. The review used data from the 2016/17 academic year (year ending 31st July 2017), which was the last year for which full data was available at the time the review as carried out (May/June 2018). A review of available historical data was also undertaken as part of the Energy Review dating back to the base-line year 2005/06, obtained from the University's HESA returns and is included in Section 3.5, Figure 12.

The purpose of the review is to accurately quantify the energy and water use and carbon emissions, and then to determine the areas of Significant Energy Use (SEU). SEUs have determined on the basis of both individual buildings (as was already standard practice at the University, in accordance with existing metering arrangements) and also by end user, i.e. heating, cooling lighting, etc. The intention of this approach is to allow insights into where energy is used across the campus and within individual buildings, and therefore where practical measures to reduce energy and water use might be applied in future. In most cases, there is no metering available to derive the breakdown of energy use by end user at building level, so an approach of using available benchmark data adjusted for individual buildings (based on knowledge gained through a campus walkover) was adopted, backed by sub-meter data where it was available.

Data has been drawn from the University's SystemsLink and Active Energy Manager data platforms. Data includes both main (fiscal) meters for gas, electricity and water, and sub-meters.

The following Figures 4-10 provide further information on the University of Salford breakdown of emissions and general performance levels of different elements of the organisation

Document Ref.: Author: Version No. and Date:EWCMP Bec Bennett/Nia Prys-Williams V4.0 03/03/2020 [11]

Top 5 buildings = over 50% of all energy use
Top 13 buildings = 90% of all energy use

In terms of Significant Energy Uses, as noted above, analysis may be made by both building and by end use.

In building terms, it is obvious and notable that the largest buildings are the highest users, with the top 5 largest buildings using around 50% of our total annual energy use, cost and emissions, and the largest half of the estate being responsible for around 90% of our energy use, cost and resultant emissions. As such these are deemed our significant energy uses (SEUs) under the ISO50001 Energy Management System, since they are likely to offer considerable potential for energy performance improvements. These are the subject of detailed energy audits as detailed in the Action Plan presented in Section 4.

The Energy Review tool also allows the user to 'slice' the data at individual building level in order to identify Significant Energy Uses by end use category, for example heating, domestic hot water, cooling, fans and pumps, lighting, data centre, office equipment, laboratories, swimming pool and so on. Whilst not absolutely accurate (due to the absence of sufficient end user sub-metering), it does serve as a useful indicator of which services are the Significant Energy Uses, and therefore where future energy and carbon emissions reduction efforts might best be targeted. It is not the intention that this

260 | *Appendix A Energy, Water and Carbon Management Plan*

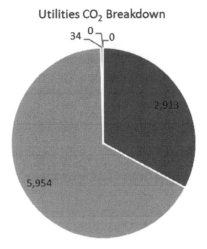

Figure 4 Breakdown of utilities by emissions.

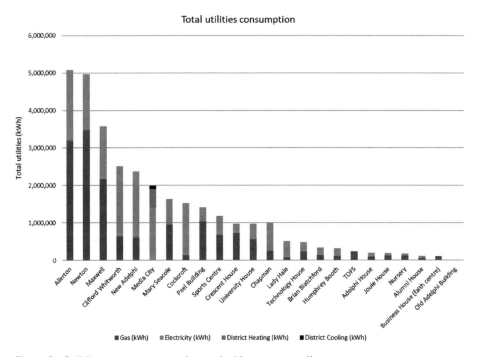

Figure 5 Building energy consumption, ranked largest to smallest.

3   Energy and Water Review | 261

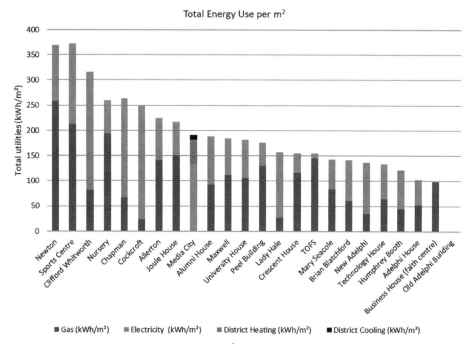

Figure 6   Building energy consumption per m² floor area, ranked largest to smallest.

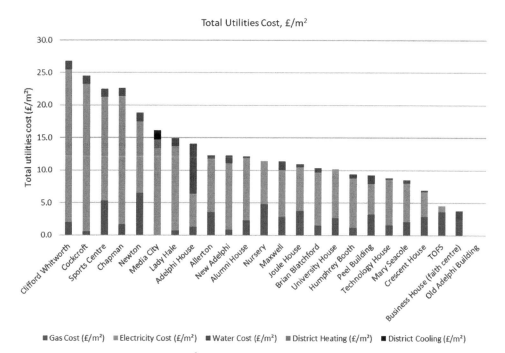

Figure 7   Building energy cost per m² floor area, ranked largest to smallest.

262 | Appendix A Energy, Water and Carbon Management Plan

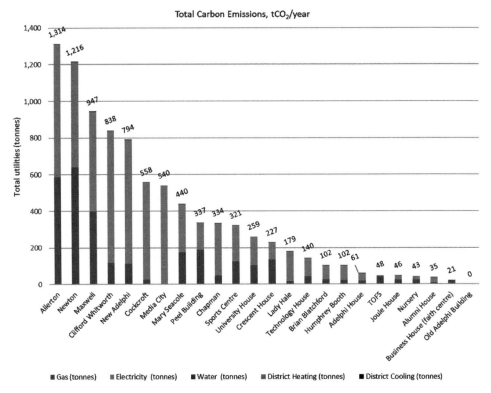

Figure 8 Building Energy-related CO_2 emissions, ranked largest to smallest.

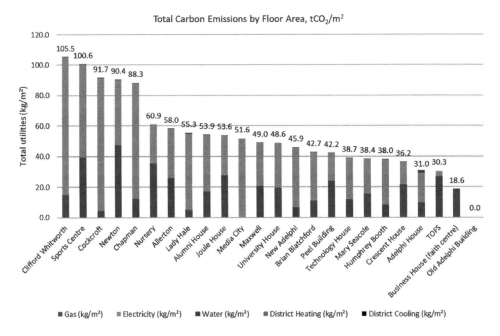

Figure 9 Building energy-related CO_2 emissions, ranked largest to smallest.

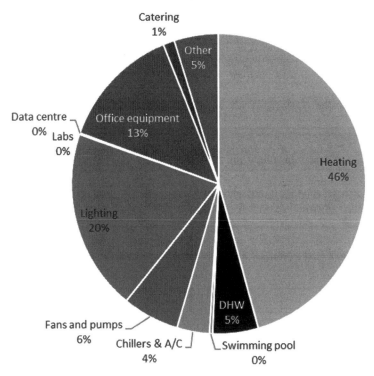

Figure 10 Breakdown of Energy use by end user across all buildings.

approach be used as a formal SEU measure in the ISO50001 system at this stage, however, this could be adopted at a later date when the level of sub-metering allows for energy used by such end-uses to be more accurately measured. For the time being, the analysis presented in the following Tables and Charts is used to inform thinking on the targeting of energy.

As can be seen, in SEU terms, the largest energy users ranked by end use category are:

End User	Breakdown of Use %
Fans and pumps	6%
Domestic Hot Water	5%
Chillers & A/C	4%
Catering	1%
Swimming pool	<1%
Data centre	<1%
Labs	<1%
Other	6%
Total	100%

The table above indicates that the SEUs in terms of end users are space heating, lighting, IT and office equipment, fans and pumps, DHW and chillers and A/C, which together account for over 90% of overall consumption. Other users can be considered insignificant at this stage at a campus level as their percentage consumption is 1% or less of the total (unless they are found to have significant potential for energy performance improvement as identified through detailed energy audits).

3.4 Energy Market Context and Risk Assessment

The current market for electricity, gas and water has been examined in the context of fuel supplies and availability, anticipated market condition and future pricing, legislative impact and taxation, together with an examination of the risks and opportunities faced by the University. The output of the Assessment is included in Appendix A for information.

One point worth noting is that the unit costs of electricity and gas are forecast to rise markedly in future. According to forecasts carried out by the Department for Business, Energy and Industrial Strategy (BEIS), the cost of electricity is forecast to rise from current (2018) levels by around 28% to 2028, before falling back to an overall rise of around 8% by 2035. Gas prices are forecast to rise by around 90% between now and 2030 before levelling off.

The forecast rises are shown in Figure 11.

3.5 Energy and Water Forecast

Following the Energy Review, historic energy and water use and carbon emission data along with data on floor area, staff and student numbers and turnover has been obtained from the University's HESA (formerly HEFCE) returns. Together with heating and cooling degree

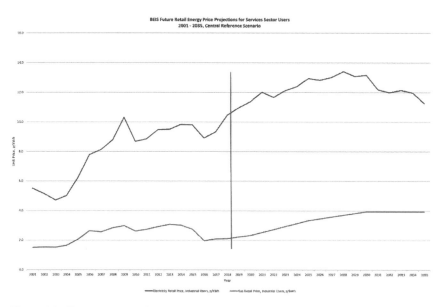

Figure 11 Forecast rise in electricity and gas unit costs to 2035 (BEIS, Services Sector, Central Reference Scenario).

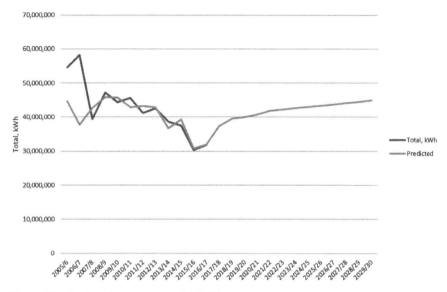

Figure 12 Predicted energy use to 2030/31.

day data, this has been used to explore the relationship between energy and water use and the various driving variables. It has been found that the strongest correlation for energy use (gas and electricity) is a combination of floor area, staff and student numbers (expressed as FTEs) and heating degree days (for water any combination tried gives a much weaker correlation though there are uncertainties in the data which make this analysis less reliable).

Our Campus Masterplan has been considered for planned new build and refurbishment works in order to obtain a reasonably accurate projection of floor area going forward to 2030/31 and forecast student numbers from 2017/18 to 2022/23 (and an estimate has been used thereafter to 2030/31) have also been considered. Degree days are difficult to predict, so an average figure has been assumed.

Using the Forecast tool an estimate of gas and electricity to 2030/31 has been generated. The resulting profile (showing total energy use) is shown in Figure 12:

This has then been overlaid with forecast gas and electricity price rises (derived from the BEIS Forecast referred to in Section 3.4, Figure 11), forecast grid electricity emission factors (again derived from BEIS) and forecast rates of Climate Change Levy (currently set only as far ahead as 2019/20, so with an RPI-based inflation factor built in thereafter).

The result is a forecast of both energy costs and Scope 1 and 2 carbon emissions between now and 2030/31.

This is then used to model various energy use and emission reduction scenarios to give a 'top down' view on what measures we will need to put in place to achieve our 81% carbon emissions reduction target by 2030/31, and also what would be a sensible interim target to be achieved by implementation of the EWCMP.

The model also provides a view of the energy and emissions 'Value-at-Stake', as discussed in Section 3.6.

3.6 Value at Stake

The Value-at-Stake analysis examines the consequences of 'doing nothing' in terms of energy and emissions management, versus the impact of 'doing something'. For the purposes of this analysis, we have taken the energy use as forecast by the methodology described in Section 3.5, overlaid with forecast energy price rises, carbon emissions taxes and forecast grid electricity factors. The impact on overall utilities costs is shown in Figure 13.

It can be seen, that without action on energy, water and emissions management, costs are forecast to increase from current levels of £2.8 million per year, to £4.6 million per year by 2030/31, an increase of 64% at today's prices.

Figure 14 present the overall impact of the University of Salford in terms of carbon emissions and general environmental performance.

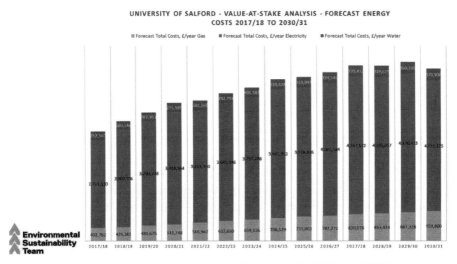

Figure 13   Value-at-Stake – Impact on overall utilities costs, 2017/18 to 2030/31.

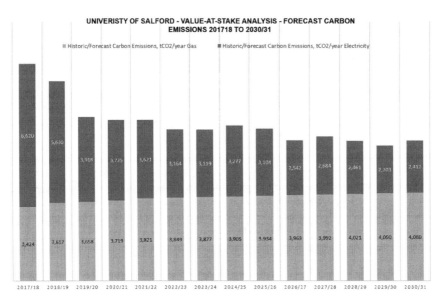

Figure 14   Value-at-Stake – Impact on overall carbon emissions, 2017/18 to 2030/31.

In the case of carbon emissions, the chart shows that, over the period to 2030/31, emissions will actually fall, driven largely by the successive decarbonisation of the electricity supply grid as forecast by BEIS, despite increasing consumption. Emissions from gas use, however, will continue to rise slightly in line with increasing consumption.

It can be seen, that without action on energy and emissions management, emissions overall are forecast to fall from current levels of around 10,000 tCO$_2$e per annum to around 6,600 tCO$_2$e per annum by 2030/31, a decrease of 35%. When compared to the 2005/06 baseline, the overall reduction is around 65%.

The question of what could be achieved through pro-active measures to reduce energy and water use and consequent costs and carbon emissions is the true Value-at-Stake and is addressed in Section 4.

4 Energy and Water Objectives, Targets and Action Plan

4.1 Development of Energy and Water Objectives and Targets

The aim of our Energy Water and Carbon Management Plan is to review our current position and to develop appropriate Objectives and Targets for the five-year period 2017/18 to 2022/23. This will help set the direction to achieving our long-term targets of reducing carbon emissions by 81% by 2030 and net zero carbon by 2038. Targets set under the EWCMP will be SMART (Specific, Measurable, Actionable, Relevant and Time-bound).

The analysis undertaken during the development of the EWCMP has resulted in two approaches to setting potential new realistic and achievable targets over the coming five-year period to 2022/23 and beyond to 2030/31. The first is a 'top-down' view of the emissions 'glidepath' required to meet the 2030/31 target, taking into account the impact on energy use of increasing student numbers and expanded floor area, together with the anticipated impact of grid decarbonisation as used in the forecasting tool described in Section 3.5.

The results of the top-down forecast are shown in Figure 15:

Gas

The UK operates a liberalised gas market, enabling consumers to change suppliers and tariffs at will (within normal contractual obligations), in order to take advantage of a competitive

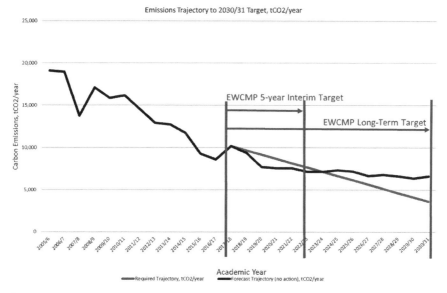

Figure 15 Historic emissions and future trajectory to 2030/31.

market. The market is dominated by a handful of larger players, with smaller operators at the margins. Competition and the availability of imported gas via transcontinental and subsea pipelines bringing gas from Russia and Central Asia, together with shipped imports of LNG from the Middle East and North Africa, have resulted in historically low gas prices over the last 15 or 20 years; however, these are starting to increase. The main factor driving the increase is competition for gas as the developing World both industrialises and decarbonises. Overall, according to BEIS forecasts, gas prices are expected to double between 2018 and 2030.Costs are made up of commodity costs (the cost of the gas supplied itself), plus non-commodity costs which cover the cost of gas storage and distribution, plus Government taxes/levies to support the development of renewable energy and reduce carbon emissions. The main tax is the Climate Change Levy, which is expected to increase dramatically in the coming few years as Government balances it up to meet that of electricity, in order to reflect the lower carbon impact of electricity now.In addition, VAT is charged on top of gas charges and levies.

Rising prices as a result of increased demand.
Potential increases in tariffs on imported gas (up to 50% of the UK's gas demand is imported through sub-sea interconnectors and shipping) as a result of Brexit. Geopolitical events causing market instability and rising prices. Potential supply interruptions as a result of local, regional, national or international demand, geopolitical events or malicious disruption. Increases to CCL to offset closure of CRC scheme. Gas will increase at a greater rate than electricity to level up the disparity between the two.

Purchase gas at lower prices
Move to all-electric campus to avoid gas costs and associated CCL. Reduce consumption through behavioural change and investment in energy efficient technologies.

Water
Water has traditionally been supplied on a regional basis, with consumers tied to regional water supply companies. The water market in England has been opened up to competition from April 2017; however, with businesses able to purchase water and waste water services from other providers on a national basis, or even elect to become a water supplier themselves ('self-supply', generally suited to organisations with large numbers of premises spread over a wide geographical area). The ultimate aim is to introduce competition between water service providers and reduce costs to consumers, although it is unclear how successful this has been to date.

Rising prices caused by the water companies' need to invest to replace aging infrastructure. Potential supply interruption or restrictions as a result of water stress resulting from climate change.

Reduce consumption through behavioural change and investment in water efficient technologies.Invest in rainwater harvesting or greywater recycling systems to reduce mains water demand. Investigate the viability of on-site abstraction from boreholes with on-site treatment.

Use water metering to identify and eliminate water leaks.

Reduce meter sizes to match actual demand to avoid meter size-related standing charges.

Look at on-site primary wastewater treatment to reduce tariff charges. Check water tariffs to ensure that these are fair and reasonable, particularly with respect to meter standing charges, surface water drainage, water not otherwise returned to sewer (e.g. evaporation), etc.Investigate options for purchasing water and waste water services from other providers on the open market.

Appendix B

Environmental and Sustainability Strategy Document

Our Mission

By pioneering exceptional industry partnerships we will lead the way in real-world experiences preparing students for life.

As a civic institution, our mission is to:

- Educate the next generation of modern industrialists, innovators, creators, entrepreneurs and leaders
- Develop the skills and knowledge needed to capitalise on the next industrial revolution
- Work in collaboration with public and private sector partners to address local and global economic and societal challenges

Our Vision for Sustainability

By embedding sustainability in all aspects of Organisation life, we will enable our Organisation community to have maximum beneficial impacts for society and environment.

1 UN Sustainable Development Goals

The Sustainable Development Goals, or Global Goals, are a call for action by all countries to promote prosperity while protecting the planet. They recognise that ending poverty must go hand-in-hand with strategies that build economic growth and address a range of social needs including education, health, social protection, and job opportunities, while tackling climate change and environmental protection.

We are aligning our Environmental Sustainability Plan to the Global Goals by mapping each area to the relevant goals. Figure 1 represents these Global Goals related to each area

2 About This Plan

This Plan provides narrative to the Organisation policies and objectives to improve environmental sustainability performance. At the Organisation we are proud of our place in the world. Situated in the heart of the vibrant Greater Manchester conurbation we play a significant role in the community. While our heart is anchored in the North West, we have a

Appendix B Environmental and Sustainability Strategy Document

powerful global reach. We focus on equipping our students, our future leaders, for the world of work.

We recognise the global climate crisis and that climate change is one of the biggest challenges facing our society across the globe. As a higher education provider we have a major role to play in enabling our students and staff to respond positively to global challenges.

We also recognise our responsibilities for the direct impact on the environment, as one of the largest organisations in the Greater Manchester region offering a place of study and work to over 18,500 students and 2,500 staff across 27 buildings and around 70 hectares. In the 2016/17 Academic Year (the baseline year selected for the majority of this Plan), energy and water costs were around £2.1 million per year. Total carbon emissions from energy use for the same period were 8,901 tCO_2e and waste from Organisation operations was 900 tonnes. We have an obligation to reduce our negative impact on our environment but also an opportunity to demonstrate environmental sustainability through our operations and using our campus environment as a living laboratory.

This plan sets out how we will demonstrate management of our environmental sustainability including how we will mitigate climate change, be resource efficient and contribute towards sustainable development through the operation of our campus and how this will contribute to education and research for sustainability.

Our Environmental Sustainability Policy sets the framework for our environmental objectives. This plan sets the key actions we will take, how we will monitor and review our progress. This plan is reviewed at least every two years and any amendments to objectives, targets and KPIs will be highlighted.

Figure 1 Our 2016/17 environmental impact.

3 Our Story so Far

We have built on our first Carbon Management Plan launched in 2011 and our Environmental Sustainability Policy, adopted in 2015, to ensure that our commitments around environmental sustainability remain current, challenging and reflect our Campus Masterplan. Most notably through the adoption of significant objectives to support the Greater Manchester science-based carbon target and ambition to become the greenest city region in the UK. Our new Energy, Water and Carbon Management Plan builds on our significant progress in carbon reduction (65% reduction since 2005/6) to ensure we can achieve our objective of an 81% reduction by 2030 and work towards net zero carbon by 2038. We have also committed to eliminate avoidable single-use plastics from our campus by 2022.

We recognise the importance of our green space on campus and have committed to maintaining these to a high standard including encouraging biodiversity. In recognition of this we received a prestigious Green Flag Award for our Peel Park and Frederick Road Campuses, well ahead of our target date. Our Campus Masterplan will also look for opportunities for new green spaces in our urban environment to help promote learning, well-being and healthy lifestyles.

In 2018 we completed development and implementation of our environmental and energy management into a system through the EcoCampus framework and are proud to have subsequently been awarded EcoCampus Platinum, ISO 14001 and ISO 50001 certifications.

We are proud of our achievements so far but we are committed to doing more. We are currently positioned 64th, a 2:2 classification, (2019) in the People and Planet Organisation League, our aim is to increase to the 1st class banding.**ronmental Management**

4 Environmental Management

An environmental management system is a way for an organisation to demonstrate that it addresses and minimises its environmental impacts, manages its legal compliance and continuously improves its environmental performance. An Environmental

Management System is a systematic approach to managing an organisation's impacts on the environment. An effective EMS will involve the following:

- An initial assessment of how the organisation's activities, products and services might affect the environment
- Making sure they comply with all relevant environmental regulations
- Gathering data, e.g. energy use, waste, water, raw materials
- Developing and communicating an environmental policy
- Procedures for controlling activities with significant environmental impacts
- Identifying risks and opportunities associated with environmental issues
- Setting targets and measuring progress
- Defining roles and responsibilities for all employees
- Training and awareness
- Periodic internal auditing
- Management Review and commitment from Leadership

The Organisation Estates and Facilities Division leads on the Environmental and Energy Management System (EEMS) which is certified to both BS EN ISO 14001: 2015 and BS EN ISO 50001:2018 standards covering all of our physical estates and operations. The EEMS was developed over a number of years using the EcoCampus approach with awards from bronze to platinum.

We will meet our environmental compliance obligations, protect the environment in which we operate our estates and facilities, and prevent pollution by reducing and eliminating pollution sources.

We will achieve this by

- Developing and maintaining an Environmental and Energy Management System externally accredited (ISO 14001:2015 and ISO 50001:2018 standards) for Estates and Facilities Division
- Maintaining a register of environmental compliance obligations and regularly evaluating our compliance
- Reducing pollution risk (emissions and discharges) to land, water and groundwater

We will monitor this by

- ISO 14001:2015 and ISO 50001:2018 certification
- Number of major non-conformities related to a breach in compliance obligations
- Number of pollution incidents

Relevant Documents

- Environmental Sustainability Policy

5 Energy, Water and Carbon

Carbon emissions are broken down into three categories by the Greenhouse Gas Protocol in order to better understand the source.

Scope 1 – All Direct Emissions from the activities of an organisation are under their control. Including fuel combustion on site such as gas boilers and fleet vehicles.

Scope 2 – Indirect Emissions from electricity purchased and used by the organisation. Emissions are created during the production of the energy and eventually used by the organisation.

Scope 3 – All Other Indirect Emissions from activities of the organisation, occurring from sources that they do not own or control. These are usually the greatest share of the carbon footprint, covering emissions associated with business travel, procurement, waste and water.

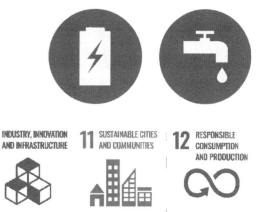

In 2011, the Organisation launched their first Carbon Management Plan with targets to reduce scope 1 and 2 carbon emissions by 43% by September 2020 compared to a 2005/06 baseline. As of end of academic year 2017/18 the Organisation had already reduced its scope 1 and 2 carbon emissions by 64%. This has been achieved partly through reductions in energy use at the Organisation through investment, partly through reductions in the grid electricity emissions factor, and also partly through the Organisation's divestment of its student accommodation in 2007/08.

Our Energy, Water and Carbon Management Plan builds on our significant progress in carbon reduction to ensure we can achieve our objective of a scope 1 and 2 81% reduction by 2030 and work towards net zero carbon by 2038 in line with the Greater Manchester carbon reduction strategy.

Scope 3 carbon emissions account for just over 80% of the Organisation's total emissions and we recognise the impact of these emissions so have committed to improving our monitoring and measuring of these and setting reduction targets where possible.

We will reduce the energy and water consumption of our buildings to meet the targets as established in the Energy, Water and Carbon Management Plan.

We will achieve this by

- Improved monitoring of energy, carbon and water data
- Eliminating or minimising energy and water waste
- Improving energy and water efficiency
- Monitoring our scope 3 carbon emissions

We will monitor this by

- Percentage reduction of scope 1 and 2 carbon emissions from baseline
- Energy consumed by gross internal floor area
- Percentage reduction in water use (per staff and student FTE) from baseline
- Publication of an annual scope 3 carbon emissions report

Relevant Documents

- Energy and Water Policy
- Energy, Water and Carbon Management Plan
- Scope 3 Carbon Emissions Report

6 Waste Management

Waste and recycling is one of the most visible areas in the sustainability agenda and can demonstrate our commitment to staff, students and visitors as well as offering an opportunity for people to easily get involved in environmental improvement. The 'waste hierarchy' ranks waste management options according to what is best for the environment. It gives top priority to preventing waste in the first place. When waste is created, it gives priority to preparing it for reuse, then recycling, then recovery, and last of all disposal (e.g. landfill).

Waste management and recycling are co-ordinated by the Organisation Estates and Facilities Division. In the academic year 2016/7 we produced approximately 900 tonnes of waste across the Organisation (excluding construction waste). We have worked with our waste contractors to divert as much waste as possible from landfill – the majority of our waste is sent for energy recovery. We operate a reuse scheme through Warp-it which facilitates reuse within the Organisation and wider. Our recycling rate in 2016/17 was 38%; we are working to increase this as well as reducing our total waste output.

Along with other universities and colleges in Greater Manchester, the Organisation will work to eradicate avoidable single-use plastics from catering, labs and stationery by 2022.

On plastics, we have already worked with our caterers, Chartwells (Salfood) and the Students' Union on a number of actions, including incentivised use of re-usable cups and non-plastic cutlery and takeaway containers. Chartwells and the Students' Union have also added a zero-waste element to one of their shops which sells dried goods. Instead of plastic packaging, customers bring their own containers or bags to fill up products and pay by weight or item.

We will apply the principles of the waste hierarchy to prioritise reduction, reuse and recycling to reduce the impacts of waste management.

We will achieve this by

- Reducing total waste sent for recycling and disposal
- Increasing waste recycled
- Eliminating avoidable single-use plastics on campus
- Encouraging contractors to apply the principles of the waste hierarchy on capital projects

We will monitor this by

- Percentage reduction in waste from baseline
- Percentage waste recycled
- Number of avoidable single use plastics on campus

Relevant Documents

- Waste Management Plan

7 Sustainable Procurement

Sustainable Procurement is the process of recognising the negative and positive environmental, social and economic impacts of purchased goods, works and services. It is a cradle to grave approach and should consider the entire supply chain rather than just the individual product or service. The UN Marrakech Task Force defines Sustainable Procurement as 'Sustainable Procurement is a process whereby organisations meet their needs for goods, services, works and utilities in a way that achieves value for money on a whole life basis in terms of generating benefits not only to the organisation, but also to society and the economy, whilst minimising damage to the environment.'

The Estates and Facilities Division at Organisation has significant spend covering everything from facilities management, minor works and regulatory maintenance and testing to major building projects and infrastructure development. With this amount of spend and potential environmental impact, the need for staff involved in purchasing to ensure that spend provides value is evident. Environmental and social responsibility is considered as part of our value for money process and appropriately considered in the procurement of supplies, services or works. For example we refer to the Energy and Water Technology Lists in relevant purchases.

Through our membership of North West Universities Purchasing Consortium, we are members of Electronics Watch, an independent monitoring organisation set up with the purpose of protecting the rights of workers within electronics supply chains. Through this membership we can collaborate to have a positive impact on workers through the electronics we purchase. We are also exploring ways to ensure transparency in supply chains of other goods.

We will develop and apply an ethical and sustainable procurement system to ensure social and environmental impacts of purchases are appropriately considered.

We will achieve this by

- Implementing a Sustainable Purchasing Policy and Plan for Estates and Facilities
- Ensuring all Estates and Facilities tender evaluations include sustainability
- Meeting Level 4 in all areas of the Flexible Framework
- Developing and implementing a Sustainable Food Policy

We will monitor this by

- Flexible Framework Self-Assessment
- Reviewing Estates and Facilities Invitation to Tenders for sustainability

Relevant Documents

- Sustainable Procurement Plan
- Sustainable Food Policy

8 Sustainable Construction

As a Future Enabler, one of the key pan-Organisation projects that will enable us to deliver the Organisation's vision, is our Campus Masterplan. Central to the plans is an energy strategy that takes a major step towards a zero carbon future, while providing high-quality spaces for residential, teaching, research and commercial uses that are cost-effective to run. Organisation buildings will undergo refurbishment to make them highly energy efficient, with maximum potential for renewable energy generation. We recognise there may be competing priorities and financial pressures, therefore we have developed a policy and design standard to guide our development teams on our sustainability principles.

We are committed to managing design, construction, refurbishment and post-completion occupancy of its buildings in order to reduce environmental impact, enhance the well-being of staff and student users of the building, minimise operating costs and comply with all relevant sustainable building legislation.

We will ensure sustainability is a core principle of the Estate Masterplan by embedding sustainable construction practices and performance specifications into plans for new buildings and refurbishments.

We will achieve this by

- Implementing our Sustainable Construction Policy
- Aiming for BREEAM Excellent for high value new builds where possible

- Considering the use of SKA methodology targeting at least Silver for lower value new builds and refurbishments
- Implementing our Energy Design Standard on all projects regardless of value

We will monitor this by

- Percentage of construction projects meeting sustainable construction policy requirements

Relevant Documents

- Sustainable Construction Policy
- Energy Design Standard

9 Travel and Transport

Our latest Travel Plan covered the period 2012–2017. We have recently reviewed this Plan and are aiming to relaunch an updated version as soon as possible.

Our travel surveys show that the majority of students travel by non-car modes (75%) and 49% of staff travel by non-car modes. Single occupancy car use has increased for both staff and students slightly overtime and is likely to be a consequence of reduced car sharing promotion and allowance of unrestricted parking for all users. The surveys also show that a high number of staff and students are unaware of the existing tools and infrastructure measures.

Our ambition is to consolidate and rationalise car-parking across campus and to reduce internal car circulation to promote a healthier, more pedestrian-friendly campus through our Campus Masterplan. The vision also promotes greater levels of cycling in and around the site. Improved pedestrian connections will be key to improving permeability across the campus – think tree-lined boulevards, public squares and a generally more well-defined network of routes and open spaces.

We will develop a travel plan which encourages alternative methods to single-occupancy car journeys and minimises the environmental impact of the Organisation fleet and business travel.

We will achieve this by

- Developing an updated Sustainable Travel Plan for 2021–2030
- Improving facilities for cyclists on campus
- Supporting the use of electric vehicles by staff and students
- Increasing the use of electric vehicles on campus by Estates and Facilities

We will monitor this by

- Published Sustainable Travel Plan
- Number of cycle parking spaces
- Number of electric vehicle charge points on campus
- Percentage of electric vehicles in Estates and Facilities fleet

Relevant Documents

- Sustainable Travel Plan

10 Biodiversity

Biodiversity, or biological diversity, is the variety of life! This includes variety in habitats (e.g. grassland and woodland) as well as diversity in species and the ecosystems in which they occur. Biodiversity is important because of the essential contribution that it makes to the functioning of our planet and because of all the benefits that it provides, from foods and medicine to climate regulation. Contact with biodiversity and the natural world has also been linked to improvements in health and emotional well-being.

Despite being just a mile and a half from the City of Manchester, our main campus in Salford is a leafy, riverside site adjacent to Peel Park and the River Irwell. The habitats on the Organisation campuses include lawned areas, memorial gardens, wildflower areas and woodlands. Many common British wildlife can be found on the campus and surrounding areas, as well as some not so common.

Our Campus Masterplan plans to fully utilise the green spaces in and around the campus, particularly Peel Park, David Lewis Playing Fields and The Meadows. Through our Masterplan, we will look for opportunities for new green spaces in our urban environment to help promote learning, well-being and healthy lifestyles.

We will protect and enhance the natural habitats and biodiversity on the Organisation estate.

We will achieve this by

- Implementing our Landscape Management Plan
- Maintaining our Green Flag Award

We will monitor this by

- Green Flag Award

Relevant Documents

- Landscape Management Plan

11 Community Engagement

Communication and engagement are important elements of our Environmental and Energy Management System (EEMS) and our Environmental Sustainability Policy.

To achieve our EEMS objectives and reduce the impact of the Organisation's activities on the environment, we need to raise awareness amongst staff, students and other interested parties to encourage and support them to take action and adopt positive behaviours as well as ensuring that any legal compliance requirements, our policies and environmental objectives are clearly communicated to staff, students and other interested parties.

Our colleagues' efforts in our Green Impact programme are a key component of our environmental sustainability aims. From energy-saving and waste reduction initiatives to greening offices with plants and Fairtrade coffee mornings our Green Impact teams lead a wide range of creative activities that reduce the environmental impact of our day to day work. Green Impact is an international scheme which brings staff and students together to increase sustainability across campus. Alongside the Green Impact teams, the scheme is facilitated by student auditors who assess project teams across our Organisation to meet their green targets.

We also work with key partners such as our Student's Union, catering teams and external partners such as TfGM and Salford Council on Go Green Salford, our four-week long annual campaign to raise awareness of the importance of sustainability and encourage positive change among our students and staff. Usually, taking place in February/March, it incorporates national campaigns such as the People and Planet, Go Green Week, Student Volunteering Week and Fairtrade Fortnight.

We recognise the importance of networking with relevant sector and professional bodies to understand best practice. Raising the external profile of the Organisation's activities and achievements in environmental sustainability and promoting any areas of good practice in environmental sustainability is also important. We have been shortlisted as finalists in the sector Green Gown Awards every year between 2011 and 2018, with a number of winning and highly commended entries.

We will embed care for the environment in the culture of our organisation as part of our goal to be a sustainable Organisation and empower and motivate the whole Organisation community to support our policy and strategy through appropriate education and communication.

We will achieve this by

- Implementing our Communications and Engagement Strategy for sustainability
- Increasing the influence of the Green Impact initiative at the Organisation
- Increasing engagement with students and staff at the Organisation with sustainability

We will monitor this by

- Number of staff/students influenced in Green Impact
- Number of staff/students engaged with sustainability campaigns and events

Relevant Documents

- Environmental Sustainability Communications and Engagement Strategy

12 Teaching, Learning and Research

One of the principles of the Organisation Industry Collaboration Strategy is to 'promote sustainability and social responsibility'. We take a broad definition of 'industry' and when we refer to industry partnerships and collaboration this includes work with private, public and third sector organisations of all sizes. From our inception, we have been committed to providing

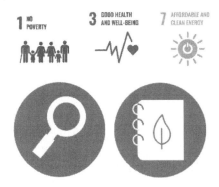

real-world solutions to tomorrow's challenges and the application of our research through partnership working and our current Research and Knowledge Exchange Strategy continues this tradition.

For example, our Applied Buildings and Energy Research Group (ABERG) is focused on establishing an evidence base to better understand and address the issues of energy consumption in buildings. The team covers a wide range of disciplines including building performance, electrical engineering, construction management, design and social sciences. ABERG is also the home of the unique Salford Energy House, a full sized two-bedroom terraced house built inside an environmental chamber that can replicate almost any weather conditions. With Energy House 2.0 coming soon we are looking to do even more in the area of Smart Living.

Whilst this strategy is focused on the environmental sustainability of our Organisation operations we aim to support and encourage teaching, learning and research through this. We aim to 'practice what we preach' by using our campus as a Living Laboratory. We are already partners in a multi-million pound European contract for a project to develop 'green' resilience to extreme climate hazards. The IGNITION project backed by £4 million from the EU's Urban Innovation Actions initiative, will see the Organisation host a 'living laboratory' for urban green infrastructure including roofs, walls and gardens. The project is led by Greater Manchester Combined Authority and supported by 11 partners, including the Environment Agency, and comes in response to increasing flooding, heat hazards and worsening air quality.

In our teaching and learning, one of our principles of a curriculum for Industry Collaboration is 'Education for ethical behaviour', which presents an approach to student education that seeks to balance human and economic well-being, knowledge, and competencies with cultural traditions, respect for other people, and for the earth and its natural resources. In practice this requires design of academic programmes with consideration for sustainability including all the Sustainable Development Goals.

We will support our Organisation Strategy in providing real-world experiences preparing students for life and supporting our research to provide real-world solutions to tomorrow's challenges.

We will achieve this by

- Continuing to share our environmental sustainability experience and challenges through guest lectures
- Offering our environmental sustainability challenges as live briefs for students
- Encourage and support the implementation of a living laboratory on the campus showcasing sustainability solutions

We will monitor this by

- Number of environmental sustainability guest lectures delivered
- Number of environmental sustainability live briefs/student projects supported
- Number of demonstrator environmental sustainability projects on campus

Relevant Documents

- Organisation Strategy 2016–2021
- Programme Design, Approval, Amendment, Review and Withdrawal Policy

We have the ability to contribute to all the Global Goals through our teaching, learning and research.

13 Our Governance

We manage and control environmental sustainability on our campus and in our operations through our Environmental and Energy Management System (EEMS) which is certified to ISO 14001:2015 and ISO 50001:2018. This provides the framework for continual improvement in our environmental performance. We will deliver, review and monitor this plan through our EEMS.

Environmental sustainability is embedded within the Organisation through high-level commitment from Organisation leadership and our policies. A number of Organisation groups provide strategic direction, support and oversight for implementation of the Organisation policies.

The Environmental Sustainability Team leads the delivery of this plan, directly reporting to the Environmental Projects Board, chaired by the Director of Estates and Facilities, who report to Organisation senior management groups via the Estate Programme Group and Sustainability Strategy Group.

The Environmental Sustainability Team manage budgets for energy management, waste management and sustainability communications and engagement. Business cases and requests for further funding are submitted via the Environmental Projects Board through the central Organisation financial process.

We have created a number of policies and detailed plans to support our continual improvement aims and progress is reported quarterly to the Environmental Projects Board and publicly annually

Appendix B Environmental and Sustainability Strategy Document

Environmental Sustainability Policy

- Energy and Water Policy
- Sustainable Construction Policy
- Estates Sustainable Procurement Policy
- Sustainable Food Policy

- Energy, Water & Carbon Management Plan
- Sustainable Procurement Plan
- Waste Management Plan
- Landscape Management Plan
- Sustainable Travel Plan
- Communications and Engagement Strategy
- Sustainable Food Plan

Appendix C

Considerate Constructor Scheme Site Registration Monitors Checklist

Site Registration Checklist

Monitors' checklist for registered sites

This Checklist should be used as a basis for discussions that complement direct observations by the Monitor. Questions highlighted in bold on the Checklist must be satisfactorily addressed for the site to achieve compliance and this indicates that a site has reached a standard beyond statutory requirements. Sites are not only assessed for compliance, but also to identify measures taken which are above and beyond these requirements and addressing the non-bolded questions on the Checklist may result in a higher score.

The Checklist contains a number of prompts that highlight specific areas the Monitor may take into consideration when reviewing that question. The prompts shown are not exhaustive and inevitably other items will need to be considered. Monitors will only look at the activities of the site during the construction phase.

Monitors will use their discretion when assessing whether questions or prompts are relevant. Where they are not, they will not be considered when assessing performance or awarding a score. Monitors will decide whether a question has been adequately addressed taking into account the size, type and location of the site. Credit will not be given for activities that are planned but have yet to be carried out.

The safety of the working site is outside of the scope of the Code of Considerate Practice and the Scheme's monitoring process.

Site details

Brief description of the work, sector, site location and context, anything 'sensitive', type of contract and other relevant matters.
Are there any constraints that restrict the site's ability to address certain areas of the Checklist?

Care about **Appearance**
Constructors should ensure sites appear professional and well managed

1.1 Does the external appearance of the site present a positive image of the industry?	Initial impressions, signage, enclosures, entrance, obstructions, mud, debris, litter, graffiti
1.2 Does the site appear well organised, clean and tidy?	Tidiness, organisation
1.3 Does the external appearance of all facilities, stored materials, vehicles and plant make a positive impression?	Screening of facilities, remote compounds, organisation, layout and tidiness, supply chain, accommodation
1.4 Does the appearance of the workforce project a positive impression?	Onsite dress code, offsite appearance, smoking and vaping
1.5 What actions are taken to keep the perimeter and surrounding areas clean, tidy and free of litter, mud and dust?	Inspection, public rubbish, surrounding roads/pavements, dust prevention
1.6 What arrangements are in place to ensure that the public and visitors see a site that is organised, clean and tidy?	Site waste, viewing points, vandalism, waste bins
1.7 How are compounds, waste and storage areas cleaned, managed, and maintained?	Supervision, procedures, checklist
1.8 How does the site encourage the workforce to contribute to cleanliness and good housekeeping?	Workforce awareness, involvement
1.9 How is all smoking and vaping managed to avoid a negative impact on the public?	Guidance, discreet areas, ashtrays
1.10 How are company values, corporate identity and a positive industry image promoted?	Communication, guidance, branding, signage, websites, social media, corporate badging, First Impressions standards

Respect the **Community**
Constructors should give utmost consideration to their impact on neighbours and the public

2.1 Are all those affected by the work identified, notified and kept informed and shown courtesy and respect?	Pre-start information, updates, advance notice of disruptive works, site, company and out of hours contact information, complaints procedures, sensitivity to neighbours, special needs, working hours
2.2 Are all reasonable efforts being made to minimise the impact of deliveries, parking and work on the public highway, cycleways and footpaths?	Routes, timings, unloading, public diversions, utility works, traffic management arrangements
2.3 Is the site contributing to and supporting the local community and businesses?	Goodwill, community liaison, local shops, trade contractors, labour, suppliers
2.4 Is the site actively promoting the benefits of registration with the Scheme?	Workforce, inductions, toolbox talks, consultants, clients, supply chain, public, newsletters, Scheme posters and banners
2.5 How does the site ensure that all those affected, including visitors, are treated with consideration, courtesy, and respect?	Local and special needs, operative conduct and behaviour, induction and training, ramps, facilities for visitors, community notice board

(Continued)

2.6 How is nuisance and intrusion minimised?	Noise, privacy, outlook, radios, phones, cameras, parking and obstruction
2.7 How are compliments, comments and complaints sought, recorded and managed?	Feedback, public surveys, regular reviews
2.8 What is being done to support and contribute to the local community including promoting local employment?	Corporate Social Responsibility action plan, schools/colleges, businesses, residents
2.9 How do company directors, senior managers, clients, consultants and the supply chain assist the site in meeting the requirements of the Code?	Scheme champions, CCS briefing and action plan, meeting agendas, management review, promoting Scheme registration
2.10 What is being done to leave a positive and lasting impression of the industry on completion of the project?	Co-operation, support, charities/organisations, legacy

Protect the **Environment**

Constructors should protect and enhance the environment

3.1 Are environmental issues identified, communicated, managed, and promoted?	Environmental policy promoted, initial environmental assessment, site environmental plan, workforce, supply chain, site specific induction
3.2 Is waste avoided and the use of resources and energy minimised?	Policy and plan, reducing, reusing and recycling, sustainable solutions
3.3 Are all reasonable efforts being made to minimise the impact of vibration and of air, light and noise pollution?	Working methods and equipment, programming, monitoring, inspections
3.4 Are all reasonable efforts being made to protect the existing ecology, the landscape and watercourses?	Birds, trees, plants and wildlife, river, hazardous substance storage, spill control
3.5 How are environmental issues identified and managed?	Management policy, training, investigation, specialist input, monitoring, incident procedures, certifications, actual vs target
3.6 How are environmental issues communicated and promoted to the workforce and the general public?	Newsletter, notice board, local groups, consultation, involvement, promoting achievements
3.7 How is the site measuring and minimising its use of natural resources including water?	Sustainable sources, policy, implementation, water/energy saving measures, harvesting rainwater, offsite construction, prefabrication
3.8 How is the site measuring and reporting its carbon footprint and what is being done to reduce it?	Carbon footprint reporting, report to company level, energy and fuel use minimised, energy efficient accommodation, signage, travel plans, green purchasing, carbon offsetting

(Continued)

(Continued)

3.9 How is vibration, and air, light and noise pollution managed to minimise impact?	*Campaigns, plant and machinery, idling vehicles, hybrid/electric plant*
3.10 How is a positive contribution being made to the natural environment?	*Awareness, goodwill work, planting, landscaping, local materials, improvements, post-completion impact, plastics and packaging*

Care about **Safety**

Constructors should attain the highest levels of occupational health and safety performance

> Note: The CCS Code of Considerate Practice expects constructors to have practices and approaches in place which demonstrate that they care about occupational health and safety for visitors and the workforce on site. In relation to the public, we expect constructors to have practices and approaches in place which demonstrate that they care about the public's safety around the site or in relation to the construction activity.

4.1 Are practices and approaches in place that care for the safety of the public, visitors and workforce?	*A&E/MIU, first aid, safety plan updated, inspections and reporting, risk information, PPE, protected and controlled access, signing in and out*
4.2 Have all risks to neighbours and the public been considered and addressed?	*Site/boundary/property security, scaffold protection, traffic management, child safety, escape routes, safety signage*
4.3 Are initiatives in place for continuous occupational health and safety improvements?	*Workforce consultation and information, training, campaigns*
4.4 Does the site encourage attitudes and behaviours that enhance occupational health and safety performance?	*Management, supervision, site-specific induction, drugs and alcohol policy, RAMS review*
4.5 How is the safety of the public outside the site addressed and monitored?	*Falling debris, roads, footpaths and diversions, barriers, lighting*
4.6 What arrangements are in place for dealing effectively with emergencies?	*Emergency procedures, visible first aiders, defibrillators, drills, injuries, visitor medical details*
4.7 What is done to ensure that the movement of vehicles and plant to and from the site is not a risk to vulnerable road users?	*Initiatives, cycle safety, CLOCS, speed limits, supervision, inspections, access/egress, FORS, vehicle enhancements, traffic marshal*
4.8 How are accidents, incidents and near misses recorded, and what is done to learn from them?	*Identification of near misses, recording, analysing, communicating, training*
4.9 How does the site provide current safety and risk information to operatives and visitors?	*Hazard board, daily briefing, driver and visitor information, language differences, visual/hearing impaired*
4.10 How does the site embed a culture of continuous positive occupational health and safety performance?	*Attitudes, behaviour, incentives, controls, supply chain engagement, drugs and alcohol testing*

Value their **Workforce**

Constructors should provide a supportive and caring working environment

5.1 Does the site demonstrate a commitment to respect, fair treatment, encouragement and support?	*Equality, diversity and inclusion, bullying, harassment, inappropriate language, management attitudes, open door policy, recognition, feedback, consultation*
5.2 Are personal development needs identified and is training promoted?	*Training – employed, self-employed, trade contractors, supply chain*
5.3 Does the site care for the health and well-being of the workforce?	*Occupational health risks assessed and addressed, posters, healthy lifestyle and mental health advice, emergency contact details, medical conditions, medications*
5.4 Are suitable, hygienic and well-maintained welfare facilities provided within a reasonable distance of the work area?	*Changing, secure storage, drying, toilets, canteen, cleaning regime, separate facilities, sanitary bins*
5.5 How does the site assess and monitor the legitimacy and competency of the workforce?	*CSCS, skills cards, illegal workers, spot checks, modern slavery*
5.6 What is the site doing to support the company in promoting construction as a career of choice?	*Campaigns, careers advice, apprenticeships, placements, disadvantaged groups, minority groups, Promoting Construction campaign*
5.7 How is the health and well-being of the workforce assessed and addressed?	*Worker fatigue, weather protection, stress, vaping, health screening, counselling services*
5.8 What is being done to help encourage more equality, diversity and inclusion on site?	*Religious considerations, cultural needs, literacy and numeracy training*
5.9 What additional facilities are available for those working onsite?	*Showers, lockers, rest, multi-use room, recreation, Wi-Fi, laundering*
5.10 What is the site doing to support the company to improve its image and the overall image of the industry?	*Qualifications, e-learning, financial advice, contribution to Scheme's Best Practice Hub, rewards, benefits*

Additional Information
Innovation

Please detail any measures implemented on the project that demonstrate innovation and original thinking.

Site-specific data

This information is used to capture key information and identify trends within the industry

What is the average number of operatives on this project?	
Of these, how many on average are women?	

(Continued)

(Continued)

Is there someone on site who champions Equality, Diversity and Inclusivity?	Y / N / Don't know
What percentage of the current workforce holds a CSCS card?	%
How many companies within the supply chain on this project are registered with the Scheme?	
Are there processes in place to ensure subcontractors (and subsequent subcontractors) are conducting 'right to work' checks?	Y / N
Are physical spot checks conducted to ensure minimum standards of 'right to work' checks are taking place within the supply chain?	Y / N
Is the company a CLOCS Champion?	Y / N / Don't know
Is this site operating to the requirements of the CLOCS Standard?	Y / N / Don't know
Has the site updated its profile on the Construction Map?	Y / N
Does this site have visible posters on display for the workforce to highlight the relevant Covid-19 Government App and/or Covid-19 Government QR code check-in (noting that one or both may exist depending on the information provided for England and each devolved UK nation and Ireland)?	Y / N

Please note that questions below refer to data **since the project first commenced**

How many compliments have been received and recorded by the site?		
How many complaints have been received and recorded by the site?		
How many reportable accidents have there been?		
How many non-reportable accidents have there been?		
Have there been any fatalities on site?		Y / N
If so, were any construction related?		Y / N
How many schools, colleges or universities have visited the site or has the site visited?		
Approximately, how many students have been spoken to about the construction industry?		

Feedback

Does the Site Manager have any comments, questions or suggestions for the Scheme?

Notes

Does the administration office need to be informed of any changes to the site details or are there any product queries?

Appendix D

COVID-19 Project Review of Site Operating Procedures

Project Name		Project Number		
Completed by		Date		
			Compliance Review	
Site Operating Procedure	**Project Arrangements**		Yes	No
Information on symptoms				
What information has the project provided to staff and supply chain to remind them of the symptoms of Coronavirus and what action they should take if they or their family (household) develop symptoms, or they are living with an at-increased risk or vulnerable person?	SHE Alert issued to all contractors explaining what the symptoms are and what to do if you or a family member are showing symptoms.			
	A copy of the SOP Document rev 3 sent to all contractors and a SHE Bulletin explaining the changes.			
	TBT completed by supervisors to all operatives on site.			
The SOP requires us to remind the workforce 'at every opportunity' of the safe operating procedures and how they are implemented at this location, paying particular attention to changes. Please detail how the project is achieving this? (briefings, toolbox talks etc.)	SOP is discussed during the daily breakfast briefing with all supervisors.			
	POWRA completed and briefed to operatives daily.			
	Weekly VOICE meetings with operatives.			
	SHE Bulletin displayed explaining the changes to the SOP document.			

(Continued)

(Continued)

Site Operating Procedure	Project Arrangements	Compliance Review	
		Yes	No
If someone falls ill whilst at work			
What are the projects arrangements for dealing with a person who falls ill whilst at work and are these detailed within the CPHSP and communicated to staff?	Temperatures of all persons entering site are taken at gatehouse using CCTV heat camera (max. 37.9 deg). Employer to be advised of high temperatures and person prevented from entering site. Person to be sent home and follow PHE self-isolation guidelines. Deep clean of welfare to be carried out. Arrangements are added to the site COVID-19 plan included in the PEP document.		
Travel to site (wherever possible workers should travel to site alone using their own transport):			
How are we informing contractors that they need to consider means of travel for staff that avoid them sharing small vans/cars where possible and if this is unavoidable that they are following the SOP guidance around sharing vehicles?	All contractors have been issued a copy of the SOP document and the SHE Bulletin detailing the requirements. Reminders given in the Daily supervisor briefings.		
What additional parking arrangements for cars and bicycles are in place?	Additional cycle storage is available.		
What arrangements are in place to allow staff/contractors to avoid public transport, and if this is not possible what measures have been taken to avoid use of public transport in peak times? (staggered shifts, not between 0545–0730 and 1600–1730)	Staggered start and finish times considered where possible.		
What hand-cleansing facilities are provided immediately adjacent entrances and exits to site? (This should be soap and water wherever possible or hand sanitiser if water is not available. Consider pop-up hand-cleaning stations.)	Mobile handwashing station provided in compound areas. Wall-mounted hand sanitiser to site entrance/exit. Bottles of hand sanitiser available in offices and welfare.		
How would somebody be taken home if they became ill? (Avoid sharing vehicles if possible, if not what additional controls are in place?)	Affected person's family to be contacted to arrange pick up of the person.		

(Continued)

Site Operating Procedure	Project Arrangements	Compliance Review	
		Yes	No
Site access points			
What action has been taken to stop non-essential visitors coming to site?	Meetings are encouraged to take place over video conferencing. Non-essential meetings/site tours have been cancelled.		
What arrangements are in place to prevent congestion at key locations around site? (Consider staggered start and finish times, limiting numbers of people.)	One-way pedestrian routes established. 2 m markings at turnstiles and pedestrian routes to maintain distances. Staggered start and finish times considered.		
• Is signage such as the below in place at all site access points:Floor markings to ensure 2 m distance is maintained • To remind workers not to attend if they have COVID-19 symptoms?	Yes Yes		
• What arrangements are in place for the following?Monitoring access points to ensure social distancing • Eliminating regular touch points such as fingerprint readers at turnstiles (these should only be put back into operation where their use is contactless, or is cleaned by an attendant between every touch) • Visitors or staff needing access to the site offices	Security staff monitor entry points for compliance. Separate access and exit points set up where possible. All biometrics at turnstiles are turned off. MSite APP in use by security staff to record attendance on and off site. Non-essential visitors attendance prohibited.		
Is signage in place requesting staff to wash hands on entry to, and exit from the project? What other arrangements have been taken to encourage handwashing and good hygiene?	Yes, signage displayed at entry points and in prominent places throughout the project. Toolbox talk issued to all contractors.		
What arrangements are in place to regularly clean common contact surfaces in reception, office, access control and delivery areas, e.g. Scanners, turnstiles, screens, telephones, desks, particularly during peak flow times?	Additional cleaners employed. Increased cleaning at regular intervals. Cleaning provisions made available in the offices.		
How is the project reducing the number of people in attendance at site inductions and holding them outdoors wherever possible?	Online Pre-registration completed by all new starters. All inductions must be booked in advance. Restricted numbers per induction, multiple induction presentations if required.		

(Continued)

(Continued)

Site Operating Procedure	Project Arrangements	Compliance Review	
		Yes	No
Drivers should remain in their vehicles if the load will allow it and must wash or clean their hands before unloading goods and materials. How is this being managed at the project?	Gatemen controlling deliveries and providing instructions to drivers. Gloves are mandatory for all persons.		
What arrangements are in place to regularly monitor compliance to the arrangements for site access detailed above?	Daily monitoring by Supervisors. Weekly SHE inspections by site management. 2 weekly inspections by visiting SHE team.		
Handwashing			
How are handwashing facilities, in addition to the usual welfare, provided at key locations around the project, especially where the project has large numbers of people, is spread out, or has remote areas? Consideration also needs to be given to plant operators. Consider the use of pop-up handwashing stations to supplement where required.	Hand pop-up washing stations have been placed within easy access locations throughout the site. Hand wipes provided to plant operators to regularly clean controls.		
• Detail how the following is being managed and monitored throughout the day: That soap and fresh water is always readily available and kept topped up. • Handwashing facilities are regularly cleaned • That suitable and sufficient rubbish bins for hand towels are located in key areas, and regularly removed and disposed of. • Adequate supply of soap, hand sanitiser and paper towels are available, and storage for additional is secure.	Cleaning team regularly replenish soaps each day along with bin changes and paper towels. Regular stock checks of soap and toilet roll carried out with Buying dept. All supplies are securely stored away when not in use.		
Is hand sanitiser provided in locations where handwashing facilities are unavailable, and is it a minimum of 60% alcohol based? Please detail locations.	Yes but washing facilities are regularly available. 60% alcohol-based sanitiser supplied as a minimum.		
Toilet facilities			
What restriction is in place on the number of people using toilet facilities at any one time. How is this being managed, and is signage in place to make people aware of these requirements, including floor markings for those queuing to use them?	Security staff employed to monitor welfare facilities. Social distancing signage displayed. Additional toilets arranged where required.		
Is signage in place to inform workforce to wash hands before and after using the toilet facilities?	Yes, signage displayed.		

(Continued)

Site Operating Procedure	Project Arrangements	Compliance Review	
		Yes	No
What is the cleaning regime for toilet facilities particularly door handles, locks and the toilet flush?	All handles, locks, taps and flushes included in the regular cleaning regime.		
Portable toilets should be avoided wherever possible, but where in use these should be cleaned and emptied more frequently. If in use, detail how this is being undertaken.	Included in the regular cleaning regime and additional arrangements made with supplier.		
Canteens and eating arrangements			
What arrangements are in place to ensure that microwaves, kettles etc. are cleaned between use?	Kettles and toasters have been removed from site. Microwaves and hydro boilers have been added to the cleaning regime.		
What arrangements are in place to reduce congestion and contact times within canteens? (staggering breaks, limiting numbers etc., consider increasing the number or size of facilities)	Staggered break times scheduled by contractor. Spare seating removed to prevent use. Additional welfare provided where needed.		
Is the maximum capacity of any welfare facility clearly identified at the entry points, and what arrangements are in place to supervise compliance? (consider attendants where appropriate)	Removed spare seating from canteen. Additional outside seating area provided. Canteen layouts displayed and signage stating where to sit. Welfare attendant in place.		
Are handwashing facilities or hand sanitisers available at, or immediately adjacent, the entrance to any room where people eat? Please detail locations.	Yes, at entry/exit points.		
To limit the need to visit local stores, the workforce should be asked to bring pre-prepared meals and refillable drinking bottles from home. Has this been communicated?	Yes, discussed in VOICE meetings and raised with supervisors to communicate to all of their staff.		
Workers should sit (2 m) apart from each other whilst eating and avoid all contact. What arrangements are in place to ensure this occurs?	Spare seating removed. Additional outside seating provided. Signage and seating plans displayed.		
Have catering/server facilities that provide hot or cold food for sale been shut down?	Yes		
What enhanced cleaning measures are in place around drinking water facilities such as dispensers and taps?	None touch drinking water stations are in use. Taps are included in the regular cleaning regime.		

(Continued)

(Continued)

Site Operating Procedure	Project Arrangements	Compliance Review Yes	No
How is the project ensuring tables are cleaned between each use?	Schedules developed for staggered brew times. Cleaning regime works around this schedule.		
Are enough bins located in canteens so that rubbish can be put straight in the bin and not left for someone else to clear up?	Yes		
What arrangements are in place for all areas used for eating to be thoroughly cleaned at the end of each break and shift? (including chairs, door handles, vending machines)	Schedules developed for staggered brew times. Cleaning regime works around this schedule.		
What arrangements/communication is in place to advise staff to use own cutlery and mugs, store within own bags at all times along with food and remove non-essential items? Consider disposable cups/cutlery as an alternative.	Staff encouraged to bring their own cutlery and mugs. Dishwasher provided for remaining items.		
Changing facilities, showers and drying rooms			
What arrangements are in place to reduce congestion and contact times within these areas? (staggering start and finishing times, limiting numbers etc.)	Staggered start and finish times considered. Additional facilities provided. Monitored by security staff.		
What arrangements for enhanced cleaning of all facilities throughout the day and at the end of each day is in place?	Additional number of cleaners in place. Increased cleaning regime in place throughout the day.		
Are facilities adequate to allow social distancing measures to be effective, and if not, could we increase the number or size of facilities available on site if possible?	Yes facilities are currently adequate but monitored as the numbers increase, additional facilities will be arranged if needed.		
Based on the size of each facility, determine how many people can use it at any one time to maintain social distancing of 2 m?	Seating plans for welfare facilities arranged. Staggered brew and start/finish times considered.		
Are enough bins located in canteens so that rubbish can be put straight in the bin and not left for someone else to clear up?	Yes		
Is the maximum capacity of any changing/drying room facility clearly identified at the entry points, and what arrangements are in place to supervise compliance? (consider attendants where appropriate)	Posters displayed showing maximum numbers.		

(Continued)

Site Operating Procedure	Project Arrangements	Compliance Review Yes	No
Avoiding close working (within 2 m)			
What project level arrangements are in place to review all tasks and identify close working so it can be avoided where possible?	All RAMS reviewed against latest SOP document. Task assessments undertaken. POWRA completed daily.		
Has work that requires working within 2 m been fully risk assessed using the Morgan Sindall Close Working Guide?	Yes This is communicated to all contractors.		
If a supply chain contractor wishes to use their own format for assessing close working, this must as a minimum meet the requirements of the Construction Leadership Council Safe Operating Procedures.			
What arrangements are in place to regularly monitor close working activities to ensure compliance?	Contractor Supervisor inspections, MS Site Manager inspections, SHE Team inspections, Leadership Engagement tours.		
• Where close work is required, ensure the following arrangements are being considered:Increase ventilation in enclosed spaces • Skin to skin contact is being avoided • Re-useable PPE is being thoroughly cleaned after use and not shared between workers • Single use PPE should be disposed of so that it cannot be reused	The MS 'Assessment of close working' document is to be used. where works cannot be completed while maintaining 2 m; this has been briefed to all contractors.		
• Stairs should be used in preference to lifts or hoists; however where lifts or hoist are used please detail the arrangements for:Lowering their capacity to reduce congestion and contact at all times • Regularly cleaning touchpoints, doors, buttons etc.	Restriction on the number of occupants in hoists/lifts. Signage displayed. Social distancing markers in waiting areas. Operator regularly cleans touch points in hoist/lift.		
Prohibit the sharing of vehicle/plant cabs wherever possible, but where essential, regularly clean the inside of vehicle cabs and between use by different operators. Please detail arrangements for this and how it has been communicated?	Where possible plant is to be used by a single operator only. Cleaning provisions to be provided to plant operators to allow regular cleaning of controls.		

(Continued)

(Continued)

Site Operating Procedure	Project Arrangements	Compliance Review Yes	No
• Detail arrangements for site meetings to ensure the following is undertaken.Attended by outside participants only when absolutely necessary. • Attendees are more than 2 m apart from one another. • Rooms are well ventilated and fresh air circulating. • Meetings held in open/external areas wherever possible.	Meeting rooms attendee numbers restricted. Non-essential attendees prohibited. Meetings take place over video conference whenever possible. Meeting rooms to be appropriately ventilated.		
What arrangements are in place to maintain 2 m separation around main access and egress routes and within offices, canteens etc. (One-way systems on stairs/corridors, thumbs-up to pass, floor markings etc.) How are we achieving good ventilation in all areas?	One-way systems in place where possible. 2 m markings on all floors at access/egress routes and waiting areas. Introduction of a 'Thumbs up' to pass system when using staircases and routes without one-way systems. Office windows left ajar during day to circulate air		
Cleaning • Detail the enhanced cleaning procedures in place across the site, particularly in communal areas and at touch points including:Taps and washing facilities • Toilet flush and seats • Door handles and push plates • Hand rails on staircases and corridors • Lift and hoist controls • Machinery and equipment controls • Food preparation and eating surfaces • Telephone equipment • Key boards, photocopiers and other office equipment	Additional cleaning staff in place. Increase cleaning regime and inclusion of all multi-touch surfaces. Operatives advised to bring food which doesn't require heating. Operatives advised to bring thermos flask for hot drinks. Prevention of the sharing of devices.		
Detail the arrangements for rubbish collection and storage points? These should be increased and emptied regularly throughout and at the end of each day.	Skips and wheelie bins provided for office and welfare wastes. Bins emptied regularly throughout the day.		
First aid Are adequate resources for first aid being maintained at the project and agreed between all parties? (number of first aiders, equipment etc.)	Yes, appropriate coverage of first aiders is regularly monitored. The MS COVID-19 guidance for 1st aiders sent to all contractors.		

(Continued)

Site Operating Procedure	Project Arrangements	Compliance Review	
		Yes	No
Consideration must be given to delayed response from emergency services, this may require additional first aid resources or rescheduling high risk work? How has this been considered and what arrangements are required?	AED defibrillator provided. Appropriate number of first aiders and first aid equipment. Blankets available to keep a casualty warm. Additional emergency arrangements added to RAMS for all tasks.		
Have first aid supplies been assessed to ensure they contain adequate supplies of items required to maintain good hygiene (Face shields, gloves etc.)	Yes		
Have first aiders been briefed on the actions to take in the event a first aid incident being required. MS Guidance for First Aiders.	Yes, guidance given to all staff regarding first aid.		
Review			
This SOP needs regular review at each project a minimum of weekly by the site team, and at regular points by senior management. What are monitoring and review arrangements for this project?	Reviewed on a weekly basis and during team meetings.		

Appendix E

Supply Chain Communication and Coronavirus Business Continuity Planning

March 2020

COVID-19 ('Coronavirus')

We appreciate that you may be concerned about the impact of Coronavirus on our people and our operations.

Please be assured that we have a robust Business Continuity Plan (BCP) in place, which is regularly reviewed and updated. We are using this to help guide our actions in response to the Coronavirus, and taking a number of specific actions as a result:

- Daily calls with a core Business Continuity Team to monitor our exposure, to ensure we stay abreast of the latest Government guidance, and to ensure this is effectively communicated to our people and our supply chain colleagues working on our sites
- Regularly reminding our people about the need to maintain good hygiene, and updating them on the latest Government guidance on self-isolation
- Liaising with supply chain partners to identify and mitigate any supply risks as part of wider project-specific business continuity assessments

Our BCP includes contingency arrangements for our back-office functions including transaction processing, finance and payroll.

What you can do to help

1) Reiterate to your people the importance of maintaining good hygiene and following Government guidelines regarding self-isolation. Do not allow operatives to attend site if they are showing symptoms that would require self-isolation.
2) Inform the Construction Project Manager immediately if an operative working on one of our projects is required to self-isolate or has a confirmed case of Coronavirus.
3) Liaise with your supply chain to understand any supply issues – please see overleaf for areas to consider.
4) Please notify us immediately of any operational issues relating to Coronavirus.

If you have any questions or require further information, please do not hesitate to speak to your Construction contact or get in touch with me directly.

Appendix F

COVID-19 Management System

Management System

Toolbox talks

What is Coronavirus?

Coronaviruses (CoV) are a large family of viruses that are common and are typically associated with mild illnesses, similar to the common cold.

A novel coronavirus (nCoV) is a new strain that has not been previously identified in humans.

How is Coronavirus spread?

Coronaviruses are zoonotic, meaning they usually begin in animals. While rare, coronaviruses can evolve to infect people. For example, investigations found that SARS-CoV was transmitted from civet cats to humans, and MERS-CoV was transmitted from dromedary camels to humans.

In some cases, the Coronavirus can evolve further and spread from person-to-person. The route of transmission is not always known, but the viruses are generally thought to spread by respiratory droplets when people are in close contact.

What are the symptoms of infection from Coronavirus?

Each Coronavirus will vary in the severity of the infection it causes. Common signs of infection include respiratory symptoms, fever, cough, shortness of breath and breathing difficulties. In more severe cases, the infection can cause pneumonia, respiratory issues, kidney failure, and sometimes death.

COVID-19 presents with fever and symptoms of lower respiratory illness (e.g., cough or difficulty breathing).

What is the recommendation to prevent the spread of Coronavirus?

There are general principles set out by Public Health England that you can follow to help prevent the spread of respiratory viruses, including:

Hand and cough hygiene

- Washing your hands more often – with soap and water for at least 20 seconds or use a hand sanitise when you get home or into work, when you blow your nose, sneeze or cough, eat or handle food.
- Avoid touching your eyes, nose, and mouth with unwashed hands.
- Avoid close contact with people who have symptoms.
- Cover your cough or sneeze with a tissue, then throw the tissue in a bin and wash your hands.
- Clean and disinfect frequently touched objects and surfaces in the home.

(Continued)

(Continued)

Social distancing

What is social distancing? Social distancing measures are steps you can take to reduce social interaction between people. This will help reduce the transmission of Coronavirus (COVID-19).

The General Principles outlined by the construction leadership council are that:

Non-essential physical work that requires close contact between workers should not be carried out.

Work requiring skin to skin contact should not be carried out.

Plan all other work to minimise contact between workers.

Re-usable PPE should be thoroughly cleaned after use and not shared between workers.

Single-use PPE should be disposed of so that it cannot be reused.

Stairs should be used in preference to lifts or hoists.

Where lifts or hoists must be used: Lower their capacity to reduce congestion and contact at all times.

Regularly clean touchpoints, doors, buttons etc.

Increase ventilation in enclosed spaces.

Regularly clean the inside of vehicle cabs and between use by different operators.

Have you got any questions?

Record questions on the attendance sheet.

Index

a

Abraham Maslow Hierarchy of Needs Theory 146, 160, 166
Abuse of power 80
Academic programmes 225
Accountability 18, **214**
Accreditation 140
Adams Theory 160
Adaptability of leadership styles **149**
Additional fees 109
Age **85**
Alderfer's Theory of Motivation 161
Alienation 159, 169
Alliances and partnerships 130
Altruism
  Altruistic goodwill 45
  Altruistic offerings 52
An Inconvenient Truth 2006
Apprenticeships and internships 48
Architects 29
Association for Project Management Body of Knowledge 138
Association of Certified Fraud Examiners 193
Associative discrimination 80
Ethical position 162
Autocratic
  Autocratic management styles 163, 169
  Autocratic style of leadership 149

b

Banwell Report (1964) 127

Behaviour 159
  Behaviour of professionals 14
  Behaviours and leadership skills 7
Beliefs 239
Belongingness 250
Beneficial impact 162
Best Companies accreditation 18–20, **70**, 73, **139–140**, 233, 238, 242
  Best Companies benchmark 71
  Best Companies Index 140
Best practice 7, 18, 88
Bid/contract rigging 5, 193
Billing fraud 193
Blacklisting 224
  Blacklisting of the supply chain 5
Blame game 123
Bolstering incomes 179
Boredom 169
Bottom-up management initiatives 169, 245
Breaches 126
  Breaches of data security 91
BREEAM 26–27, 30, 37, 145
Bribery **192, 216,** 249
  Bribery Act 192, 248–249
  Bribery conviction 5
British Research Establishment (BRE) 26, 30
Build standards 3
Buildability 113, 123, 128
Building Information Modelling (BIM) 129
Building performance 224
Building technologies 235
Built environment 1–4, 6–8, 10–11, 13, 17

Bullying   80, 227
Business
    Business ethics   17, 191
    Business ethics model   51
    Business organisations   3
    Business population estimates   130
    Business relationships   10

c

Calculations   200
Carbon
    Carbon emissions   29
    Carbon friendly building   235
    Carbon generation   42, 145, 234
    Carbon management   23, 28
    Carbon Management Plan   36
    Carbon management reports   51
    Carbon management strategy   32
Career development   170
Case studies   3, 6, 11, 23, 48, 115, 172
Construction Design Management Regulations
        CDM (2015)   95, 101, 103
Change order manipulation   193
Chartered Management Institute (CMI)   138
    Chartered Management Institute Code of
        Ethics and Professional Conduct   24
Chartered institute of Building   4, 195, 212
Claims conscious   123
Client   2, 4–5, 7, 29
    Client satisfaction   73, 129, 131, 225
Climate change   23, 25, 27, 42, 234
Codes
    Codes of conduct   15, 95, 189–190
    Code of Considerate Practice   52–53, 236
    Code of ethics   148, 190
    Codes of practice   10–11, 155
Coding and judgement   **200**
Coercion   80
Collaboration   4, 9, 140
    Collaborating stakeholders   2
    Collaboration and partnering skills   41
    Collaboration toolkit   36, 39–40
    Collaborative alliances   126
    Collaborative and partnering
        strategies   148

Collaborative approaches   71
Collaborative innovation capacity   126
Collaborative model of procurement   36
Collaborative relationships   109
Collaborative working   2, 6, 10, 117
Collaborative working approaches   17,
    19, 114
Collaborative workshops   114
Collusion   5, 209, **215**
    Collusion on tenders   224
Common goals   165
Common Good Approach   **215**
Communication   19, 169
    Communication processes   81
    Communication routes   93
    Communication, as a motivation
        factor   **159**
    Communication, the importance
        of   **148**
    Communications strategies   71, 159, 168,
        244
Community
    Community education   43
    Community engagement   43, 235
    Community relations   39
Company profits   46
Conditions of work   166
Conflict   159, 168
    Conflict of interest   18, 209
Confrontational practices   3
Conservation   23
Considerate Constructors Scheme   9, 53–54,
    61, 226, 236
Construction
    Construction Industry Council   161, 213
    Construction 2015   4
    Construction and engineering sectors   9
    Construction companies   5
    Construction industry   2, 4–5
    Construction Leadership Identity   137
    Construction phase plan   102
    Construction practices   4
    Construction Products Association   130
    Construction professionals   7
    Construction quality   68

Construction regulatory bodies 230
Construction Sector Deal 59
Construction Skills Certification Scheme 95, 98, 104, 227
Construction, 2025 Industry Strategy: Government and Industry in Partnership 129
Construction management 121
   Construction management curriculum programmes 126
   Construction management teams 164
Contamination 234
Conscientious refusal 195
Continued professional development (CPD) 52, 77, 79, 219, 223, 229, 249–251
Contracting organisations 4
Contractor management skills 128
Contractual
   Contractual implications **179**
   Contractual routes 128
Contravention 163
Coronavirus Job Retention Scheme 178
Corporate
   Corporate codes **56**
   Corporate financial performance 3
   Corporate responsibility 48
   Corporate social responsibility (CSR) 2–3, 6, 8–9, 10, 23, 25, 31, 39, 45–46, 223, 234–236
   Corporate Social Responsibility Model 54
Correct Action Plan 88, 89, 94, 240
Corruption 70, 73, 167, 192
   Corruption and bribery 67
Competitive tendering 130
Cost 157
   Cost and time predictability 128, 129
   Cost predictability 121
COVID 19 Coronavirus pandemic 10, 104, 143, 172, 176–177
Criminal
   Criminal charges 166
   Criminal law 67
Critical incident 237
Cross pollination of ideas 2

Culture 9, 163–164, 166
   Cultural barriers 246
   Cultural change 31, 35, 82, 129, 147, 251
   Culture of professional ethics 10, 155
   Culture, importance of 20
Curriculum 126

d

Data
   Data Controller 92
   Data Processors 92
   Data protection **91**, 227
   Data Protection Act 2018 91
Decision-making 65, 70, 139, 217
   Decision making processes 17
Deep Interdependence 115, 127
Deepwater Horizon oil spill 2010 47, 60, 235
Democratic styles of management 170
Demoralisation 166
   Demoralised employees 169
Department of Trade and Industry (DTI) 26, 42, 59, 234
Dependent variable 120
Depletion of natural resources 234
Dereliction of duty 9
Design
   Design consultants 7
   Design and build procurement 123
   Design and construction (gap between) 123
   Design modifications 131
   Design process 28
   Design team consultants 29
Direct discrimination 80
Disability **83**
   Disability Confident 83
   Disability Discrimination Act (1995) 76
   Disability equality **78**
Disciplinary action 163
Discrimination 9, 79
Dispute
   Dispute avoidance 128
   Dispute resolution 127
   Disputes and litigation 237

Diversity 9
  Diversity and inclusivity   50, 60, **83**
  Diversity impacts   81
Domestic subcontractors   66
Downward cycle   160
Downward cycle of trust   127
Duty
  Duty and moral obligations   15
  Duty to warn   237
Dynamic environment   169

e

Economic
  Economic climate   168, 244
  Economic goals   3
  Economic perspective   45
  Economic responsibility   51
Education   13, 19, 232
  Education of construction staff   162
  Educational institutes   35
Embodied carbon   28–29, 36, 39
Emmerson Report (1962)   127
Empathy   140
Employees productivity and commitment   13
Employment relations   19, 233
Energy   23, 28
  Energy and carbon footprint   57
  Energy and water policy statements for environmental performance   58
  Energy consumption   25, 42
  Energy creation   28
  Energy efficiency   42
  Energy efficient construction techniques   61
  Energy Performance Certificate   32
  Energy, Water and Carbon Management Plan   36
Environmental
  Environmental ethics   5–6, 26, 8–9, **22–24**, 28, 41, 54, **233**
  Environment issues   **215**
  Environmental accreditations   37
  Environmental agenda   28
  Environmental and energy audits   51
  Environmental and sustainable building technologies   29
  Environmental awareness   28
  Environmental benefits   33
  Environmental conservation   234
  Environmental considerations   9, 45
  Environmental enhancement   36
  Environmental ethical behaviour   22
  Environmental impacts of development   38
  Environmental issues   33
  Environmental management   23
  Environmental Management System   **38, 43**
  Environmental performance criteria for buildings   26
  Environmental problems   42
  Environmental responsibility   22
  Environmental social responsibility   23, 27, 42
  Environmental standards   61
  Environmental sustainability   22, 38, 40, 43
  Environmental sustainability credentials for organisations   **225**
  Environmental sustainability performance   38
  Environmental sustainability projects   38
  Environmental sustainability strategies   3
  Environmentally sustainable building technologies   39
  Environmentally beneficial outcomes   41
  Environmentally sustainable policies   42
Equality   9
  Equality Act (2010)   76, **78**, 79, 81, 94, 227, 239
  Equality and diversity   75
  Equality assessments   79
  Equality issues   94
  Equality, diversity and inclusion   82
Ethical
  Ethical 'game playing'   247
  Ethical adherence   73
  Ethical and legal aspects   9
  ethical approach   191

Ethical behaviours   6–7, 10, 148, 174
Ethical behaviours and practices   108
Ethical bodies   218
Ethical challenges   **196**
Ethical codes of conduct   71
Ethical compliance   16, 19, 88
Ethical considerations around trust and collaborative working   **108**
Ethical debate   64, 72
Ethical decision-making   4
Ethical dilemmas   16–17, 29, 85–86
Ethical duty   67
Ethical factors   4
Ethical initiatives   3, 11, 13
Ethical issues   66–67
Ethical issues around culture   **81**
Ethical leadership   10, **137–140**, 144, 182, 228, 241
  Ethical leadership models   63
  Ethical leadership theory
Ethical management practices and models   8
Ethical motives   15
Ethical perspective   10
Ethical policies and procedures   139
ethical policing   11, 189
Ethical position   162
Ethical practice   3, 6, 13
Ethical principles   6, 17
Ethical principles, adherence to   19
Ethical procurement   3
Ethical procurement processes   43
Ethical question   85
Ethical standards   3–4, 6–7, 11, 16, 17, 19, 66, 68
Ethical statements   **56**
Ethical supply chain management   **85**, 94, 240
Ethical support procedures   218
Ethical tools   218, 248
Ethical usage   218
Ethical values   6, 17–19
Ethics   1, 9
  Ethics and compliance   22
  Ethics and leadership   **70**
  Ethics and quality   63
  Ethics and reputation   64
  Ethics and trust   **70**
  Ethics around equality and inclusion   **77**
  Ethics around health and safety   86
European Union (EU)   26, 91, 84, 94, 240
  European Union Directive 2006/12/EC   30
Experience   109
Experiential model   143
Exploiting staff   165
Extrinsic measures   169
  Extrinsic motivational measures   169

f

Failures   2
Fair reward   18
Fairness   19, 64, 167, 238
Faith   239
False representation   193
Fictitious vendors   193
Financial
  Financial claims   67, 72
  Financial considerations   33
  Financial damage   66
  Financial gain   85, 166
  Financial incentives   165–169
  Financial resilience   46
  Financial resources   66
  Financial rewards   166
  Financial risk   66
  Financial viability   33
Fire
  Fire dampening system   175
  Fire safety   104
Fitness for purpose   128
Fraud   70, 73, 167, 193
  Fraudulent performance bonds   199
  Fraud and corruption   64
  Fraudulent acts   179, 238
  Fraudulent billing   198
Freedom of speech   77, 227
Front load problems   131
Frustration of construction staff   168
Furlough Scheme (see also Government Job Retention Scheme)   172, 178, 246, 248

g

Gender **84**
General Data Protection Regulations (GDPR) 9, 75, 91, 240
Global
  Global and national agenda on sustainability 234
  Global cases of unethical behaviours 167
  Global climate change 42
  Global institutions 2
  Global warming 234
Governance 251
  Governance and regulation 5, 142
  Governing bodies 212
Government financial aid 179
Government funding opportunities 226, 234
Government Job Retention Scheme (see also *Furlough Scheme*) 246
Greed 179
Green
  Green credentials 39
  Green Deal 23
  Green issues 24
  Green travel plan 34
Greenhouse
Greenhouse gas emissions 41, 233
Greenhouse gases 28
Greening initiatives 28, 42, 234
Groupthink 76

h

Harassment 227
Health and safety 11, 69, **95**, 142, 250
  Health and safety compliance 107
  Health and Safety Executive (HSE) 9, 106
  Health and safety files 107
  Health and safety expertise 128
  Procedures and documentation, around health and safety 105
Highest rank 166
Honesty 16
Hospitality 249
Human resources 2, 233
  Human resources initiatives 227
Human resources management 3, 6, 9, 224
Human relations policies 19–20
Human resource considerations 11
Human resources development 162
Human resources policies and procedures 17

i

Institute of Civil Engineers 212
Illegal
  Illegal conduct 5
  Illegal immigrants 165
Improved outcomes 4
Improvement measures 123
Inappropriate behaviours 81
Incentive based motivational measures 164
Inclusivity 9, 75
  Inclusivity, equality and diversity 93
Increased productivity 11
Independent variable 120
Indirect discrimination 80
Industry accreditation 170, 246
Information Commissioners Office (ICO) 92
Infrastructure Cost Review 2010 129
Innovation 130
  Innovative initiatives 1
Institution 1
  Institution of Occupational Safety (IOSH) 95, 99, 106
Institutional
  Institutional codes of conduct 4
  Institutional membership 109
Insubordination 209
Integration
  Integrated project teams 169
  Integration of contractors 128
Integrity 16, 18, 64, 140, 238
Intergovernmental Panel on Climate Change 25
Interview
  Interview design **199**
  Interview questions and aims **202**
Intimidation 80

Intrinsic factors   166
ISO 14001 Energy Management System certification   58, 226
ISO 50001:2015 Environmental Management System certification   58, 226

j

Job
  Job rotation   162, 169
  Job satisfaction   11, 141, 165
  Job training   170

k

Key performance indicators (KPI's)   51, 61

l

Lack of recognition   164
Landfill Tax Escalator 2010   30
Law enforcement   89
Leadership   2–3, 6, 9–10, 63, 72
  Leadership around ethics   7
  Leadership balance   **214**
  Leadership challenges and dilemmas   137, **147**
  Leadership Identity Model   142, 228, 242
  Leadership roles   23
  Leadership skills   10
  Leadership styles   **147**
  Leadership, poor   10
Learning and Skills Council   30
Learning pathway   143
Legal frameworks   116
Legal perspective   45
Lessons learnt   143, 168
Liability for discrimination   **79**
Life cycle   2, 28, 41, 233
  Life cycle of buildings   2, 28
  Life cycle around carbon dioxide emissions   29
Litigation proceedings   3
Local Transport White Paper 2010   25
Loss of client confidence   111
Low Carbon Construction Final Report 2010   5
Low Carbon Final Report (MH Government 2010)   120
Low Carbon Final Report 2010   129
Low esteem   159, 168
Low productivity   168, 233
Low-carbon technology   32
Lower energy consumption   225
Lowest capital costs   235

m

Macho culture   157
Maintenance
  Maintaining standards   156
  Maintenance costs   128
Management
  Management processes   126
  Management styles   159
  Management strategies   168
Manufacturers   2
Marie Curie   57
Marketing   226
Masterplanning projects   30
Maximisation of profits   123
Micro ethics   15, 16
Mistrust   111, 123
Mobilisation of staff   169
Model of good practice   1
Modern Day Slavery   146
  Modern Slavery Act 2015   9, **86**, 75, 94
  Modern slavery legislation   88
  Modern slavery statements   89
  Identifying modern slavery   **88**
Money   166, 168
Monitoring Checklist   105, 107
Monotony   169
Moral
  Moral codes   18
  Moral compass   146
  Moral considerations   3
  Moral dictum   4
  Moral principles   14, 15, 191
  Moral responsibilities   1, 45
  Moral values   14
Morale   159, 162
Motivation   9, 63, 72

Motivational management   2, 6, 10–11,
    156, 16, 169
Motivating factors   164
Factors which affect motivation   158
Motivation initiatives   160
Motivation levels   7
Motivation of employees   40
Motivational management tool   156
Motivational measures   140
Multi-disciplinary   177

n

National Infrastructure Plan 2013   129
National Health Service (NHS)   174, 178
National Procurement Plan   34
*National Social Value Measurement
    Framework*   48
National Tender Board   198
*National Themes, Outcome and Measures
    (TOMS)*   48
Natural resources   25
Negative publicity   70
Negativity   168
Nepotism   207, **216**
New Engineering Contract 3 (NEC3)   67, 113,
    126–127, 179

o

Off-off nature of projects   163
OJEU process   113
Open communication routes   77
Operation
    Operation costs (Opex)   145
    Operational efficiency   234, 226
Organisational
    Organisational behaviour   148
    Organisational change   **35**
    Organisational culture   13, 18, 20, 159, 165
    Organisational culture and values   140
    Organisational ethics   1, 2, 4, 6–7, 9–10, **13**

p

Pain-share/gain share   115
Partnering
    Partnering arrangements   6
    Partnering based approaches   132

Partnering organisations   127
Partnering philosophies   114
Partnering toolkit   36
People and Planet ranking (of Universities'
    environmental sustainability)   28
Personal
    Personal development   13, 19, 232
    Personal protective equipment (PPE)   144
    Personal satisfaction   164
Philanthropy   50, 60
    Philanthropic efforts   51
    Philanthropic perspective   45
Photovoltaic technologies   28, 42, 234
Political
    Political efforts   234
    Political influence   233
Poor
    Poor ethical compliance   168
    Poor leadership   7, 10, 152
    Poor management systems   163
    Poor materials   224
    Poor morale   169
    Poor productivity   167
    Poor quality   72
    Poor quality of design   224
    Poor working practices and
        behaviours   224
    Poor workmanship   224
Potts Report (1967)   127
Practitioner framework   5
Pregnancy, maternity, and carer
    responsibilities   **84**
Preliminaries   114
Prequalification questionnaires   107
Principal Contractors   102
Principal Designers   102
Procurement strategies   129
Productivity   9, 13, 17, 19, 159–160, 166
    Productivity on site   249
Professional
    Professional ethics   4–6, 10, 14–15, 108
    Professional practice   10
    Professional practice   7, 11, 69
    professional bodies   142, 189, 217, 230
    Professional codes of conduct   70

Professional development   40
Professional ethics related to data protection   **91**
   Breaches of professional ethics   4
Professional governing bodies   218
Professional judgement   4
Profit
   Profit levels   235
   Profitability   19
   Profits and overheads   114
Programme disruption   174
Project
   Project based environment   157
   Project controls   71
   Project failure   **7**, 150
   Project outcomes   3
Protected characteristics   79, 94
Public
Public perception   226
   Public relations   23, 30
   Public relations benefits   141
Punishment for non-compliance   251

q

Qualifications   109, 170
Quality   9, 72, 157, 159, 166, 209, 216
   Quality of construction   9
   Quality of service   69
Qualitative analysis for data collection   **199**

r

Recognition   161
   Recognition initiatives   170
Recycling   42
Reduced waste   225
Regulations   163
   Breaches of regulations   9
Regulatory standards   11
Relationships   159
Reliability   19
Religion, belief and race   **84**
Renewable
   Renewable sources of heat and power   226
   Renewable technologies   36
Repetitive work   159

Reputation   9, 13, 39, 45, 63
Reputation of organisations   16
Reputational
   Reputational benefits   46, 226
   Reputational damage   2–3, 5, 9, 16, 64, 66, 71, 112, 235
Resource availability   29
Responsibility   18, 64
Retention monies   175
Royal Institute of British Architects (RIBA)   *126*   212
   Royal Institute of British Architects (RIBA) Plan of Work 2013   121, 125
   Royal Institute of British Architects (RIBA) Plan of Work 2013: Strategic Definition   122
   Royal Institute of British Architects (RIBA) Plan of Work 2013: Concept Design   122
   Royal Institute of British Architects (RIBA) Plan of Work 2013: Developed Design   122
   Royal Institute of British Architects (RIBA) Plan of Work 2013: Technical Design   122
   Royal Institute of British Architects (RIBA) Plan of Work 2013: Construction, Handover and Close Out   122
   Royal Institute of British Architects (RIBA) Plan of Work 2013: In Use   122
Royal institution of chartered surveyors (RICS)   4, 130, 212, 229
RIDDOR (Reporting of Injuries Diseases and Dangerous Occurrences Regulations)   96
Rio Declaration on Environment and Development   25
Risk   9, 63, 72
   Risk and ethics   63
   Risk assessment   64, 88
   risk assessments and method statements   106
   Risk materialisation   237
   Risk workshop   66, 237
Roles and responsibilities   126

s

Safe working environments 76
Safety 42, 163
  Safety and time predictability 132
  Safety programmes 159
SARS outbreak 172
Scandals
Scope of services 109
Self-motivation levels 160
Senior management 168, 244
Sensitive and confidential information 112
Sexual orientation 239
Shared perceptions approach 165
Shareholders 46
Short-term
  Short-term economics 157
  Short-term fixes 169
Simon Report (1944) 127
Single use plastics 235
Site inductions 102
Site Registration Monitors Checklist 54
Skills 42
  Skills and knowledge 110
Social
Social agenda 46
Social and environmental
    responsibilities 24
Social class protected characteristics 239
Social facilitators 4
Social impact 146
Social performance 3
Social responsibility process 88
Social value 6, 45, 48
  Social Value Portal (2020) 48, 60, 236
Solar power 28, 42, 234
Stability 19
Staff development 140, 142, 162
  Staff development and training 141
Stakeholders 46
Standards
  Standards and values 155
  Standards of conduct 14
Statement of ethical principles 198
Stereotyping 81

Subcontractor
  Subcontractor teams 161
  Subcontractor team of the month 170
Success
  Success of projects 6
  Successful outcomes 46, 226
  Successful project outcomes 11
Supply chain 5, 28, 57, 88, 120, 228
  Supply chain, blacklisting 5
  Supply chain, knowledge 128
Sustainability 2, 9, 22, 24–25, 30–31, 34, 50
  Sustainability credentials 39
  Sustainability initiatives 23, 34
  Sustainability measures 145
  Sustainability principles 26
  Sustainability strategy 35
  Sustainability weightings 39
  Sustainability, global and national
    agenda **26**
  Sustainable and ethical considerations 46
  Sustainable construction 43
  Sustainable development 33
  Sustainable efficiencies 5
  Sustainable environment 1
  Sustainable procurement 39, 48
  Sustainable resource management 32
  Sustainable resources 33
  Sustainable solutions 5
  Sustainable transport policies 32, 36, 42,
    225, 234

t

Teaching 19, 232
Team spirit 140
Teamwork 17, 19, 233
Tender price 128
The Carbon Plan (2011) 25–26
The Considerate Contractor Scheme 45
*The Farmer Review of the UK Construction
    Labour Model* 58
*The Office for Government and Commerce
    (OGC)* 34
*The Pyramid of Corporate Social
    Responsibility* 50

The RICS Professional Ethics Working Party
The United Nations Brundtland Commission 25
Theft or substitution of materials 193
Time 55
  Time certainty 121
  Timescales 166
Town and Country Planning Act (1990) 27, 42, 234
Traditional procurement 121
Training 159, 162, 238
  Training and development 170, 245
  Training and development programmes 88
  Training and education 19, 52, 155
  Training and education initiatives 2
  Training institutions 126
  Training to improve skills 18
Transparency 16, 19, 215
Transparency International 192
Transport 23, 28
Trust 6, 9, 63, 72, 167
  Breach of trust 112
  Degrees of trust 113
  Trust and collaboration 64
  Trust building mechanisms 127
  Trust creation 18
  Downward cycle of trust 115
  Reputational Trust 112
  Unconditional trust 115
  Trust, as a collective necessity **125**
  Cognitive based trust 109
  Conditional or unconditional nature of trust 116
  Degradation of trust 167
  Duties of trust 10
  Fragile and dynamic nature of trust 112
  Fragility and robustness of trust **110**
  Importance of trust for collaboration and professional ethics 112
  Lack of trust 4, 11
  Status of trust 111
  Violations of impersonal trust 112
  Trustworthiness **213**
  Trustworthy 109

u
UK construction 129
  UK construction industry 4, 52, 84, 108, 120–121, 127, 130–131, 241, 217, 246
UK Government 27, 41–42, 59, 233
UK Law and Care Commission 56
UK local authorities 176
UK recession 177
Unethical
  Unethical and unfair practices 7
  Unethical behaviour 17, 108, **176**
  Unethical conduct 7, 13, 16
  Unethical environment 197
  Unethical practice 7, 68, 224
  Unethical practices and behaviours 4, 111
  Unethical tender practices 5
  Unethical practice 10
Unity 167
University of Salford 60
University of Salford Ethical Statement 57
Unlawful discrimination 81, 240
Unprofessional practices 108
Upward cycle of trust 126

v
Value
  Value engineering 113
  Value for money 128
  Value management 123
Victimisation 227
Vroom's Expectancy theory 161

w
Wages 23, 28, 42, 166
Waste
  Waste management 34
  Waste management strategy 30
  Waste recycling 28
  Waste reduction initiatives 43
Welfare facilities 102
Well-being of employees and stakeholders 19
Whistle blowing processes 82, 190

Win
    Win and lose outcomes   121
    Win-win scenarios   245
    Win-win outcomes   148, 242
Wind
    Wind power   28, 234
    Wind turbines   234
Work-based scenarios   249
Workers' rights   209
Working practices   4
Workmanship   158
Workshops   127

z

Zero carbon
    Zero carbon homes   23
    Zero carbon targets   233